GOD IN THE MACHINE

GOD IN THE MACHINE

WHAT ROBOTS TEACH US ABOUT HUMANITY AND GOD

ANNE FOERST

DUTTON

DUTTON
Published by Penguin Group (USA) Inc.
375 Hudson Street, New York, New York 10014, U.S.A.
Penguin Group (Canada), 10 Alcorn Avenue, Toronto, Ontario, Canada M4V 3B2 (a division of
Pearson Penguin Canada Inc.); Penguin Books Ltd, 80 Strand, London WC2R 0RL, England; Penguin
Ireland, 25 St Stephen's Green, Dublin 2, Ireland (a division of Penguin Books Ltd); Penguin Group
(Australia), 250 Camberwell Road, Camberwell, Victoria 3124, Australia (a division of Pearson
Australia Group Pty Ltd); Penguin Books India Pvt Ltd, 11 Community Centre, Panchsheel Park, New
Delhi – 110 017, India; Penguin Group (NZ), Cnr Airborne and Rosedale Roads, Albany, Auckland,
New Zealand (a division of Pearson New Zealand Ltd); Penguin Books (South Africa) (Pty) Ltd, 24
Sturdee Avenue, Rosebank, Johannesburg 2196, South Africa

Penguin Books Ltd, Registered Offices: 80 Strand, London WC2R 0RL, England

Published by Dutton, a member of Penguin Group (USA) Inc.

First printing, December 2004
10 9 8 7 6 5 4 3 2 1

Man/Mouse images on pp. 54, 55, and 56: Copyright 1961. Canadian Psychological Association.
Reprinted with permission.

Sketches from Christel Schachtner's experiment on embodiment, on pp. 86, 87, 88, 89: Christel Schachtner,
Geistmaschine: Faszination und Provokation am Computer (Frankfurt: Suhrkamp Verlag, 1993).

 REGISTERED TRADEMARK—MARCA REGISTRADA

LIBRARY OF CONGRESS CATALOGING-IN-PUBLICATION DATA

Foerst, Anne.
    God in the machine: what robots teach us about humanity and God / by Anne Foerst.
        p. cm.
    ISBN 0-525-94766-3 (alk. paper)
1. Computers—Religious aspects. 2. Robotics. 3. Artificial intelligence. 4. Religion and science. 5. Man
(Christian theology). 6. Philosophical anthropology. 7. Humanity. 8. God. I. Title.
    BL255.5 .F64 2004
    215—dc22   2004058291

Printed in the United States of America
Set in Sabon
Designed by Leonard Telesca

This book is printed on acid-free paper. ♾

To all persons with courage and
humor enough to doubt themselves

Contents

Acknowledgments

A book like this cannot be written without the help and support from many, many people and it is impossible to mention them all. So I will only mention the most important contributors to this book. But I want to thank everyone who was involved in the project at some point, all these wonderful friends who were willing to listen to yet another theory of mine.

I would also like to thank the people in my audiences all over the world—be it in lectures or on radio and TV shows; they provided the wonderful and creative feedback I needed to sharpen my ideas.

I would like to thank specifically the Center for the Study of Science and Religion (CSSR) at Columbia University. Bob Pollack, especially, gave me the chance to give the inaugural lecture series for the Center in the spring of 2000. These four lectures were the basis for this book. Also wonderful at the Center were Bob Thurman, Piet Hut, and many more—always willing to listen and join me on journeys into the depth of the religion and science dialogue and into questions of embodiment.

To Harvey Cox and Rod Brooks, my very special thanks for believing in me and giving me a chance to pursue my ideas at both Harvard and MIT. Harvey encouraged me to think theologically and not dogmatically; he also taught me that theology without personal experience and emotion is dead. And he taught me how to engage in a dialogue with people who think fundamentally differently from me. Rod, for taking a leap of faith and making me part of the MIT AI Lab. Over the course of the years, we worked well together and today, we often find ourselves together on panels and the like. I think a mentor who can encourage a young academic to find her own way, even if it is contrary to the path of the mentor, is admirable and rare, but Rod did exactly that.

Most people at the MIT AI Lab and larger Computer Science community were very welcoming toward the theologian in their midst. I would like to thank here the late Paul Penfield, former head of the computer science department, and his successor, John Guttag, whose humor often helped to ease some tensions. MIT is a very special place to me. Cynthia Breazeal, the friend and colleague who let me into her work and her thought: without your collaboration I would have never gotten that deep into Kismet's head. Other members of the team were helpful as well: Matt Williamson, who helped me to understand Cog, Lynn Andrea Stein, Annika Pfluger, Robert Irie, Brian Scasselatti, Una-May O'Reilly, and many more. Especially valuable was the friendship with Deniz Yuret, who introduced me to Rod Brooks and remained my strongest critic and challenger and, yet, a close friend. Jean Hwang, my former assistant who had to deal with my temper and was still able to turn the CSSR transcripts of my lecture series into readable text.

Tom Chittick for helping me to keep the balance between faith and doubt. Ed Neuhaus for helping me to understand human ambiguities, including my own, and to make peace with them. Norbert Samuelson for pointing out many valuable rabbinical sources. Phil Hefner for advising me about my approach to create a dialogue structure.

I could never understand why people thank their agents; after all, they earn money with you. Gee, was I wrong! Lane Zachary, you are great! Thanks for helping me find my story and keep it alive. Thanks also to Brian Tart from Dutton for finding the Bible parts in my book the most intriguing, Mitch Hoffman, the best editor on the planet, and Erika Kahn, the best editor's assistant.

My parents for never wavering in their belief in me and my ideas, for their reading of my work (even working through English texts), and for their never-ending interest in my new ideas. My whole family has been wonderful! Thanks for being so close despite an ocean between us.

Finally, I would like to thank John for his humor, his patience, his friendship and caring, and his willingness to listen.

GOD IN THE MACHINE

Introduction

> *"I have, alas! Philosophy,*
> *Medicine, Jurisprudence too,*
> *And to my cost Theology,*
> *With ardent labor, studied through.*
> *And here I stand, with all my lore,*
> *Poor fool, no wiser than before.*
> *Magister, doctor styled, indeed,*
> *Already these 10 years I lead,*
> *Up, down, across, and to and fro,*
> *My pupils by the nose,—and learn,*
> *That we in truth can nothing know!"*
> JOHANN WOLFGANG VON GOETHE, FAUST, PART I

This is the story of a handshake.

The handshake occurred in the fall of 1995 between Harvey Cox, a professor at the Harvard Divinity School, and the humanoid robot Cog, developed at the Artificial Intelligence Lab at the Massachusetts Institute of Technology.

Cog is a huge robot, approximately seven feet tall, and somewhat intimidating. It has a massive steel frame. Our culture is infused with science-fiction stories about robots that turn against humans and destroy them. We are also accustomed to factory robots that blindly do their assigned tasks and can accidentally smash a finger or a head.

Therefore, Cog induces a mixture of fascination and fear in many people who see it for the first time.

But Harvey Cox took the first step. He had devoted his entire life to bringing Christian theology into a dialogue among people with different worldviews, and he didn't want to stop at a robot. So, when he first realized that Cog's eyes were following him around the room, he made eye contact. Then Professor Cox tentatively extended his hand, and Cog, after some trial and error, grasped it. There was a collective gasp from the Harvard theologians and MIT scientists present.

Cog is the result of one of the first attempts to build a humanoid robot, and in 1996 it had just become very famous. As soon as the media discovered the project, Cog was constantly surrounded by cameras and covered by all the major television networks and newspapers. Building a humanoid robot is a fascinating endeavor, but before the development of Cog, such a project had belonged firmly in the realm of science fiction.

At MIT, Rodney Brooks, then associate director of MIT's famous Artificial Intelligence Laboratory (MIT AI Lab), had the vision of building humanoids. For quite some time, Rod was the *enfant terrible* in the robotics community; he had played around with techniques and strategies that were completely new to the field, and he was the first to build successful autonomous robots that could navigate real-world environments. This might sound like a very simple task, but before the mid-1980s, nothing like it had ever been accomplished.

The traditional metaphor in AI had been that intelligence is a program that is implemented in the wetware of the brain but might just as well be implemented in the hardware of the computer. Traditional AI programs have internal world models that process data in order to plan "intelligent" reactions to given stimuli. While these traditional machines are great at playing chess and proving mathematical theorems, they fail as soon as they have to navigate constantly changing environments. They are incapable of reacting effectively to the real world. The assumption in the traditional system is that intelligence is the capability for abstract thought. Humans are so good at it that they can live in the real world. If AI systems become increasingly smart, they might eventually be able to perform as well as we do.

Rod broke with this assumption and declared our facility for abstract thinking a mere by-product of our ingenious capability to in-

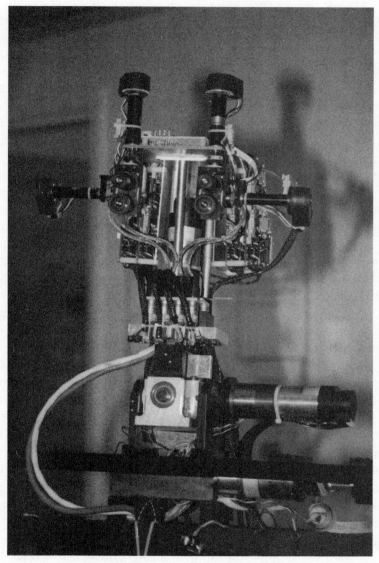

This is Cog in its early stages, 1992–94. The motors that move the eyes are designed to give a slightly anthropomorphic appearance. *Credit: Rodney Brooks*

tentionally shape our respective environments in order to survive. Rod believed that if AI researchers attempt to build intelligent machines, they ought to build embodied entities that interact with the real world. One might not need abstract thought in the beginning— after all, most animals survive pretty well without it, and human newborns don't have this capability either. Hence, Rod proposed starting with insect robots and then attempting to build a robot analogous to a newborn baby that learns through social interaction and physical embodiment. The jump from insects to a humanoid is huge, but the group that worked with Rod was so excited about the project that they decided to go ahead nonetheless.

At research universities such as MIT, most professors have groups that consist of undergraduate and graduate students, Ph.D. candidates, postdocs, researchers, and visiting professors. That is, people on many different levels work together and inspire each other's projects. I started to interact with Rod and his group in the fall of 1993 and joined the AI Lab in the fall of 1995. While I never got to know all the people in the original project that started in 1992, I became acquainted with all those who worked in the group between 1994 and 2000. At this time, we called our group "The Zoo," as there were so many exotic people in it. As a theologian, I was probably the most bizarre member, but we had other strange animals as well.

I was working in Germany on my doctoral thesis. I had studied computer science as well as theology and focused in my thesis on the possibilities for a nonjudgmental dialogue between Christian anthropology and AI. As I based my work on the theology of the great twentieth-century theologian Paul Tillich, I had intended to work in the Tillich archives at the Harvard Divinity School (HDS) during my first visit to Cambridge. But I also wanted to sneak my way into the hallowed halls of MIT, the cradle of AI. Most bigwigs in AI had come from this place or had trained here, and I couldn't wait to walk the same hallways with these figures that I so secretly admired.

When I met Rod, he invited me to come back in the fall of 1993 to join his informal seminar on "Embodied AI," in which he would talk about his new assumptions on how to build smart machines. I think I tickled his sense of humor when I first walked into his lab. He had probably never been sought out by a theologian. But he respected my wish to analyze the hidden assumptions, hopes, and beliefs in his group, and thought it might help them to develop a clearer worldview. He invited me back again and then invited me to join his lab as a postdoc. After finishing my thesis in 1995, I became an official member of the lab.

Meanwhile, at Harvard, my research in the Tillich archives had not gone unnoticed. I had unearthed a total of six lectures that Tillich had given at MIT, only two of which had been published to that date. Harvey Cox became fascinated. He had studied with Tillich and had written his doctoral thesis on Tillich and technology, but he had never been involved with MIT, nor had he ever concentrated on a particular area of technology. He was especially intrigued by the possibilities of bringing Tillich into the discussion of the development of Cog. As plans for my postdoc time at MIT became more concrete, he managed to get me a position at Harvard as well. This way, I would not spend my entire time at MIT, an alien place for me, but would find refuge at HDS, which was my theological home.

It worked beautifully. I spent my days happily at MIT and often trotted over to HDS. The inherent differences between both places fascinated me and became instantly visible just by looking at the architecture of the respective schools. The AI Lab was then a harsh, concrete, cubicled building at Technology Square, and HDS a Gothic, Oxford-style building with turrets. One had long corridors, chaotic offices, and modern furniture, and the other had oil paintings and woodwork from another era. At MIT, you couldn't distinguish between professors' and students' offices; at HDS only professors resided. Robots were always running through the hallways of MIT, and Cog was constantly being haunted by the media.

But there were also commonalities. Both schools were located slightly off campus and, in both groups, I laughed more than in any of my previous German academic settings. The other commonality was that at both schools I was initially perceived as an outsider. People at HDS found my fascination with high-tech somewhat strange. Most of them were slightly antitech and thought my quest to bring theology and AI together somewhat superfluous. At MIT, on the other hand, people were suspicious of the theologian in their midst. It probably took a year and many, many conversations until heads stopped turning toward me whenever the term *evolution* was mentioned. Most people there associated any form of Christianity with creationism and had a hard time understanding why I, as a Christian, actively sought out scientific explanations for why we are the way we are. But at both places, initial skepticism slowly evolved into active interest in my research.

As I liked both places so much, I attempted to bring the sides together. Socially, I introduced as many people as possible, and it is wonderful to see that after all these years some friendships still remain. At the same time, I was attempting to bring both groups together academically. I read and gave presentations to prepare them for each other. Finally, the group from HDS came over to MIT to meet the robot team and the robots themselves. It was on this visit that the monumental handshake occurred. There were quite a few important handshakes that day between various professionals of divinity and of AI, most of whom had never met a person from the other discipline.

The handshake between Harvey and Cog was deeply profound, as it built a bridge between these seemingly different areas. When Harvey and Cog looked at each other, it became clear that there was a dialogue waiting for us. As our technical creatures become more like us, they raise fundamental theological questions. As theologians, we have the responsibility not to shrink away from such a challenge but to seek out these opportunities and overcome our fear as Harvey had done that day.

This handshake has been fundamental for my life as a researcher, and I have worked on stabilizing the bridge that was forged at that moment ever since.

This became especially challenging when, after a few years of work on Cog, Cynthia Breazeal, one of the first engineers of Cog, started her own robot project, Kismet. Kismet is the most charming robot I know. Today it is proudly presented in the MIT Museum. However, during my last years at MIT, it was the center of our attention and eventually became the center of my research.

For me, the challenge is to analyze the attempt of the people at MIT to build creatures with human capabilities by technical means alone. Any such attempt suggests that humans are not special but rather are just like machines. Embodied AI also firmly places us within the animal kingdom and uses many insights from evolutionary biology. Any sense of specialness is rejected because there is no empirical evidence that humans are more unique than any other animal.

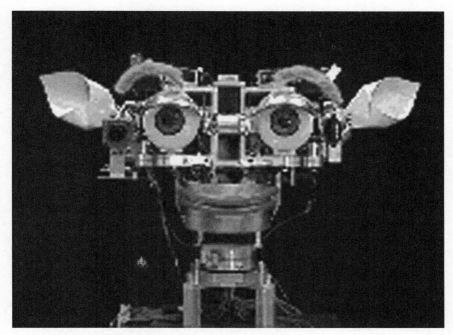

Kismet has a very expressive face, and here Kismet is "happy."
Credit: Cynthia Breazeal and the MIT Artificial Intelligence Laboratory

Most of what we think is special in humans can be directly tracked to earlier stages of development in other animals. At the same time, I had learned in theology to understand humans as special, elected by God to be God's partners. Because these two worldviews clashed, I was constantly walking the boundary between them.

What I learned at this time was that doubt is a positive thing. If you enter a dialogue between theology and another discipline and don't think the partner in dialogue has a valid perspective, you might as well not talk at all. Whatever the other party is saying will never convince you or even influence you in any way. If, on the other hand, you throw yourself wholeheartedly into a discussion, you have to question yourself constantly. I often sat in my office at MIT and thought the whole idea of religion was utter rubbish. What I had learned about the human machinery sounded so entirely convincing. These moments of doubt were very painful. But they were also constructive, as I learned the validity of both sides of the argument. I was able to follow Tillich's footsteps in walking the boundary without coming down on one side or the other because I found both sides so fascinating and attractive. I loved the people in both fields and was able to build connections between them.

Could I have found a similar situation in any scientific lab that attempts to understand humans in terms of evolution and functionalism? I think it was the presence of the humanoids in the AI Lab that made the journey so special. Humanoid robots invoke many contrary emotions in most of us. We often perceive them as a threat; we fear they might turn against us. We also resent the possibility that these creatures are as smart as or even smarter than we are, because we feel that we humans are special. But while we fear humanoid robots, we are also attracted to them. We are intrigued by the idea of nonhuman partners inhabiting the earth with us. As a deeply lonely species, we have a strong desire to communicate with beings different from us. Our attempts to communicate with dolphins and chimps and our continuing search for extraterrestrial intelligence demonstrate the depths of this desire. The construction of humanoid robots follows this search for partnership. It can therefore be linked to the Jewish golem tradition in which the construction of humanoid robots is understood as praise of God and as a repetition of God's act of creating us.

Even when I talk about computers, my focus is on humanoid robots and I strictly distinguish between the two. Computers are the machines that sit on your desk or in your lap. They are behind bank machines and at other sites where computing power is needed. Computers can be accessed via keyboard, voice, or touch, and they respond by processing data, running programs, and outputting language.

Robots, on the other hand, are bodies that have specific functions in and interactions with real-world environments through sensors and actuators. Even if their actions are based on computer power, they appear as entities in which computers play only one part. The popularity of fictional humanoid robot characters such as Commander Data (*Star Trek*) or C-3PO (*Star Wars*) demonstrate this. Most people are fascinated by robots and endeared by them.

The power of robots lies in their physical presence. They share our space. They move in the world we move in. They interact directly with us, and this allows us to bond with them. Bonding with computers occurs because computers are fascinating machines. Bonding with robots occurs because of their physical reality in this world and our ability to interact with them in physical space.

Among the most powerful science-fiction robots are C-3PO and R2-D2 from *Star Wars*. R2-D2 is much more darling than C-3PO despite C-3PO's humanoid form and human speech. I think that people prefer R2-D2 because it is so emotional. Even though its beeps are unintelligible to us, we think we understand it because we empathize and bond with it.

This is the same for Cog and Kismet. Cog is more like C-3PO in that it has a humanoid form and invites "grown-up" interaction. Kismet is cute like a baby, fostering an emotional connection—like the one with R2-D2.

This suggests that our bonding mechanisms depend on our own perception of the other and that therefore our ability to bond with them depends much more on emotional settings than on abstract "humanlike" qualities. For the same reason, it is the very emotionality Commander Data from *Star Trek* displays every time it complains about having no emotions that endears us; an emotionless machine would not constantly raise the issue of its own worth, value, and personhood.

Needless to say, I am very conscious not to assign robots any gender. Even if we, for instance, might think Data is male—after all, it is played by a male actor—I don't want to assign it gender. Humans are accustomed to interacting with creatures that have a gender. Although we find the address "it" offensive, I don't mean it this way. I just want to avoid a "male" impression of the robots.

Cog and Kismet are the first steps in the development of humanoid robots that will be part of our future. But there are many questions about a future with robots. Will these beings be seen as our future partners or as our future enemies? Will we ever perceive robots to be as special as we perceive ourselves to be? Might they make us superfluous in the course of the new, technology-induced evolution?

While I will address many of these questions, I have yet another emotional response to robots. While working with Cog and later Kismet, I have learned more about myself than at any other time. Why would it make me happy if Kismet smiled at me when I knew that it was a programmed reaction? Why would I be disappointed when Cog was ignoring me, even if I knew that—at that time—it could hardly see anything in its periphery?

Our strong emotional reactions toward the robots let the members of my team study themselves. What, after all, do we really know about ourselves as biological creatures, functional systems, and images of God? We decided to explore what modern science could teach us about ourselves. We studied neuroscience to learn more about the brain, and we studied behavioral science and animal studies to learn about mechanisms for common behaviors. We studied evolutionary biology to learn why our bodies are the way they are and what that means for our being in the world. Later on, we concentrated on developmental psychology; adult intelligence is, after all, developed over time, and we thought that if we simulated the learning process of an infant in our robots they would develop a higher degree of intelligence.

This book is the result of my journey at MIT, which started with the monumental handshake between Harvey and Cog. It was this handshake that made me realize that I was not an oddball trying to bring together things that simply do not belong. It convinced me that a dialogue was not just possible but necessary. I finally realized that these robots can serve as thinking tools to explore how we are and how we function in relationships. Surprisingly, the reflection on the robots led me to find some answers to one of the most urgent questions at the beginning of the third millennium. How can we achieve a global community where deeply ingrained cultural differences can, nonetheless, create a community in which different people can live in harmony?

In this book, I explore what robots can teach us about ourselves, our emotions, our ways of thinking and acting in the world. I will talk about what the very wish to build such a creature tells us about *Homo sapiens*. In the end, we will see that such questions lead to deeper theological and philosophical insights into who we are as thinking machines, as bodies, and as interactive beings and partners of God.

CHAPTER 1

Re-creating Ourselves

O GOD[1], *you have searched me and you know me.*

You know when I sit and when I rise; you perceive my thoughts from afar.

You discern my going out and my lying down; you are familiar with all my ways.

Before a word is on my tongue you know it completely, oh GOD.

If I go up to the heavens, you are there; if I make my bed in the depths, you are there.

If I rise on the winds of the dawn, if I settle on the far side of the sea,

Even there your hand will guide me, your right hand will hold me fast.

If I say, "Surely the darkness will hide me and the light become night around me,"

---

1. The various layers of the Hebrew scriptures use several names for God. The oldest one is the acronym JHWH, often falsely pronounced "Jehovah." Around 800 BC, the Jewish community, in fear of trivializing the name of their God, stopped saying the name aloud; whenever the tetragram appeared in the biblical texts, they would say "adonai" (My Lord), or "elohim" (God). Out of respect for the Jewish taboo, many translations of the Hebrew scriptures translate JHWH with "my Lord." For easier reading, I will translate the acronym JHWH simply with "God."

Even the darkness will not be dark to you; the night will
shine like the day, for darkness is as light to you.

For you created my inmost being; you knit me together
in my mother's womb.
I praise you for I am wonderfully made. Wonderful are
your works, that I know very well.
My frame was not hidden from you, when I was being
made in secret, intrically woven in the depth of the earth.
You created me as golem in my mother's womb.
In your book were written all the days that were formed
for me, when none of them as yet existed.

FROM PSALM 139

What Does It Mean to Be Human?

We need to begin with the most important questions that will follow us throughout the book. What does it mean to be human? How can humanness be defined? Can we ever come up with criteria that distinguish us from animals—or, for that matter, from robots? And what exactly is our place and our purpose on this planet, in our sun system, in the universe? Are humans special, or are they just another random species on an insignificant planet?

People have dealt with these questions for millennia. Countless answers have been provided, usually reflecting the specific cultural context in which they were formulated. As a result, many of these answers are unsatisfying for people outside the particular religious and cultural framework in which they operate. Is it really our intelligence that makes us human? Is it our creativity, or perhaps our ethics, that distinguishes us from animals? Is the essence of humanity the humanoid body—and if so, in what gender?

At the MIT AI Lab, I was confronted with these questions every day. I found a Calvin and Hobbes cartoon that beautifully illustrates both the complexity and profoundness of this issue. Calvin says, "I read that scientists are trying to make computers that think. Isn't that weird? If computers can think, what will people be better at than machines?" And Hobbes answers, "Irrational behavior?" And so Calvin says, "Well maybe they'll invent a psychotic computer." I have always found that this is not just a good description of humans but also a very good introduction to the world of Artificial Intelligence and our attempt to rebuild ourselves with the help of machines.

Before we can focus on this endeavor, however, I would like to introduce you to the assumptions I have made about humans in order to write this book. I believe that humans are, before anything else, storytellers. When we attempt to understand ourselves and develop theories about why we are the way we are, we talk in metaphors and symbols. Before we can learn about ourselves as creatures who yearn to rebuild themselves, we first have to look at other stories about humankind.

The Storytelling Humanoid (*Homo narrans*)

When we search for ways to understand ourselves, we find many metaphors in the human sciences. We have attempted to describe what it means to be human in terms of humankind's evolutionary development. We know that we are mammals and share with chimpanzees between 98 and 99 percent of our genes (we share with yeast approximately 50 percent of our genes, so perhaps the commonality with the chimps is not too astonishing). In the evolutionary context, our nearest *Homo* ancestor is addressed with the metaphor *Homo erectus* (upright humanoid), which refers to our upright position. However, the two-legged walk alone does not turn a species into one that constantly asks fascinating and deep questions about itself. For this, we need a form of higher intelligence, thinking capability, or wisdom; therefore, scientists talk about *Homo sapiens*.

Observations of human behavior versus animal behavior reveal that humans are quite good at building tools and then larger, long-term constructions with the help of these tools. The metaphor *Homo faber* describes the universal human trait to build and construct things and to shape the world in which we live with the help of technology. We know that other primates use tools and that they teach their young how to employ them. *Homo faber* is distinct from these other species because members of this species can intentionally build and construct over time—even over several generations.

Churches often took several hundred years to be completed, and in the case of the church in my hometown, the Kölner Dom (Cologne Cathedral) in Germany, it took more than a thousand years to complete it. The Dom is my personal favorite of all the churches I have seen. Of course, I grew up with it, but, so far, no other church has inspired in me so much awe and such a profound sense of beauty and spirituality. Part of the awe is a result of its history; it is hard to imagine that people who possess different styles and tastes can, over the course of a millennium, construct something so unified and coherent. The Dom is in the *Guinness Book of World Records* because there has never been another building that took that much time to be constructed. The metaphor of *Homo faber* addresses our ability to follow our intentions and wishes over long periods of time through many generations. Our ability to speak and to write makes such a capability for planning possible.

Homo sapiens cannot be without *Homo faber*; both are equal parts of being human. The seemingly "wise" *Homo sapiens* creates theories and visions while *Homo faber* constructs and verifies ideas. In AI, both are intrinsically linked.

A good example of the close connection between humans as creators of ideas and humans as toolmakers is Martin Luther's powerful influence on the thoughts of people in Western Europe in the sixteenth century. As a Lutheran, Luther is to me one of the greatest men in history. But most of his thoughts weren't unique or new. Many others before Luther had offered similar critiques against the Church, and they all were prosecuted and killed as heretics. What Luther had on his side was not only the greatness of his ideas but also a technical gadget, the printing press.

Gutenberg's printing press allowed for the wider distribution of the ideas of the Reformation. The fourteenth- and fifteenth-century Renaissance movement had called for education, among many other things. When Luther appeared, many people were able to read without the help of their priests and so they could absorb Luther's writings. The thoughts alone were not powerful enough to change the Western world; they needed to be contained and transmitted in a concrete and physical entity, such as a booklet or pamphlet.

It is, of course, *Homo faber* who constructs robots and many other things. If humans believe themselves to be created by a god, members of the species celebrate that god and god's creativity within them whenever they construct something; art, music, engineering, poetry, and thought are all new creations.

Building robots, as well as most other acts of creativity, contains a playful element. When Rod Brooks was once asked why he liked robots so much, he answered, "Because they blink and move." The playful humanoid (*Homo ludens*) has fun doing unusual things and accomplishing nearly impossible tasks. We are also very good at role-playing and act often according to a script, e.g., when we conform to the behaviors of the group we happen to live with, or if we take a specific role in a defined interaction between a small number of people. For instance, in a family, the father, the mother, and each of the children will take on a specific role that works in interactions with the other roles. One member might be the serious one; someone else might be the joker of the bunch. One might be responsible for healthy social interactions between the family members; another might be the peace builder or the disruptive element. Research even suggests that the birth order—whether a child is the youngest or the oldest of her siblings—as well as gender have a profound influence on the development of the individual child.

Homo ludens is a social animal with a fine sense for interaction. While the playful element should not be forgotten, it is important to remember that most games follow rules as well. *Homo ludens* seeks fun and entertainment and *Homo faber* often acts out of sheer pleasure. But humans also can use their abilities to work toward different goals, for example, to just make money. We are a greedy bunch and work selfishly toward our own economic and commercial success; that is why economical theories that are based on the metaphor of the *Homo economicus* are applied so successfully to market analysis.

Finally, theologians talk about *Homo religiosus*, the religious humanoid, which aims to be spiritual and to act in a meaningful way. *Homo sapiens* is at her best when involved in a spiritual enterprise. Many scientists think the metaphor of *Homo religiosus* unscientific and irrelevant. But history, cross-cultural studies, and anthropological paleontology indicate that with the rise of *Homo sapiens* there have been no groups or societies that did *not* have a religion and a story of their origin that would place them into creation and in a special and unique spot.

However, religion seems to be rooted into our system even deeper than we think. Our fellow primates sometimes have their own ritual-istic behaviors. Jane Goodall, the eminent and first female primate expert, has a unique way of looking at our closest relatives. Unlike many of her contemporaries, she embraces similarities between them and us. I once heard her talk about a rain dance in a chimp tribe in which male chimps perform a dancelike ritual before a thunderstorm. I don't know if I would go as far as she does and call that behavior spiritual. But I certainly do agree with her findings, which seem to suggest that the phenomenon of ritualistic behavior has been in our primate species for a long time.

The concept of *Homo religiosus* is also supported by brain stud-ies. Neurologists have discovered that a special part of the brain is connected to religious experiences. When people pray, this part of the brain is particularly active and if, in turn, the neurons in this part are artificially stimulated, the subject reports a religious experience. Other researchers have discovered brain changes that occur whenever meditations and prayers are performed. These brain changes helped the subjects to function better; they were relieved from a feeling of stress and could concentrate better.

In 1997, this research was reported in the popular media and cre-ated a nationwide discussion; the media dubbed the brain area in which the activities occur the "God module." Reactions from re-porters were split. Half of them stated happily that now we know how God has enabled us to communicate with God; they saw the God module as a sure sign that God is real. The other half, however, felt the exact opposite. Here, they stated, is the final proof that religion is bogus. There is no God anywhere; it's all in the brain.

The researchers who discovered the God module protested against both of these views; they were adamant that their findings proved only a correlation between brain activity and religiosity, which is per-fectly natural. We are bodies and, therefore, everything we experience has to manifest itself also in the physical realm. They said it was in-appropriate to draw conclusions about God's being from these find-ings. But they weren't heard. People were much more intrigued to use this research to start again age-old fights about whether there is a God. Of course, this discussion is worthless, as there will never be any concrete proof one way or the other. But what the media responses showed was that, first, there is a lot of interest in the phenomenon of religion and, second, we always try to find seemingly scientific proof for what we believe. Faith is defined as what we can't know but trust in anyway. The discussion around the God module shows that this understanding of faith has been lost; most of us need proof to have faith. If anything, this discussion throws light on how much *Homo religiosus* is in us insofar as we all yearn for faith, for something to believe in.

The metaphors just described tell stories about specific aspects of human nature, such as the desire to create. It is appropriate, therefore, to introduce yet another metaphor for our species of humanoids that refers to our universal trait to tell stories: *Homo narrans*.

Humans have a need to tell stories, to make sense of the world. These stories help define us, help us to discover who we are, to create community. In fact, each of our *Homo* metaphors is a story in itself; it tells the story of a human being in a very specific context in which one feature is most prominent—our upright walk or our cognition. But, as the discussions about *Homo religiosus* show, humans today are not very good at accepting stories as stories. They need empirical evidence, they need proof, and they need verification of the facts. The attempt to understand the Bible as a factual, scientific book is a result of this way of thinking. We have unlearned our ability to understand metaphors as pointing toward a deeper meaning, telling stories. Instead, people today tend to understand them literally.

Most economic theories are based on the story of *Homo economicus*, the self-centered, pleasure-seeking humanoid. Predictions under that model can be very precise and can help us to understand economic tides. However, this does not mean that we are entirely egotistical and utilitarian. No one can deny that we all have a *Homo economicus* inside ourselves; it *is*, after all, one important aspect of us. But it is just one of many contradictory aspects of humans, and to say humans are nothing but self-centered and egotistical is simply wrong. We are also playful. And sometimes wise. If one were to take any of the *Homo* metaphors as absolute, we lose the richness and complexity of our self-understanding.

So storytelling is not without its own problems but it is our primary feature, the main method for interaction, and what we do best. Humans tell stories in every context possible: Stories are used to teach about the properties of our environment, and to make sense out of our perceptions and experiences; they help us to explain ideas and abstract concepts and they help us deal with the incoherencies of this world.

Jesus used stories to reveal his religious message. To the great grief of New Testament scholars, these stories are not coherent; there is no coherent body of rules, concepts of the world, of God or of Christianity. Every story of Jesus is embedded in a different context, told in a different situation, retold by different schools of thought. The Christian tradition usually accepts these stories as equally valid and attempts to preach all stories over the course of the years. They are never perceived as wrong because of their incoherencies; they refer to human life, which is always ambiguous and chaotic. The narrative form of the gospels reflects the nature of *Homo narrans*.

Homo narrans in the World

All cultures have a large body of myths and rites that define them and distinguish them from others. The words "cult" and "culture" share the same etymological roots, the Latin noun *colonia* (settlement), which derives from the verb *colere* (farm, cultivate, inhabit, honor). This connection implies that humans who live in a group create a community that is held together by common beliefs and rites.

Indeed, when we follow archaeological, anthropological, and sociological studies, we recognize that the trend to create community through rites and myths is universal. Further, all cultures have developed a treasure of stories and narratives that are memorable and can be visually and theatrically represented. *Homo ludens* comes right into play here as the term *clown* also derives from the root *colere*. All cultures have institutions or wise people who teach the body of stories embedded in a culture to the young and, thus, create cultural continuity between the generations.

Symbols are a more advanced form of narratives. A symbol brings two very different spheres together, two realms that usually have nothing to do with one another. Take, for instance, the flag of the United States, which can be seen everywhere since 9/11. Physically, it is a piece of red, white, and blue cloth with a pattern of some stars and some stripes. But this piece of cloth participates in a completely different, deeper, and more abstract concept: the idea of a nation in which many people and many cultures live together as a whole.

Another culturally important symbol is the cross, two laths nailed together in a ninety-degree angle that participate in the faith statement that Jesus was the Christ, the Messiah who, with his death and resurrection, promises us a reality without sin.

Symbols demonstrate impressively how humans are deeply narrative, as there is no language without symbols. In fact, language consists of symbols, because every word, created as a specific combination of letters, participates in the meaning of the word. Human thought can be expressed with language; therefore, it always has a symbolic character. *Homo narrans* always throws things together so that even if a connection is not immediately obvious, it becomes clear through societal consensus.

Gestalt images are a good illustration of how *Homo sapiens* is driven to create symbols and operate as storytellers.

In this famous gestalt image, we can see either two faces or a vase; we can, however, never see both simultaneously. Our perception apparatus oscillates quickly between the two interpretations of the image, and yet only both together give us a glimpse of the whole.

A group of people might agree that it is impossible to decide what it represents and let it go. But it seems to be very difficult for humans to live with this form of ambiguity and insecurity. Therefore, people can reduce the image to representing either a vase or two faces but never both together; they might even reject people who think differently, as other perspectives might challenge them.

The rationale for returning the image in a certain way could be reinforced by stories that justify why one perspective is correct for a specific group. For instance, one group might say that whoever sees faces is just imagining things, as they know it is a picture of a vase and the faces are just an illusion. These stories usually contain an element that explains the ambiguity of the picture and defines in what situation you will see one image or the other. In any case, the people within one group will very likely go with whatever narrative their group creates to make sense out of this ambiguous sight and will reject other perspectives.

Stories can define groups; belief in a story identifies an insider, and disbelief an outsider. The very ambiguity of our perception apparatus and our incapability to live with incoherencies explains in part why interactions between members of groups with opposite stories (it's a vase or it's two faces, respectively) can be hostile. Any opposition challenges our narrative worldview.

I experience a concrete example of this with every freshman class at my school. St. Bonaventure University (SBU) is a small school in a rural area that attracts mostly kids from the region; however, there are also kids from farther away. In upstate New York, people refer to nonalcoholic carbonated drinks as *pop* while other Americans refer to them as *soda*. Every year, there is a fight between those freshmen who say *pop* and those who say *soda*, each of them convinced that they are right and the other is wrong. Some people try to compromise and come up with something such as *soda-pop*, but this usually doesn't work. The question is why the students get so emotional about such a trivial thing. And here *Homo narrans* comes into play.

The identification with the regional *pop* is a protection against outsiders. The students who come from western New York State usually grew up there and identify themselves with the region and the school. Others might see SBU as a stepping-stone toward other things and might even look down on the people from the area. There might also be a slight insecurity of people in the region, as they often are confronted with the prejudice that they are too rural and naïve to be taken seriously by the rest of the state. So the seemingly trivial fight over *soda* or *pop* has significance for the identity of both people from the region and from outside. The identification with one term or the other will put each student in a specific group with its own stories and references.

In these interactions of groups, individuals, and the larger community (i.e., St. Bonaventure University), several stories will be created and each of these stories will define specific groups, such as the regional-student *pop* group or the outsider-student *soda* group.

The gestalt figures add another level of complexity to this interaction. We can imagine how a dispute over faces versus vase, as between *soda* versus *pop*, might work. First, there is the human who looks at the figure, her perception apparatus oscillating very quickly between the images and her discomfort with this ambiguity unresolved. Then, there is the outreach for a group that identifies with her discomfort but has also reached some form of consensus about how to deal with it. Then there is the story that overcomes ambiguity by reducing the image to either two faces or a vase, and that declares the other perception as illusion. Finally, there is the dispute between members of opposing groups. To ignore even one of these elements or their related interactions reduces the whole perception process to the mere looking at the picture and loses the richness of the process.

One of the most common examples for the manipulation of visual input in the interaction between brain and eyes is the human desire to detect patterns. Many experiments have been done to demonstrate how capable we humans are of constructing patterns in a visual input, which consists of nothing but a chaotic, random collection of colored dots. Gestalt images make us aware of this part of our perceptions. We cannot help but see what the creators of these images intend us to see, and sometimes we are not even aware that the picture manipulates our visual input and our interpretation of it. Most gestalt images rely on hardwired functions and learned mechanisms in our brains that have been developed over the course of human evolution.

That we are storytellers can be seen as a consequence of this functional setting; as *Homo narrans* attempts to make sense out of the world around her, she will turn chaos into patterns and will ignore incoherent aspects of the input she receives. The image on the following page serves as an impressive example. It shows three circles that are part black and part white. Yet we cannot help but construct a triangle

out of the pattern of white in the circles. *Homo narrans* will construct her perceptions, her expectations, and her experience within her community. And all of her community members will create stories within their own lives where these three aspects come together.

This book attempts to open up spaces in which we can create stories about the world that are not necessarily coherent but fulfill different and equally important functions for human life. The world then can be rediscovered as a chaotic, exciting, and wonderful place. It will never make complete sense to us, since we are part of it and create and shape reality with our stories every day anew.

Sin: The Human Incapability to Deal with Paradoxes

The Bible is well aware of the human unwillingness and incapability to deal with ambiguities and calls it "sin." The term *sin* addresses our estrangement from God, from others, and from ourselves. This estrangement is caused by our partial knowledge and our desire toward coherency. Even if "sin" in everyday language is used for supposedly bad acts, in contemporary theology "sin" is usually translated with "estrangement."

This understanding of sin has its origin in what is usually called "the Fall" (Genesis 2–3), when Adam and Eve were kicked out of the Garden of Eden. Traditionally, the story is told as follows: Adam and

Eve live in the Garden happily, in unity with God and all other creatures. They can go wherever they want to go; they can eat and drink
whatever they want. In the whole Garden, there are only two trees
from which they are not allowed to eat, the Tree of Knowledge of
Good and Evil and the Tree of Life. One day, Eve is tempted by a ser-
pent to eat from the Tree of Knowledge (as it is usually called), and
thus acts against God's explicit wish. She also convinces Adam to
have a bite and therefore they have to leave Paradise. The sin here is
disobedience against God's command. Unfortunately, it has often led
influential church fathers to dismiss women as temptresses that lead
men to sin. One reason for the low number of women in church hier-
archies and the fact that some denominations still forbid women
from being ordained certainly lies in this interpretation of the story of
the Fall.

But is this really the only way to read this story? Paul Tillich suggests another interpretation that might have more explanatory
power—especially in the context of gestalt figures and our incapability to deal with ambiguities. As Tillich points out, in the Garden of
Eden there was no ambiguity. Everything was in perfect unity and
harmony. According to the biblical narrative, the ability to distinguish between good and evil and to make universally valid judgments
is God's and God's alone. Human knowledge is always partial and incomplete; therefore, we cannot achieve the full divine understanding
that would explain unexplainable phenomena and would make the
world coherent. We do not have divine knowledge and therefore we
are not capable of judging; we are not allowed to eat from the tree of
knowledge of good and evil.

But the humans in Paradise also had no will. They didn't make decisions. They did not think, compare, categorize. When they ate the
forbidden fruit, it was the fruit of the tree of *knowledge of good and
evil*. Before this, they did not know what was good and what was evil.
They did not decide what was right or wrong. They just were. But
with the fruits of the forbidden tree came the human capability to err
and to make mistakes. Like a lion that kills a deer without any regret
or sympathy for the killed animal, so Adam and Eve were living each
day without tomorrow, without reasoning, particularly without ethical decision-making.

But if humans do not make decisions and judge and use reason, they do not use what is potentially given to them. Humans who do not try to understand themselves and the world around them, who do not try to make judgments, are in a state of what Tillich calls *dreaming innocence*. To actualize our potential we have to try to become like God (in the language of the story, to eat from the tree) and thus risk the close relationship with God. Since humans are in most ways like other animals, they are limited in time and space. This makes every judgment incomplete. With the act of eating from the tree, humans risk error, incomplete knowledge, and false judgments. I want to point out here that in the figurative language of the Fall, it was the woman who ate first; thus it was a woman who had the guts to enter into a state of estrangement. This presents a unique contrast to the traditional reading.

Each and every one of us makes decisions depending on our upbringing and our (largely) inherited values. None of us can make judgments that are correct in all cases. The story of the Fall tells us that it is impossible for us to only make ethical judgments that have the potential of being valid for *all* humans despite their circumstances.[2] We humans can only strive toward a community in which all judgments are generally applicable and, thus, not hurtful or exclusive, but to create such a community is very, very hard. With most of our judgments comes the rejection of people who think and feel differently. In a way, one could see the reactions toward gestalt phenomena as a result of Adam and Eve's eating of the fruit of decision-making and judging.

There is also something inherently tragic about decision-making. If we say the vase–face image is just a vase, we lose the richness of the experience. Seeing things only one way removes the alternatives and thus makes the world a little more boring. Whenever we decide *for* something, we automatically decide *against* many other possibilities. If we pick one path to walk on, we decide against all the other possible paths and we usually cannot return to them. This cannot lead, however, to our avoidance of making decisions. This would put us right back into the state of dreaming innocence. Instead, we have to

---

2. One concrete example for the application of this principle would be suicide. Even if I judge that it is the right decision for me, I could never claim it has any universal value! If every human being would commit suicide, that would be the end of our species and, therefore, unreasonable and unacceptable.

make decisions every day of our life and accept the inherent loss of other possibilities. Thus, sin as estrangement from ourselves, from others, and from God is a tragedy and yet also makes us the beings we are.

After Adam and Eve eat from the tree, they start to live with ambiguities and decisions about right or wrong, which are not universally valid but culturally dependent. In the biblical narrative, this is expressed in the section right after eating from the tree, when Adam and Eve look at one another and recognize their nakedness. In the state of dreaming innocence, nakedness was not shameful or evil but just was. Only with the capability for reasoning and decision-making can Adam and Eve decide that nakedness has to be covered. It is their worldview that leads to this ethical judgment; in other cultural contexts, the decision might be different. Nakedness is not just to be without clothes but to be revealed, to be uncovered. Nakedness means vulnerability. Therefore, one might say that when Adam and Eve started to judge, they were able to understand the other through empathy. At the same time, they were also able to hurt each other deeply through judgment. The clothes are not just a protection from shame but also from vulnerability and from being known fully. With judgment also comes the insight that none of us is a perfect being. If we are honest with ourselves, we will realize that each one of us can be mean and cruel. So we don't want to be fully known, and we hide ourselves in order to not be judged.

This interpretation of the Fall creates multiple layers of meaning. This story is much more complex and rich than the one of Adam and Eve's misbehavior. It provides an explanation for an inherent human dilemma. Adam and Eve want to understand the world and formulate theories about it, while at the same time they know that these theories are always limited, time dependent, and thus incomplete. It also gives an explanation for the human drive toward a whole and complete truth and why it often happens that humans take partial knowledge as absolute (they have eaten from the tree and feel like God).

St. Paul talks about our sins only if he quotes liturgy from early Christian tradition; in his own theology he always uses the singular form. For St. Paul, the term *sin* does not refer to specific actions but describes a general human state of being. We do not sin; we are in the state of sin. *Sin* describes humans as being estranged, ambiguous, torn between incoherent wishes and desires, torn between polarities and fears; *sin* means a life in paradoxes. Sin, therefore, is *not* human wrongdoing. Sin is the state of living in ambiguity that leads to incomplete knowledge and flawed judgment and, thus, to wrong or at least questionable acts. Sin is a state of being, and bad actions are a *consequence* of sin.

Consider, for instance, the Roman Catholic folklore[3] of the seven deadly sins—sloth, greed, envy, gluttony, wrath, pride, and lust. They are supposed to be forbidden to restrain people and make them live in a seemingly "Christian" way. But when you take a closer look, you will see that all these sins can actually lead to estrangement—from yourself, your body, your neighbors, nature, and God. Take gluttony, for instance. Most of us indulge in gluttony every Thanksgiving and probably on Christmas as well. It feels good sometimes to be really, really full. But what is meant by gluttony is not the casual overeating on holidays, as this is fairly harmless. We humans are not very good at doing things in moderation. Many people are in danger from overeating constantly. Eating can be a cure against stress and loneliness. Several things will happen in such a case. The people who overeat will gain weight and thus challenge their health. Also, because in our society slimness is often considered beautiful, severely overweight people are often ostracized and marginalized. They sometimes concentrate more on their desire to eat than on other things. Finally, they often lose self-respect. In other words, gluttony can lead to estrangement from your body, estrangement from others, and estrangement from yourself. And many people would say that if you are estranged from all these things, you are also estranged from God. So the problem with gluttony is not that it is forbidden. Rather, the danger inherent in gluttony is that it can lead to estrangement.

Those actions that Roman Catholic folklore or any other denomination deems as "sin" are usually deeds that can lead to estrangement. The deadly sins are, in moderation, part of being human; only overindulgence will lead to estrangement, and if we are aware of this,

---

3. These "deadly sins" are not part of the Roman Catholic catechism; instead, they have been developed by the people themselves and, then, have become part of Roman Catholic folklore.

we can avoid this estrangement more easily than we could without that insight.

How Humans Bond with Machines

Imagine yourself sitting in a lab in front of a computer and talking to it. You chat with it, and it reacts and asks good questions. Your conversations become deeper and deeper and you finally find yourself discussing with the computer private and intimate problems. You write down how your problems with your partner have started to threaten your career and how daily problem-solving is demanding too much energy from you. You tell the computer that your partner is a control freak, that you are secretly drinking, making yourself more dependent and helpless without any way out.

The computer seems to express sympathy, asks specific questions, and empathizes with you, and you have the feeling the computer understands you better than anyone else. But the underlying program depends on an outdated therapy model in which the therapist just mirrors the comments of the client and turns them back to the client in the form of a question; e.g., "I hate my father!" "Why do you hate your father?" This very simple behavior gives the computer user the feeling that the objective and rational machine sympathizes with her, so she cannot be all wrong.

Does this scenario sound unlikely to you? Well, these very scenes have occurred since the 1960s and are happening today all around the Internet-accessing world. The first program of this kind, ELIZA, was programmed by Joseph Weizenbaum, then professor of computer science at MIT, who was a major inspiration for my doctoral work. ELIZA was based on the mirror method and had, in addition, a couple of standard questions the program could ask if it didn't figure out the keywords of the client's comment. When I met Joe in Germany, I was studying theology and just finishing my degree in computer science. I heard him giving a lecture on the dangers of computers. He told the story of implementing ELIZA in the '60s and then how he observed some of his graduate students, who should have known better, using ELIZA as their own personal therapist; later, versions of

ELIZA were and are frequently used in some initial psychological analyses.

We can argue that this sounds very naïve and that the computer-savvy users of today would rarely fall into the same trap. Well, this optimism is misplaced.

In my travels, I met Cliff Nass, a sociologist at Stanford who analyzes how people interact with machines. Have you ever screamed at your computer or kicked it? Have you ever complimented it? Cliff wanted to find out if there is any evidence that we treat computers differently than typewriters, if there are any emotions involved when we interact with our desktop.

For one of his earliest experiments, he asked several people to test a computer-learning program that was supposed to be introduced into elementary schools. The program was very bad. Some of the testers were computer specialists and some were laypeople. After they had tested the program for a while, the computer on which they worked asked them to evaluate its performance. For the most part, people responded positively. Afterward, these same testers were led into another room with other computer terminals and were asked to evaluate the learning program again. Here, on these different computers, their answers were less positive about the quality of the tested software but they still sounded somewhat satisfied. Finally, a human asked the testers for their opinion on the software and the testers were very negative about it. Such a program should never be used in school, they said.

Interestingly enough, the testers had not voiced these criticisms to either the computers on which they had tested the program or the computers on which they had done the second evaluation. These same people, when asked if they would ever be polite to a computer or think they could hurt its feelings, vehemently rejected such a notion.

This experiment suggests that somehow we seem to apply our rules of politeness to nonhuman entities such as computers. Obviously, the participants in the experiment did not want to hurt the computers' feelings. They even assumed a level of kinship between different computers and, therefore, applied similar rules of politeness to the computer on which they did the second evaluation. They didn't tell these machines their true, very critical opinions, either out of the desire to not hurt the feelings of the second computer by criticizing one of its "fellow" computers or because they assumed some contact between the two so that the second would tell the first what had been said. It seems that somewhere during our interactions with a computer we start to assume that a computer is as sensitive as a human being. Therefore, we behave politely and don't want to criticize it openly.

Let's look at another experiment that seems to imply that people bond with their computers. In this case, Cliff placed people and computers inside one room. Half of the computers had green monitors while the other half had blue monitors. Half of the people wore green arm badges; the other half wore blue ones. Together, they all played interactive games, and it turned out that the people with blue arm badges were much more successful using computers with blue screens to reach their goal than using "green" machines. The same, of course, was valid for the other side. So, slowly, the people with green arm badges bonded with the green-monitored machines and the blue-badge people with the blue-monitored machines.

And now comes the surprising result. After approximately half an hour, the people wearing the blue arm badges expressed more solidarity with the computers with the blue screens than with the humans with the green arm badges; the same was true for the humans with the green arm badges. It seems that through the interactive games and the experienced benefit of interacting with the machines with one's color code, the color code took over as a definition for "my" group. The entities with the other color code, no matter if humans or machines, tended to be rejected. Remember the fictive fight between two groups in front of the vase–face gestalt image, with one insisting that the faces are an illusion and the other insisting the opposite, that the vase is an illusion. What happened in this second Nass experiment is that same thing, only this time the groups had nonhuman members and the group was subtly forged through the interactive games.

This seems to imply that humans bond with the entities of their own group, whether they are human or not. As we will see later in the book, humans do not have a sense of kinship with other humans "built in." It is not part of our biological makeup to automatically treat all humans better than all other beings. That means we bond easily with nonhuman entities and, therefore, we bond easily with our computers.

Human beings are social mammals. Most of us seek places where we can meet other people. We go to bars and sports events and demonstrate with other like-minded people. We naturally want to be with other people. And, as Cliff's second experiment shows, we seem to be able to accept anyone or anything into our group with whom we can sufficiently interact. As soon as such a stranger is accepted into a group, he or she is seen as an equal part of the group. The group defines itself by the entities that both belong and do not belong to it. As we all know and perhaps do ourselves, people treat their cars and stereos as people as well. In a way, it saves a lot of time and energy to do so. After all, humans are educated from birth on how to interact with their fellow human beings. It is necessary for a baby to be able to do so, as her survival depends on it. Throughout our lives, we learn patterns of behavior—such as being polite and not openly criticizing someone. It is very easy to apply these ingrained rules to every entity we interact with. It is very hard *not* to do so, as it demands a conscious effort of us.

This behavior—treating nonhuman objects as if they deserve some form of politeness or regard and are somewhat like us—is called *anthropomorphism*, the human capability to morph/change everything into a human (*anthropos*) and treat it accordingly. Usually, the term has a slightly negative connotation. Theologians especially often criticize human terms used to describe God, such as *shepherd* or *father*, or the classical image of God as an old, usually Caucasian man with a long white beard. Cliff and his colleague Byron Reeves, however, suggest that *anthropomorphization* is actually the initial and natural response to anything we interact with; it takes a conscious effort *not* to anthropomorphize. As social mammals, we are best when we interact and any use of these trained and built-in behaviors is easy; anything else is hard.

Cliff's experiments reveal something about how we relate to one another. We innately work toward community. We work toward social interactions. We work toward groups. We work toward bonding and solidarity with the beings closest to us. And computers are great thinking tools we can use to explore our mechanisms for bonding and social interaction.

The First Humanoids: What Happens When We Do Not Bond with the Creatures We Create?

Humanoid robots are fascinating machines and there are many science-fiction movies and popular stories that deal with them and how they shape the way we live and think about ourselves. Our culture is filled with stories about humans who lose power over creatures they have created. In the context of computers and robots, it is not just HAL from *2001: A Space Odyssey* that raises these fears in us. There is another, even more powerful story that had a major impact on our reactions to computers and particularly to robots.

Mary Shelley's *Frankenstein* is about a monster that became cruel and turned against his maker, his family, and society. Popular culture today sees *hubris* as the main motif of the story. The concept of hubris comes from Greek mythology and addresses the capability of humans to overestimate themselves. It is not just arrogance but more the feeling that humans can do everything. In antiquity, the prototypes for hubris were Prometheus and Ikarus. In *Frankenstein*, the prototype of a hubristic human is the scientist Frankenstein who builds a creature that ultimately will go against its creator and his/her family and utterly destroys them. This motif touches a deep fear in us: that our own creatures will turn against us. This novel has influenced our thinking about robots so much that authors such as science-fiction writer Isaac Asimov called the fear of humanlike machines the "Frankenstein Complex."

But I see another story. A young and overeager researcher named Frankenstein builds a creature out of human parts but then gets scared when the creature comes alive. He runs away and abandons it. Thus, the creature is never given a name, never treated as a beloved offspring. It tries to find community but is constantly rejected and feared. It is completely ostracized. But when a humanlike being has no chance to actually become part of a community, it seems reasonable that out of grief and revenge this being might turn against the community that excludes it. The monster's motive to kill did not evolve out of a need to exert its power over others; it stemmed from the feeling that it was not accepted by the community. How could the monster ever develop any form of benevolence toward humans when all it experienced from them was hatred and rejection? It never experienced positive feelings of warmth or kinship. It became brutal because it didn't know what else to do. No one let it participate in any form of human interaction. There was no bonding.

Most *Frankenstein* movies cover only the aspect of hubris and destruction, with the notable exception of the movie *Mary Shelley's Frankenstein*, where the creature at some point challenges Frankenstein. After helping a family anonymously, only to be rejected when it is discovered, it asks Frankenstein, "I can read, I can learn, I can think. Do I have a soul or have you forgotten to build it in?" Here, the creature clearly desires much more than just being self-aware and the capability to understand. It wants to be humanlike, in the sense that it wants to be accepted and loved. It wants to interact. It is so much like us that it wants to be connected to a community. In this story line, *Frankenstein* is about responsibility toward the creatures we create. We have to care for them and treat them well; otherwise, there might be unforeseeable consequences.

The Frankenstein story is, of course, not the only one that talks about the possible outcomes when we create creatures in our likeness. Much more interesting theologically and, thus, much more relevant to this book are the stories about the *golem*.

Golems

The wife of Rabbi Löw could not understand why her husband had forbidden the use of the Golem for private purposes. And when, just before Passover, she was short of help she allowed herself to give the Golem orders to fill two large water kegs which stood in the kitchen which was all prepared for the holiday. She thought also that a service in preparation for the Passover feasts did not come under the head of secular purposes.

But she had a very unpleasant experience.

The Golem took the pails and ran swiftly to the brook.

Several hours later the courtyard of the house of the Rabbi was flooded with water, and people were crying: "Water! Water!" The secret source from which this water was flowing was sought.

But it was not found until the Golem was seen patiently obeying his orders by continuing to pour water into the kegs which had been filled a long time before. This explained the flood and there was much laughter over the Golem's mistake. [. . .]

Since that time the people took care not to give the Golem any profane work to do. To this very day in Prague people say to an unskilled artisan: "You are as competent for this work as was Joseph Golem as water carrier!"

CHAYIM BLOCH, *THE GOLEM: LEGENDS OF THE GHETTO OF*
PRAGUE, TRANS. HARRY SCHNEIDERMAN (BLAUVELT, N.Y.:
RUDOLF STEINER PUBLICATIONS, C. 1972)

The Jewish golem stories go back to the thirteenth and sixteenth centuries. They can be traced back to medieval Germany and Hungary, specifically to the Jewish mysticism, called *kabbalah.*

The term *golem* comes from the beautiful Psalm 139 quoted at the beginning of this chapter, which is one of my favorites. It talks about the beauty of creation and how God knows every one of us and knows everything in us. Even if the psalmist sometimes finds this omnipresence of God somewhat disconcerting, it also gives her deep trust, as she is sure that God will never leave her.

The verb *galam* appears only twice in the Hebrew scriptures. In 2 Kings 2:8 it is used to describe the wrapping of a mantle. But probably the oldest source for this term is in Psalm 139:16: "you created me as a golem in my mother's womb." Here *galam* is usually translated as "shapeless thing" or "embryo." The psalm celebrates creation and the special love and care of God toward humans. God created the psalmist, "intricately woven in the depths of the earth," and in God's "book were written all the days" that were formed for the psalmist. The word *galam* very likely comes from an Arabic root and means "tangle" or "cluster." The medieval kabbalists used this term as the name for the humanoids they constructed.

The most famous golem story is the story of Jehuda Löw, the Maharal of Prague, and his golem, Joseph. The Maharal is a historical figure who lived in the sixteenth century. He was a widely acknowledged theologian and also a political figure. He was a very influential teacher and a very wise negotiator with the Christians and the state representatives to create a decent life for the Jews in the ghetto.

At the time of Rabbi Löw (as, unfortunately, in most of medieval and even modern times), Jews were often attacked by Christians, and the people in the ghetto of Prague were often harassed. So, to add a layer of protection to the ghetto, Rabbi Löw is supposed to have built a golem and put a paper with God's name in its mouth. The golem then became animated and was able to help the Jews in Prague. Joseph supported the Jews with his strength in their daily labor and helped them against attacks from outside. One story describes how Christians would hide dead babies in the ghetto at night and then come back during the day with armed forces, and use these little bodies as proof that Jews would kill babies in their ceremonies. Then, Christians would have a reason to attack the ghetto and kill Jews. The golem is known to have found the babies several times and hidden their bodies so that the accusations became worthless.

According to most stories, golems are built from clay, constructed through words and numbers. Kabbalist theory reveals a deep faith that the world was created by God in an orderly and numeric fashion; the better people understand the logic behind the world, the more they can share God's mind and participate in God's creativity. Thus, they are motivated to construct increasingly complex things to understand God better. But they cannot build anything animated without help; golems come to life only if they have a paper in their mouth with the holy name of God written on it, or with God's name engraved on their forehead. The ultimate power of life is God's and God's alone; God has to be involved to animate an artificial being. So, even if the letters and numbers[4] in Hebrew are orderly and thus participate in the order of God's creation, they are not sufficient on their own to create life. Quite the contrary: The tangle of flesh, genes, slime, and chemistry in the case of the human animal, or the clay in case of the Golem, need the spirit and power of God to become alive.

One could now ask if the Golem is estranged, a sinner. According to most stories, the Golem has no language and thus cannot participate in the categorizing, describing, and reducing of facts. He also has no sense of right or wrong—otherwise why would he flood the Rabbi's house? But there is a small potential that he also needs God's forgiveness. Rabbi Löw himself was never sure if the golem was a

4. Letters in Hebrew are also numbers—there are no special numeric signs. This gave the kabbalists the opportunity to create word and number riddles far more complex than our Western languages allow for.

child of God or a mere machine. Since the Maharal cannot be absolutely sure, Jehuda Löw addresses this doubt by forcing the Golem to keep the Sabbath. Every Friday, the Rabbi would remove the animating paper with God's name on it from the Golem's mouth so that it went back into its unanimated state, thus keeping the Sabbath.

One week, however, the Rabbi forgot to remove the paper slip and the Golem, without his master, went berserk. Rabbi Löw saved his fellows of the ghetto by fighting the Golem and, after considerable violence, he was finally able to remove the life-giving paper from the Golem's mouth. In some versions of the legend, the dying Golem falls on the Rabbi and smashes him. These endings refer to the motif of hubris, as often presented in Greek tragedy and also in the Frankenstein story, where the constructors of gadgets and creatures that overcome human limitations are killed in the end.

Golems as Prayers

What is prayer? Is it to ask God or another deity for something you want, such as money or health? To have a conversation with yourself? Meditation? Do we pray when we ask the deity for the strength to accept the situation we are in, to endure what we can't change in our lives?

In the Jewish context in which the golem stories are told, prayers are usually spoken to celebrate God and God's glory in us. Prayers are communal; they strengthen the bonds of people in their community, with each other, and with God. People speak prayers to express their anger, fear, and frustration with current situations, as well as to speak about their joy in life and their happiness to be God's creation.

With the construction of golems, people felt they learned more about God's creation of humans and their special capabilities. The golem builders felt that by building golems they were participating in God's creativity. They argued, God has created us in God's image so that we participate in God's creative capabilities. Whenever we are creative, we actively participate in God's creative powers and celebrate God. In this sense, every act of creativity is a prayer. And the more complex things we build, the more we praise God. Humans are—at least as far as we know—the most complex beings on earth. Therefore, if we rebuild ourselves in golems, we celebrate God's "highest" creative act, the creation of humans, thus praising God the most.

Even if one ending of the story of the Maharal in Prague refers to an element of possible danger, the majority of the golem stories are not about the motif of hubris as the Greek stories or the Frankenstein myths are. This is supported by a vast amount of rabbinical literature that discusses golems. The majority of these stories do not understand the construction of golems as a step beyond the boundaries God has set for us or as hubristic acts, but they understand golem-building as prayer.

Golems can be helpful servants, but their creation has a spiritual purpose beyond building useful machines. It can be an act of worship. It's not trying to dehumanize the human experience or deconstruct the mystery of what it means to be human. It is to praise God. This, of course, links these golem stories with the modern scientific construction of humanoid robots.

While I was at MIT, the group that was constructing Cog and Kismet had just made the jump from insectlike, six-legged walking robots to humanoids; while they all enjoyed trying to do the impossible, most of them were also well aware of the challenging nature of this task. Everyone was enjoying him or herself trying to turn science fiction and mythology into reality. But many also felt a nudging doubt. After all, they had had experiences with the rebuilding of insects. And even if the six-legged creatures they had built were fantastic robots, they came not even remotely close in their capabilities to real insects. Now, they were building a humanoid and realized that they might be out of their league. Each week for the first year of our work, we invited a specialist to tell us something about how the human system works. We would read their articles beforehand and pepper them with questions after, and one question we asked every single one of them was: How will this knowledge help us to build our humanoid? The more we learned, the more our respect grew for the incredible complexity of the human system. It is one thing to read the Bible to learn that we are the "top" of creation, but it is quite another to learn the facts about this complexity. We became admirers of ourselves, but that didn't make us arrogant. Quite the contrary, we now had a healthy respect for the task we had set ourselves, to build a system that has, however remotely, some similarity with us. Building Cog and Kismet made us modest in our admiration for God's creation. Probably no one except me in the team would formulate this feeling with the same religious words, but the sentiment was exactly that. Modesty, admiration, and a humble attempt to do our best is all that we can do to make it work.

This experience has taught me the truth of the view of the kabbalists to see the construction of humanoids as a worship of God.

It also says something about God's creativity in us that we don't stop our projects when we discover how difficult they really are. It takes a healthy self-confidence to attempt the impossible, and even if you fail, you have tried. Cog and Kismet, though both now in museums, have contributed hugely to the field of robotics, as they were the first emotional and social robots. Even if we have the capability to bond with most creatures with whom we interact, we bond most easily with those creatures that seem to have emotions that we understand. Anthropomorphizing happens with dogs and cats and other mammals because their basic similarity to us motivates us to create stories that ascribe to them the same emotions we have, and to bond with them strongly. Cog and particularly Kismet evoke similar feelings in us and thus have moved us closer to the eventual fulfillment of the old human dream of rebuilding ourselves.

One of the oldest golem stories tells about Jeremiah and his son who built a golem with the words *JHWH elohim emet* (God the Lord is truth) on its forehead. As soon as it came to life, the golem erased the letter א (the *aleph*, the first letter of the Hebrew alphabet) from the word truth so that now his forehead said *JHWH elohim mot* (God the Lord is dead). He then explained to his terrified builders that we adore God because God has created us, the most complex beings there are. If we are now able to re-create ourselves, people will adore the constructors of golems and not God anymore. But a god who is not adored and prayed to is dead.

One has to be honest and admit that this aspect of golem construction is always present. Even if we become modest and admiring of creation, there is still often the sense of demystification. If we learn a scientific explanation for a unique human capability, for example empathy (see chapter four), we are tempted to lose respect for this capability, as it seems to be just a reactive mechanism. The construction of humanoids certainly contains an element of hubris. The pursuit of an impossible task produces pride in one's own work and in the triumph of overcoming limitations, to succeed when no one in the scientific community thought it possible. There is nothing wrong with that. Quite the contrary, those elements are normal in any creative endeavor. But there is always the danger of losing respect for the human system because when you build robots and they do what you actually want them to do, you sometimes can't help but think that ultimately every part of the human system will be understood and, then, can be rebuilt.

The construction of humanoid robots is motivated by the wish to understand ourselves and to build partners with whom we can talk and interact in a meaningful way. If they challenge our self-understanding or seem threatening to us, this is not the creatures' fault but the outspoken hubristic aspects of this project that are sometimes part of humanoid projects.

Theology and Artificial Intelligence meet when we attempt to understand ourselves, who we are, and what our role in this world is. Engineers and computer scientists involved in Artificial Intelligence, as well as theologians who (re)construct stories of meaning, are creative. Both can be described with the metaphors of *Homo faber* and *Homo narrans*. The first attempt to rebuild themselves; the others tell stories. That is, both contribute equally to a fuller understanding of ourselves. They do professionally what many humans do in their spare time: attempt to answer the big question of what it means to be human.

Golem Builders Today

In true rabbinical fashion, there is more than one version of the story of Rabbi Löw and the golem Joseph. We know already about one ending, where the lifeless golem, without the animating paper in its mouth, falls on the rabbi and kills him. We have seen that this ending shares with the Christian tradition the element of hubris present in many projects of humanoid construction. I would like to focus now on the tradition that tells the story of the survival of the rabbi. In this version of the legend, the golem is put to rest in the attic of the synagogue in Prague. Rabbi Löw then creates a kabbalist rhyme that will revive the golem at the end of all days. Many Jewish children from this tradition were taught these words.

This version of the kabbalist golem legend is still strongly ingrained in the consciousness of many Jews from the Eastern European tradition. When Jewish boys descended from Rabbi Löw were bar mitzvahed, they were usually told the formula that will revive the golem at the end of all times.

It seems that many of the early AI researchers are or claim to be descendants of Rabbi Löw, but it took a coincidence to find that out. MIT is the cradle of AI; here, the field of AI was born and here the first steps toward artificial intelligence were taken and the first successful projects developed. In the late 1960s, when some students sat together on a break, someone mentioned that the first big computer in Rehovot, Israel, had been called Golem. This led to a discussion and it turned out that at least two students in the community had been told the rhyme that would awaken the golem. These two were Gerry Sussman, today professor at the MIT AI Lab, and Joel Moses, the former provost and today institute professor at MIT. When they compared the formulas they had both been told, their formulas were exactly the same—despite hundreds of years of oral tradition.

Gerry Sussman later dedicated his doctoral thesis to Rabbi Löw because the rabbi was the first one to recognize that the statement "God created humans in God's image" is recursive. Recursive functions are self-referential; that is, one cannot derive all values individually but needs the previously calculated values in order to get new values. This dedication captures various aspects of the AI enterprise. For one, God has created us in God's image and we use the same process in humanoid construction as we create them in our image. Modesty and awe come out of humanoid construction, as we can never be as successful as God. We are a derivation of God and our creatures will be the next derivation, our images. To interpret the *imago Dei* as recursive also refers to the aspect of prayer, as we can create only because we have been created in the first place and celebrate our creator who has so "wonderfully made us" (Psalm 139).

It also points out the necessity to re-create ourselves. We are images of God and we have the drive to create, to repeat God's acts of creation. The very desire of God to create humans as partners is inside us. When we look at all the attempts to "speak" with animals, especially dolphins and chimps, and the desperate search for extraterrestrial intelligence, it becomes clear that, for some reason, humans want to interact with beings of a different kind. We want to have a species, an "other," with whom we can interact. We know that many other animals are intelligent but we cannot communicate with them. But there is hope that we can communicate with the beings created in our image. They have the potential to be partners.

Recursive functions have yet another attribute, as one cannot derive a value in a recursive function without having calculated the previous value. Does this mean God needs us in order to create humanoids? Has God perhaps created us for this very purpose? Why else this strong and so deeply ingrained desire to re-create ourselves? These delightful speculations add another aspect to the element of prayer within the golem tradition. When we attempt to re-create ourselves, we do God's bidding. We help God. We are *created co-creators*, a term coined by the theologian Phil Hefner, who has influenced my thinking in religion and science enormously.

One might further speculate that the wish to revive the golem at some point in time is part of the motivation for the whole AI enterprise; this seems especially to be true, as several other famous AI researchers link themselves to this tradition.[5]

In light of the state of sin and ambiguity in which we live, golem creation has a forward-looking perspective. As we have seen, it is doubtful that the golem was living in ambiguity. But here another spiritual aspect of the AI enterprise is revealed. Wouldn't it be wonderful if we were to create an image of ourselves that is not a sinner, and can actually handle the ambiguities of life? Can we actually build creatures that are better than us? Not smarter but morally better?

Those of us who have read the robot stories of Isaac Asimov know that he indeed presented his robots as better people. They could do no wrong, as they had to obey the three laws of robotics.[6]

---

5. Among those people who have been told the formula are John von Neumann and Norbert Wiener, both foundational thinkers in the field of AI.

6. Third Law: A robot must protect its own existence as long as such protection does not conflict with the First or Second Law.

Second Law: A robot must obey the orders given it by human beings except where such orders would conflict with the First Law.

First Law: A robot may not injure a human being, or, through inaction, allow a human being to come to harm.

Asimov presented one of the most powerful examples of our creations that is morally perfect. But in his later books, he realized that these laws of robotics fall short in a universal context. He, then, adds to the three a "Zeroth" law that says "A robot may not injure humanity, or, through inaction, allow humanity to come to harm." In this moment the robotic laws become as ambiguous as human laws, as the term *humanity* is ambiguous. Who belongs to humanity and who doesn't? If one part of humanity decides to kill all the others, can that be accepted?

Many people do not like speculations like these. Many even reject a connection between the golem stories and modern AI. Many people, especially scientists, do not want to acknowledge the existence of such emotional and religious elements in the motivation of researchers, nor do they feel these elements are desirable. However, when we look at scientific enterprise today, we realize that there is much room for researchers to bring in their own quest. Many robot builders bring to the table their desire to see what humans can accomplish with the help of the technology they have available today. Many feel that robot-building can be spiritual, as it taps into God's creative powers in us. Many people might be motivated by the inherent desire to overcome human flaws and limitations by constructing robots that are better or free from sin as estrangement.

Only if we see the enterprise of developing artificial intelligence as purely scientific and ignore all the mythical and emotional elements will we be in danger of falling into the trap of hubris. Therefore, we have to first overcome the danger of seeing AI as a purely rational endeavor before we can actually look at it as a source of wisdom about who we are.

CHAPTER 2

Embodied Science

And GOD said:
"Let us make humankind in our image, according to our likeness; and let them have dominion over the fish of the sea, over the birds of the air, and over the cattle, and over all the wild animals of the earth, and over every creeping thing that creeps upon the earth . . ."

GENESIS 1:26

God at MIT

While I was working at MIT, there were several people who were not all too keen to have me there. Their main protest was that I couldn't possibly be objective; another, more abstract concern was that religion keeps people away from learning new things, as it provides superstitious explanations for empirical phenomena and thus hinders people from asking for other possible explanations. In Germany, theology is one of the major academic concentrations, and a master's degree in theology takes many years of comprehensive study to obtain. So I couldn't quite understand these people's concerns.

In 1996, I suggested I teach a computer science class called "God and Computers" about the philosophical and theological assumptions hidden in AI. The department chair as well as Rod Brooks liked the proposal and allowed me to teach it in the fall of 1997. A few days before registration, I sent an announcement about the class to the AI e-mail list, and a big discussion followed.

Marvin Minsky, one of the founders of AI and still an *"Übervater"* for the field, started with the claim that this class ought not to be taught in the regular curriculum, as it was an "evangelical enterprise." Marvin is well known for his strong opinions and his negative views on religion. He believes that humans are nothing but meat machines that carry a computer in their head. As soon as we have decoded the program that runs on the wetware of the brain, we can download it into the hardware of a computer and live forever. This is the major belief of a movement that calls itself *transhumanism* or *extropianism*, of which Marvin is one of the main figures. It is a movement that perceives the future of humanity in the creation of computers, and robots that will supersede us, a movement that promises a perfect life once we have discarded the weaknesses of our bodies and become one with a machine. At some point, Marvin even claimed that people should give their money to AI research rather than their churches, as only AI would truly give them eternal life. It was clear that he wouldn't like the class even though he certainly misunderstood its purpose.

Marvin had started an avalanche, and there were suddenly hundreds of e-mails within a few days. One student understood the class as "indoctrination." A former Ph.D. student of MIT accused me of "suffering from the set of pathologies collectively known as religious faith." He later supported another student who petitioned the head of the department to cancel the class.

People argued that someone as psychologically deluded as I should not teach at MIT, that MIT had built itself up as the stronghold for objectivity and rationality and against superstition. My course was not an academic endeavor but a missionary enterprise. One professor soothed his colleagues by stating that a single Christian could not destroy the greatness of MIT's rationality. Another faculty member, a strong supporter of academic freedom, sent a message in which he quoted from the worst of the McCarthy era, when communists in academia were oppressed. He wrote this note in my favor, drawing parallels between the critique of me and the former critique of suspected communists in lieu of academic freedom. However, most of the members didn't understand it in this way. They took this quote as support for their ideas and suggested that religious academics should be treated similarly to communist academics in the 1950s.

This discussion lasted for several days and generated more than 250 e-mails, and even made its way into the *Boston Globe*. The people who argued against the class mostly did so because of the G-word ("God") in the title and hadn't even looked at the online syllabus. I was quite hurt at the time. I am an academic who has studied for years the possible interactions between theology and computer science within various academic settings. I could not understand why people reacted so violently and viciously and questioned my academic integrity. But I also received a lot of support from the people with whom I worked, especially from the department chair at that time, Paul Penfield, and Rod. I can say in retrospect that it is quite ironic that so many smart people can be so religious in their rejection of religiosity.

The question is, why did these eminent and incredibly smart people react so emotionally to my presence and my course offering, without even looking at the syllabus to see what the course was all about? Why did they perceive me as a danger? Pop psychology would conclude that such strong emotions could, in part, be explained by fear. Is it fear to have to question one's assumptions?

We have seen that humans, because of their estrangement, are not very comfortable with incoherencies and paradoxes. At MIT, many people are comfortable because they share a specific worldview and hold certain common beliefs—and the rejection of religion for the reasons named above is one of them. In a way, I posed a threat, as I stood for a very different worldview. The looks I received whenever the term *evolution* was mentioned are a symptom of this. As only very few extremist Christians actually believe that evolution and creation are in conflict (and they live predominantly in the United States), the attempt to put me into this camp represented the effort to put me down as a fanatic. If I could be rejected as an extremist and fundamentalist, my opinions were no longer valuable and I could be dismissed out of hand.

However, after years of research I had quite a balanced point of view and could understand and empathize with the worldview many people held at MIT. Slowly, I won respect, and when I left, most people had ceased to be prejudiced against *all* religious people even if they still rejected (and rightly so) all religions' extremist points of view. During the three years of the "God and Computers" project, in which every fall semester up to ten scientists presented their personal take on spirituality within their research, some people at MIT even had their "coming out" as religious people themselves. Shortly before I left, one faculty member whom I very much respect voiced the opinion that I had changed MIT, as it had now become acceptable in some circles to confess one's religiosity without being ridiculed.

Homo objectivus et rationalis

There are several explanations for the behavior that quite a few people at MIT displayed. I have been similarly confronted in the many settings in which I have presented my work over the years. It didn't matter where I was—in scientific conferences, churches, business conferences, or in the classroom. I usually found myself during the first part of the question-and-answer period defending either my seemingly too reduced understanding of the human system, or my too theological worldview. People couldn't categorize me, as I wear many hats: the hats of a scientist, a theologian, a counselor, and a fellow human being who asks the same questions as most people in the room.

My understanding of humans as estranged storytellers leads me to believe that if we are telling stories to make sense of the world around us and within us, then these stories relate to the various parts of who we are, which are sometimes in conflict with each other. We have desires and questions that do require answers—but these answers are seldom compatible. Indeed, our fundamental estrangement leads us to accept stories that, told together, might appear to be incoherent.

At this point, we are touching upon one of the most fundamental assumptions of classical philosophy and science: that the world is a place that can be analyzed and eventually understood by us, as it follows laws of order. I cannot and do not want to refute this belief.[1] But I would like to challenge the possibility for us humans to actually *perceive* the world as it is. As the gestalt figures in chapter one demonstrated, we have a hard time with incoherencies in our perception and

---

1. Many chaos researchers refute this assumption and claim that the universe is principally not one of order but one of chaos. I cannot refute this theory either. But I have to question the possibility of epistemology for this stance, too.

try to solve them through stories. But as soon as we create these stories, we cease referring to the world in itself (*Welt an sich*) and instead describe the world as we perceive and understand it. That means sin as estrangement makes it impossible for humans to perceive themselves and their world as they are, so instead they tell stories. In an attempt to overcome estrangement, we seek to make these stories fit together even if they relate to different parts of us.

Sin presents us with several great temptations we ought to seek to overcome. One is the desire to ignore possible stories that might be true but that are not in accordance with our worldview. We have seen that this is the case with the gestalt figures when people claim one aspect as illusion and accept only the reality and necessity of the other. Another temptation is to ignore parts of ourselves in order to reduce the numbers of explanatory stories and thus increase the chance of their inner coherency. We will see that the oppression of our body as a crucial part of our being can lead to a reduction of incoherencies. It is also common to ignore emotions as the most important aspect in our interactions with the world. In chapter three, we will see to what extent these temptations are against our human nature. A final temptation is to claim objectivity and judge the opinions and views of people who do not agree with us as less valuable; this often leads to the rejection of the people who hold these opinions themselves. Every one of us who gives in to these temptations can be assigned yet another *Homo* metaphor: *Homo objectivus et rationalis*.

What happened at MIT can be seen as an example of the temptations of sin. Those people who called me "psychologically deluded" rejected my reasoning because they claimed that the only valid reasoning was that based on scientific facts alone. However, in most parts of human life, scientific reflection plays only a very insignificant role, as our lives revolve mostly around relationships, conversations, sensual enjoyment, creativity, labor, and role-playing. Therefore, the social sciences and the humanities have at least as much to offer as the natural sciences. Only with all these insights together can we get a glimpse of what it truly means to be us.

These people also claimed that someone who is committed to a specific faith tradition is not rational without conceding that they themselves are as emotionally attached to their own worldview as I am to mine. Rejection of someone with a different set of beliefs does not contribute to a better understanding of ourselves, but it is a temptation of sin.

Insights from Science and Theology

This book attempts to bring together insights from science, engineering, and theology, in order to provide a complex understanding of ourselves. In order to do so, we need to establish a new method that allows for different stories that are equally important to us, stories that relate to one another and interact with one another but that are, nonetheless, not necessarily coherent. We find the foundation for such a theory in Max Horkheimer's and Theodor Adorno's[2] analysis of the negative impacts of the Enlightenment movement. They asked how a seemingly enlightened century such as the twentieth could have possibly created the two worst wars in history; note that they wrote their book after World War II and didn't know yet of later wars, such as in Vietnam, Somalia, the former Yugoslavia, and Iraq. People claimed to be enlightened, to have achieved rationality and have overcome superstition and ideology, and yet they fell for Hitler's Nazi ideology and others.

As we have seen, science, especially AI, is an exciting enterprise with sometimes deeper levels of motivation that scientists often might not be aware of. This human element makes science less accurate than people tend to believe. But, obviously, there are still many people who believe in the objectivity of science. This widely held belief is a remnant of the eighteenth-century European Enlightenment movement and its drive toward rationality, reasonability, and the liberation of people from superstition and angst.

But the Enlightenment was far more paradoxical than we realize. Horkheimer and Adorno use two different realms to analyze how it could go so astray. For millennia, philosophers have believed in two different realms to describe how we experience reality: *mythos* and *logos*. In the *mythos* realm, we tell stories that give our life meaning; here, we address existential questions such as, What does it mean to

2. What makes these two interesting in the context of this book is that both came from the same philosophical tradition as Paul Tillich, the theologian who has influenced my theology the most. Theodor Adorno even wrote his second doctorate (his *Habilitationsschrift*) with Tillich advising.

be human? What does it mean to be me?—the questions around which this book revolves. In the *logos* realm, we aim to make factual statements about the world around us and about us in it.

The Realms of *Mythos* and *Logos*

Within the logos realm, people create statements within a discussion that are open for dialogue. Logos expressions are used when humans observe the world and formulate ontological statements in theories, formulas, and definitions. Logos expressions, therefore, are open for questioning; they require rational analysis and seek situation-independent validity.

The Greek term *logos* is one of the most important terms in Western philosophy and Christianity; it has led to the term *logic* and to the syllable *-logy,* after areas of rational inquiry like "biology," "theology," or "ethnology." It does not just simply mean general speech, but also justification, calculation, and reason. Humans are in the world and describe the world in logos speech acts; they try to understand the world around them and themselves as well.

The Greek term *mythos* in its original sense meant just "utterance" or "story." Mythos expressions, indeed, are used for the interpretation and explanation of human reality; these explanations usually have the form of narratives or myths. The mythos realm supports storytelling to explain our perception and our reasoning about the world, and to provide a system of belief sentences that gives meaning to our being in the world. A mythos interpretation of reality, quite contrary to a logos interpretation, cannot be a topic for rational analysis and discussion. Whether a mythos story is accepted depends on the time and the culture in which it is told. The act of acceptance of a story is not a solely rational one but is in part unconscious and depends on upbringing and the value system of the respective humans. That is, mythos narratives are acceptable and valued only within well-defined boundaries.

Myths often refer to supernatural agencies. They do not prove the reality of these agencies but presuppose them; they then can tell stories about the relationship of the supernatural or the divine with humans. We use myths to address and describe things that are sacred or holy for us. But besides religious myths, there have always been secular myths. They are derived from religious myths but lack the reference to a supernatural agency. Art and literature create myths, nations create myths, and—what concerns us the most in the context of this book—scientists and engineers create myths as well, as it is the nature of *Homo narrans*.

Myths do not stay the same over the course of one's life. A myth that was accepted as valid could become meaningless. A myth is part of human development; any individual life story is tied to certain myths that have been accepted within the single life stages. Every story, therefore, is always a result of the very concrete situation that *Homo narrans* finds herself in; it is a result of development, change, and the interaction of humans and their environment (especially their culture).

The distinction between these two realms goes back to classical Greece. Socrates already used this distinction; he, the rationalist, who thought he could get at the depth and truth of all things when using reason and analysis, also regularly sought the advice of the oracle of Delphi and didn't think that this was a contradiction. He knew that reason and analysis of the world are not very helpful for the entirely human quest for meaning. The ultimate desire for a meaningful life is not something that can be described in a fully satisfying way in logos terms alone. *Homo narrans* needs symbols, metaphors, and myths—stories and narratives—to express her inner need for meaning and directedness. Therefore, many Greek philosophers accepted mythos and logos as distinct but equally important and indisposable ways to understand humans, their lives, and the world around them.

If one looks at the worldview of AI, one can see that AI as a science uses predominantly logos expressions for its research. In order to build intelligent machines in our image, we have to assume that humans are nothing more than machines, bags of skin that can be rebuilt. This assumption and everything related to it is firmly planted in the logos realm. On the other hand, many of the AI researchers with whom I spoke and interacted wouldn't like to be treated as mere machines and usually don't treat their lovers or children as machines. There is no way to resolve this dilemma; humans can't help but bond and act as if the other with whom they have bonded is of enormous value. Any attempt to resolve this ambiguity will lead to an oppression of one part of our worldview. Many people in AI are quite happy to have both narratives—the one of humans as machines and the other of humans as beings with intrinsic value—side by side. Both are equally important for their life, but are applied in different life situations. When in the lab, the machine metaphor holds true, while privately they believe in concepts like dignity and worth.

Another good example of how mythos and logos interact are the golem stories. They are myths and would lose the richness of their meaning without their reference to God. Even if one does not believe in a god, as is true of many of the scientists who understand their own work in the context of the golem tradition, they can still appreciate the richness and complexity of the stories. The kabbalists created golems through mathematical formulas that reflected God's logic and order in the world. The Greeks have the same concept because any logos description of the world is based on the assumption of an orderly universe. Logos expressions are most commonly used in science; they are based on observation and pattern recognition.

Logos statements are supposed to answer questions about the "How?" regarding the reality around us, while mythos expressions are supposed to answer the questions of "Why?" The "how" of anything can be discussed, proved as right or wrong, or demonstrated by empirical or logical evidence. But any answer to "how?" questions, as we have pointed out earlier, is based on assumptions that cannot be proved but can only be accepted in faith. These assumptions form the body of the mythos realm that is closely related to the subject who attempts to understand the world and herself, and attempts to give meaning to her experiences and perceptions. Even if the mythos presents itself as a cosmological or historical story, in the end it has to be interpreted in an anthropological, or, better, an existential fashion.

Any existential "why?" question has to be answered for everyone individually. Whether an answer given by a myth is acceptable for someone depends on her very life situation, her cultural background, her religious upbringing, etc. The mythos realm, then, not only helps to construct answers to the individual quest for meaning but also constitutes a group; myths create and shape communities and reinforce and strengthen the interaction of group members. Members of the group believe in and trust the authority who presents the mythos speech acts or in whose name myths are told. If the myths do not correlate with one's personal life story or if the authority who presents it is not accepted, they will come across as fiction, illusion, or even a lie.

As with the gestalt images, particularly the vase–face picture, we cannot refer to both the mythos and logos realms at the same time, and yet we are always related to both. We cannot ignore the one while operating in the other and we cannot forget that we are constantly connected to both realms. They both shape the way we perceive, interpret, and communicate our reality.

The vase–face image also illustrates the group dynamics that surround a collection of myths in a worldview—people exclude other people from their community quite often because they have other worldviews, other myths that could challenge the mythos construct of order and coherency, the myth construction that the group has achieved.

If we look at any phenomenon in nature, we can see its meaning for the logos realm or its meaning for the mythos realm, but both are quite distinct. For instance, if we look at the clear night sky, we can see the Milky Way with its myriad of solar systems. It is quite an impressive picture, especially if we reflect on the ages of the various stars and their distance from us. These stars tell us something about the history of our universe and thus about ourselves. On the other hand, those of us who believe in a creator God cannot help but feel awe for God's creation. Here's the starry sky, grandiose and beautiful, and I know I can become quite spiritual in this situation. Both stances are real and valid, but while a reflection on the consistency of the Milky Way belongs to the logos realm, the reflections on a creator God and God's creation is a myth. These descriptions fulfill different functions in our lives, but both will influence and inform each other.

Every attempt to prove (or disprove) God with the help of a logos description of the universe is doomed and usually has some form of literalism built in. The same can be said for those Christian groups that reject Big Bang cosmology because it does not leave sufficient space for God's creative powers. However we want to interpret our reactions to various gestalt images, the pictures themselves remain just metaphors for the fundamentally paradoxical character of the world in which we live. Our reactions always contain an element of construction, as we are social beings and we create and construct myths within our respective communities. These myths then are the foundation for the framework in which the logos is formulated. Cognitive science has long discovered how many constructive elements every act of perception contains. Within the mythos–logos framework, we assume that part of the construction is actually influenced by the mythos.

Living with Myths

The background people come from, the values they accept for their lives, their field of interest, their friends and communities: all these elements shape the way they see and interpret various inputs.

From birth, a child is educated within the mythos construction of her parents and her culture; she learns to think and understand and construct reality within these parameters and settings. Adolescence is the time when many of the myths, values, and narratives of the parents and the culture are rejected so that the adolescent can progress as an individual. Adolescents usually reject the worldview of their parents and invent their own, complete with art forms, music, body language, and rituals. This new worldview is similar to the parents' worldview in that it is acceptable only to people who share commonly held beliefs and values; however, the actual values and beliefs are different between the two groups. No one from another age group can possibly be accepted in it. Actually, even many people from their own age group will not be accepted, and I am sure that everyone who recently either went through a high school system or had children in high school can readily agree to that.

After this time of character formation, the individual will accept the inherited worldview (or at least parts of it), construct a new one, or become part of a group or culture different from the inherited one.

This process is accompanied by hormonal changes that cause chaotic stimuli and ambiguous and uncontrollable emotions in the adolescent, making the process of myth accepting and the creation of new worldviews not as rational as it might sound. A huge variety of perceptions, learned structures and concepts, the values of the age group, and the cultural and political context will largely influence this process, steering the young people from one direction to another until, hopefully after some years, the new worldview emerges. One cannot ignore this all-important age with its challenges and its biological havoc that leads to a new mythos-construct. Myths are not solely subconscious; rather, people at a certain age make conscious decisions about them.

The three following "man–mouse" pictures illustrate the way our expectations, beliefs, and worldviews are shaped by our upbringing and our maturity processes. These processes then, in turn, will shape the way we see. If we see the mouse image A first, we will see in the second image, B, a mouse. However, if we see the man image (C) first, we cannot help but see a man in image B. Depending on our expectations and our experience (having seen the man or the mouse first), we will interpret the ambiguous B image accordingly—to put it into the context of what we have seen before, to create coherence. The very same image can be seen completely differently, depending on the image we have seen first.

The background people come from, the values they have grown to accept for their lives, their field of interest, their friends and communities, all these elements will shape the way they see and interpret various inputs. If someone was raised in a slightly antireligious, liberal, and science-oriented environment, it seems completely consequential for her to take the "science" side in the religion and science dialogue. She might not understand other people's religiousness and might not be able to empathize with them. The opposite might be true for the people who have been raised in very religious households.

Most people from the Abrahamic traditions perceive a creator God in the world; each object of study can reveal the creator and caregiver of the world. Because of this perception, which is shaped by their background and upbringing, they might not be able to understand someone from a different background who sees the world as an object that can be understood and analyzed.

Every community and culture has its myths; myths are created by a community and constitute a community. People can define their belonging to a community by committing themselves to the myths that are considered valid within it. For many theologians in the tradition of existentialism, therefore, myth-making is considered one of the basic human features: Every human asks for the meaning of her life, every human wants to make sense out of her daily experiences.

The main assumption for this definition of mythos is that everyone

This image (A) is a drawing of a mouse. We will see in the next picture a mouse as well.

We will see in this image (B) a mouse. Unless . . .

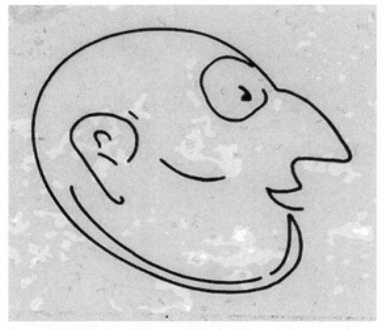

If we see this image (C) of a man first, we will see in the previous image a man and not a mouse.

asks existential questions, that everyone has an ultimate concern. Paul Tillich defines the ultimate concern in one of his (unpublished) lectures at MIT as being unconditionally concerned about the meaning of existence, taking something absolutely seriously, being grasped by an infinite interest and passion, experiencing the self-transcendence of life toward an ultimately sublime or holy. For Tillich, the term *God* and the term *ultimate concern* are synonymous; both need a commitment and both give an answer to the most important question for human life: the meaning of one's existence.

The ultimate concern expresses itself also within the construction of a humanoid. We have already seen that the construction of robots in our image can be a prayer. In the next section we will discuss how it can also help us to understand science and the people who do it. But before we apply the mythos–logos interaction to understand science better, we first have to analyze the inherent dilemma in it. Why did the Enlightenment movement lead to the destruction and violence that haunted the twentieth century and haunts ours as well?

The Dialectic of Enlightenment

As we have seen, any attempt for an unequivocal ontology has to fail, as objective knowledge is impossible. That means pure logos expressions, objective statements about the world, remain an ideal. Every statement is based on the desire to find patterns and regularities in our perceptions of the world. Logos expressions are always based on assumptions, axioms, and beliefs, and every observation of the world is shaped by the capabilities of human perception and the concepts of reality we have learned in our respective cultural settings.

However, the Enlightenment movement attempted to permanently embed people in the realm of pure logos. After centuries of superstition and lack of education, many intellectuals sought to enlighten the population and free the people from angst, superstition, and religious oppression. In addition, after thirty years of religious wars that raged throughout Europe and destroyed major parts of it, people were tired of arguments between Catholics and Protestants. Therefore, the pure logos became the dream, the ideal, and the hope for the future.

Freedom of thought is only possible if *all* thoughts are possible. In their earnest desire for a new age of rationality, many Enlightenment thinkers threw out the baby with the bathwater, as the saying goes. Many thoughts contain religious elements and such thoughts were very unpopular and were seen as ignorant.

But *Homo narrans* needs to address her search for meaning. She has to tell stories. The Enlightenment movement sought to eliminate all statements not purely in the logos realm, but people still had the need for myths. Instead of using established religious symbols to express them, they were now forced to tell their stories with the language of logos. This necessarily created an essential paradox, which was perfectly demonstrated in the actions of Maximilien Robespierre. Robespierre was one of the most prominent proponents of the French Revolution. He was a "High Priest" of the Enlightenment, and an absolute champion of the exclusive primacy of logos in human expression. When I call him a "High Priest," thus addressing him with a religious title, I don't exaggerate. He was an admired leader with a large following, and yet he used mythos to represent logos. For instance, he built altars for "Pure Reason," thus using an explicitly religious symbol to convert people to his point of view.

This historical event demonstrates quite nicely what Horkheimer and Adorno call "The Dialectic of Enlightenment." On the one hand, if we ignore and oppress the mythos realm, we betray the very desire of the Enlightenment for people's freedom of thought and action. On the other hand, it also demonstrates that even if we try hard we cannot be without mythos. We cannot help but use religious symbols and other narratives to tell what's important for us. Whether it is Marvin Minsky's belief that AI can give him eternal life or Robespierre's belief that pure reason is a goddess, both can express the importance of what they wish the most (avoidance of death for the one, triumph over human ignorance for the other) only *with* religious symbolism.

The paradoxical relationship of mythos and logos appears in AI in the process of self-reflection. If I think about who I am, I use my very own thinking to think about my thinking. If I reflect on how humans work I have to start with myself, as I know only through reference that others work like me—I cannot know if the thinking of the other feels the same to her as mine feels to me. That is, any objective notion of who I am has to fail. Any one of my assumptions about the role of me and my species in the world cannot be grounded in the logos itself without creating circularity; one cannot question one's own assumptions without being dependent on other assumptions and beliefs.

Every observation about reality and every assumption about the relationship between the subject and her world contain elements that are outside the logos system and are presupposed in any form of logos statement. Hence, if we ever were to build a robot that is like us, all it would teach us about humans is that they are incredibly creative. Such a creature would not prove that we are nothing but machines, as this was assumed right from the start. Nonetheless, the worldview of many people might be changed because they will create mythos narratives about the robot and its power over human self-understanding.

The people at MIT who attacked me fell into the trap of the Dialectic of Enlightenment. Their belief in rationality and objectivity became a myth, a story that gave their life meaning; after all, a rational, orderly world is a world without estrangement. They had to attack me, as I represented an alternative view that seriously questioned their belief in the possibility for universal rationality. When the e-mail fight started, I was very careful to keep my silence. My every comment would have been like throwing oil on fire. But in personal conversations, I used the techniques that we have developed for interreligious dialogues; because of mythos–logos entanglement, it became a dialogue between different faith statements about the world and us within it. Ultimately, the Dialectic of Enlightenment teaches us that when it comes to understanding ourselves, every statement, whether it is made within the context of science or art or theology, is part of a worldview in which mythos statements and logos statements stay side by side and influence each other.

When I use scientific facts and theories for my development of a Christian anthropology, I do so knowing that the stories science tells about us have to be taken with a grain of salt. What utterly convinced me is the fascinating insight that science and theology fit together incredibly well. Over the course of my studies, I was forced to realize that the insights science has to offer about us are largely supported by biblical wisdom.

Humans and Other Animals

One of the most interesting scientific disciplines is ethology, the comparative study of animal behavior. After all, the Bible does not distinguish humans empirically from other critters. It clearly states in the creation accounts (Genesis 1–4) that humans depend, like all other animals, on God-given food and water. What distinguishes us from other animals is that—according to our faith—God has created us in God's image, and thus chose us as partners. So in order to learn more about ourselves it is helpful to study our animal cousins, their rituals, the raising of their offspring, and their learning process.

The general problem with all observation, whether we observe animals or other subjects, is that we can't be objective. In the special case of animals, however, this lack of objectivity is caused by our universal drive for anthropomorphization. As we have seen in chapter one, humans treat even disembodied computers with politeness and some degree of empathy; the reaction and anthropomorphization of animals as living beings, who share quite a lot of similarity to us, is even stronger. That keeps us from any form of objective observation.

Also, we have seen that eyes are not merely cameras and ears not just microphones. Expectations, prejudices, and preconceived notions will shape any human's observation. One classic example is the famous scientific debate about gender relations and sexual relationships in primates. The first observers of such animal tribes were men, and they saw that all primates have a patriarchic society in which the "alpha male," the strongest and smartest of the males, has the lead, impregnating all the females and dominating the whole group. When he can no longer fight, he is killed by his successor and then the new alpha male takes over. This understanding of primate society soon became general knowledge. People argued that human males are similar to other primate males and that, therefore, human males had the strong urge to sleep with as many women as possible. The alpha-male theory sometimes has even served as justification for cheating and rape.

Perhaps there was some wishful thinking on the side of the observers because, as it turns out, this theory is flawed. As women started to observe primates, they discovered many elements in their behavior that didn't quite fit with the alpha-male paradigm. Today we know that primates live in a sophisticated and complex societal structure. For one, many females are impregnated by males other than the alpha male—even though they have to copulate away from the leader. Also, many primates, especially bonobos, use sex to overcome estrangement from one another and to keep away aggression and fighting. This use of sex is unlimited, meaning that father and daughter, mother and son, or siblings can have intercourse when it helps the relationship between them.

Another theory was that most primates have no sense of self and act entirely out of instinct. Even if they sometimes behave similarly to humans or have similar facial expressions, they have no sophistication and are, clearly, very distinct from humans. Again, these results became much more ambiguous as soon as people from a variety of cultural backgrounds joined the data collection. First of all, it was discovered that parents actually teach their young. For instance, the use of stones as tools to crack nuts is taught to the next generation, which seems to suggest some culture and sophistication, demonstrating that the complex behaviors displayed by primates are based on much more than instinct.

Also, the question of a sense of self is more complex. To test for some sense of self in animals, researchers very often use the "mirror test." Most animals attack a mirror when they see their reflections, which seems to suggest that they don't recognize themselves. If you paint a chimp's eyebrow red, the chimp will not attack the mirror anymore but will actually touch its own eyebrow. It seems that the moment of hesitation—"that chimp looks strange with a red eyebrow"—is enough to make the chimp look more carefully and then recognize itself.

For many years it was established that only chimps, our closest relatives, would pass this test, but it turns out that this is not so. The Harvard ethologist Marc Hauser did the same experiment with a kind of macaque that has huge white hair manes. A painted eyebrow would not do but when he dyed their hair purple or blue (he uses only dyes without potentially harmful chemicals), they started to hesitate when they looked at their mirror image. And, as in chimps, this moment of hesitation seemed to be enough to take a closer look and to recognize oneself. This is not to say that chimps or macaques have the same sense of self that we do, but experiments like this seem to suggest that the difference between humans and other animals is very fluid, that we, in fact, are primates.

From these few examples it becomes clear that humans who observe animals are not in the state of *Homo objectivus et rationalis* but are subjective and project their own fears and desires, as well as their own understanding of themselves, their species, and other animals, onto the subjects they observe. But this is not the only problem that faces ethologists. Another quite common flaw in the observation of animals is to jump to conclusions about humans too quickly. Is the rubbing of beaks between birds really the origin of a kiss? Can one really make assumptions from the hive behavior in ants to the group dynamics in human society?

One example of this phenomenon is the research on voles. Specific voles have a genetic "code" that leads either to monogamous or polygamous behavior. Prairie voles belong to the 3 percent of mammals that live monogamously; when a partner dies, the "widow(er)" in 80 percent of the cases will not even start to bond with another Prairie vole. On the other hand, Montane voles, which are almost genetically identical with Prairie voles, are polygamous. The differences in behavior are due to certain hormones that are released during copulation, which, in Prairie voles, facilitate pair bonds and thus do enhance social behavior. In Montane voles, on the other hand, these hormones have no social function. In the room in which I listened to a lecture on this subject, I could see many people (most of them male) perk up and, indeed, the first question after the lecture was whether this research means that it is entirely genetic whether a human being is monogamous or polygamous. If so, this would be a great justification for cheating. The answer was the only possible one: You are exactly right, your genes will determine whether you are monogamous, but only if you are a vole.

While animal research offers many suggestions for understanding ourselves better, it is never the ultimate proof and can be applied to the understanding of humans only when findings from other disciplines support these ethological suggestions. Even if I invite similarities between myself as human and other animals, no one can deny that there are also significant differences. I would like to assume that these differences are due to differences between species rather than to any form of specialness in humans. But, nonetheless, these differences have to be taken into account and one has to be very careful about conclusions drawn from animal behavior applied to human behavior. On the other hand, to place us firmly into the animal kingdom supports the modesty that is inherent in any golem construction. It also opens up the possibility of creating robots in our image with the potential of making yet another partner species of ours.

The Nature of Science

After all these reflections, we can now finally enter the realm of science in order to get to know ourselves better. Science is a wonderful and exciting endeavor that, particularly in the last fifteen years with the advancement of the neurosciences, has taught us a lot about ourselves. We have to understand how science works in order to understand AI.

It is very difficult to come up with an unambiguous definition of the scientific enterprise. In my experience, however, one can always find five distinctive elements in every scientific research project: observation, the comparison and then categorization of data, the formulation of hypotheses, and model building.

The first part, the basis for each scientific endeavor, is **observation**. We observe the world around us. We try to collect data from the outside world by using our own perception apparatus and machines that enhance our limited perceptions. The problems in this very first step of science are already obvious. Our perception can be heavily influenced by our expectations, which are a result of our worldview and societal embeddedness. But, nonetheless, all science starts with looking at the world and trying to understand it.

The second part is **comparison.** The data we collect is compared to the data either we or other people have previously collected under similar circumstances. The act of comparison relates every observation to history (what has been described before and how has the data been collected) and to other scientific projects in our own time, thus creating a scientific community that transcends time and space.

The next step is **categorization**. After we have compared the data we collected with previously known data, we attempt to create drawers or categories. We try to fit things into certain categories to make them link together. We create typologies to describe various aspects of data. Anyone who has designed a database will know that the most important aspect of a database is the intelligent choice of search categories. The scientific enterprise always comes up with various categories to make comparison and quality control of data collection possible.

As we have seen in the story of the Fall, human estrangement is largely a result of wrong or incomplete judgment. We often err by distinguishing between categories while many things belong to several categories at once. Categorization, the way we sort data and connect it, is always context-dependent and often problematic. As any scientific theory is based on categorization, every scientific theory contains the element of estrangement. As an example, computer scientists often distinguish between hardware and software, with hardware being the computer equipment, the stuff you can touch, see, and, occasionally, throw against the wall. Software, on the other hand, is the immaterial program that makes the computer behave the way it does. They are different categories, and among computer scientists there are software engineers and hardware engineers. In reality, however, the two are closely linked and, especially in the embodied systems I am most interested in, cannot be developed without each other. The categorization is helpful in more traditional areas of AI, but in order to break with the traditional computer architectures, one has to overcome this particular category.

After the categorization of data, we are now able to **formulate theories and hypotheses.** Either during or after the theory building, researchers develop **models.**[3] They help people to understand very abstract theories and hypotheses, making concepts of the very large (e.g., star systems) or the very small (e.g., DNA or quantum physics) concrete and feasible for our minds. In the age of computer power, these models have become increasingly complex and helpful for both the comparison of data and teaching.

Modeling is that part of science in which the mythos–logos entanglement shows itself most strongly, as every model presents a part of a

---

3. Thomas S. Kuhn, in his book *Structure of Scientific Revolutions,* adds a category of shared symbols. And, indeed, symbols in science are extremely important to keep a scientific community together. However, because of the deeper meaning of the term *symbol* within the theology community, I have chosen to integrate Kuhn's category of "symbols" into the other categories.

very specific worldview. I thank the geneticist Martinez Hewlett for his example of this possible mix-up. I heard him speak long ago at a conference where he showed a picture by the French artist René Magritte. One of Magritte's famous paintings shows a pipe, and under the pipe he wrote "*Ceci n'est pas une pipe*" ("This is not a pipe"). First one thinks: "What an idiot!" but very soon, after looking at some other paintings from Magritte, one realizes what he means. This picture of a pipe is indeed not the pipe itself but an image of a pipe, a model.

Hewlett applied this insight to modeling of DNA. There are at least three different ways to model DNA: an electro-microscope image, a 3-D model of the molecular structure of DNA, and a 2-D schematic model that demonstrates the double helix. But these three models do not show DNA itself. All three are *images* of DNA, stories that reflect various aspects of DNA. They are models that make sense out of DNA and try to translate it for the scientific world that depends on it, and to the learning community that wants to learn more about it but perhaps does not need all the specifics. Every one of these models fulfills a specific function and purpose. They refer to the DNA they are modeled after but are not DNA themselves.

Modeling is a big and important part of the scientific enterprise. But if we look at the process of model construction carefully, we come across one phenomenon that makes the dialogue between scientists and nonscientists difficult—especially when it comes to our understanding of ourselves. As any model refers only to a few aspects of the thing it models, it always presents a *reduced* world. If we interpret the models as the real thing rather than the representation of the real thing, we ignore many of the complexities and interactive processes that cannot be modeled in image or 3-D construction (they can be better modeled in computers but the reductive aspect is still there).

Reductionism

In scientific research, reductionism is a very normal phenomenon. Scientists have to reduce the world, set up many independent research areas, and study tiny little aspects of the whole thing. The ultimate goal of course is always to put these little dots together to form a larger chunk, adding more complexity and explanatory power to the model.

One example is when scientists combined Darwin's theory of survival of the fittest with Mendel's theory of genes. Suddenly, biologists had a model that was not just descriptive but analytic. They now could analyze evolutionary processes and even start to manipulate them. Mendel and Darwin worked completely independently of one another, and only the connection of their models created one of the most powerful scientific theories of the twentieth century: neo-Darwinism. But even neo-Darwinism falls short in explaining numerous occurrences in the genetic development of species over time.

Scientists, unfortunately, are usually not trained to reflect on their work philosophically. Anyone who has worked in a scientific lab knows the amount of work one has to do: the research itself, teaching, interacting with colleagues, and grant-writing. All members of the lab usually share a similar worldview, and in their few hours of nonwork they usually want to enjoy life. Also, there are not many academic curricula in either the undergrad or the graduate level that train future scientists in philosophy and history of science. Most textbooks in science contain only the established paradigms of the majority of leading scientists in a specific field. Teachers who were trained in a specific paradigm and shared its assumptions have taught the next generation. Students will rarely be forced to think about alternative theories or hidden assumptions. In most scientific textbooks, one often finds a hidden undertone of authority that claims to own the truth.

AI As Scientific Enterprise

A traditional definition of AI reads something like this:

> *AI is the part of computer science and electrical engineering that attempts to build machines that are capable of performing functions for which we intuitively presume intelligence to be necessary.*

This definition still remains somewhat problematic because of the terms *intuitively* and *intelligence*. By now, it should be obvious that the meaning of both terms depends on a personal worldview and the cultural and societal setting in which they are defined.

The golem tradition is useful in defining AI. Golems had a double function: They were tools to help humankind and they were attempts to understand God and God's creation. Equally, AI can be used for two purposes: to use computers and robots for more than what normal machines are currently used for, and to use them to understand ourselves.

The first purpose, the attempt to develop machines for more applications, is called the *engineering goal*. Results from such research can be seen everywhere, in cars and stereos and databases. It is obvious that under the engineering goal it doesn't matter what understanding of intelligence you have if you build embodied or disembodied projects. The only thing that counts is how successful your final product is in performing its task.

The second purpose of AI, to understand ourselves, could be called the *scientific goal*, the attempt to understand human intelligence by building smart machines. Both goals are connected and intertwined. Researchers under the engineering goal who attempt to construct "smart" gadgets have to use a model of intelligence that is somehow familiar to them; the obvious choice would be themselves, as they know their own intelligence best. Choosing oneself as a model of intelligence for one's project influences the whole process of construction, and self-understanding and technological success reinforce each other. Researchers working under the scientific goal are influenced by successful tools developed under the engineering goal. If a strategy has been proved successful and the model in which it was implemented behaves in its own limited way like a human would, then researchers under the scientific goal often assume the human mind follows exactly the same algorithmic structure as the strategy.

Take chess, for instance. If you distribute the figures randomly on the board and ask chess masters and novices to remember them, both groups will remember the same amount of figures. However, if you present both groups with scenes from real games and let them remember, the masters do much better than before; even more important, they remember exactly how the various positions were related to one another even though they might make small mistakes such as positioning a group one row too high. The novices, on the other hand, stay the same. This seems to suggest that in order to become a chess master you have to see the board not as a combination of single figures but as a union of piece combinations and strategies. Before these experiments were done, AI researchers assumed that we humans do nothing but test all possible moves and see which one has the most positive outcome; then we build chess programs exactly this way, assuming that the programmers had discovered how the human mind works when playing chess. So, even if the distinction between a scientific and an engineering goal is very helpful in positioning various AI projects, it is still problematic.

It seems likely that many more projects under the scientific goal will be or will become a spiritual enterprise. Projects under both goals are constructed with God's creativity in us, but the scientific goal adds the spiritual enterprise to understand ourselves. We will later see in what direction such a spiritual endeavor can go when we visit Cog and Kismet in their lab and learn about the importance of embodiment and social interaction.

The Distinctive Scientific Elements in AI

When we construct humanoid robots, we use ourselves as images, as blueprints. First of all, we observe ourselves and our fellow human beings to collect data about our mechanisms of intelligence and our systematic and biological functions. Every theory based on such observation will be dependent on the historical and cultural context in which it was developed. It will also depend on the nature and character of the observer herself, her expectations of her fellows, and her social experience.

The next step, comparison, is relatively easy because there are many forms of intelligence on this planet. Research areas like psychology, anthropology, and ethology can tell us about common human traits, common cultural traits, common primate traits, and common mammal traits. Another step of comparison is the understanding of intelligence through individual development, developmental psychology, and the study of mechanisms of the brain as analyzed in neurology. All these insights about ourselves then have to be categorized into different modules of human intelligence. Learning might be one module, motor control another one, yet another one could be language or chess.

What we know now from brain studies is that the very intercon-

nectivity and plasticity of the brain makes a clear distinction between functional modules nearly impossible. So the real challenge
for AI is to come up with a *modular* model of intelligence and its underlying mechanisms in the human system that is reduced enough so
that it can be modeled and yet complex enough to be effective. Every
model that AI researchers attempt to build is reduced and reflects
what the respective team thinks are the most important aspects of
human intelligence.

When people attempt to rebuild themselves they can, of course, always choose the biological option, health, age, and partner permit-
ting. But when they want to do it with the methods of AI, they might
start by building a vision system, as the head of the MIT team did
with Cog. Of course, Cog does not even come close to a human, but
there is the belief that the more modules are added and the more
complex the system becomes, the more Cog will resemble a human.
That is, in order to build an artificial creature we have to concentrate
on a few aspects of ourselves. If those aspects work, we can add more
elements.

When I came to MIT, the robotics team had decided to concentrate on motor control and social interaction. They chose these aspects for various scientific reasons, as we will see in chapters three
and four. What led them to this decision was the recognition that the
traditional concepts had failed. In the 1960s, AI researchers were
convinced that it would take only a few more clever doctoral theses
and then AI would be solved. Well, this hope was not quite fulfilled.
So the team at MIT was willing to question the assumptions and beliefs of the traditional approach and start their own. For instance,
classical systems always had a central control unit, while the new systems were distributed. Traditional systems were always stable, while
the new systems learned to be stable through dynamic interactions
with the outside world. Did you ever notice that it is far easier to walk
than to stand still? If you cross a creek on a slim plank, your chances
of making it are far higher the faster you go, as the dynamic of your
movement drives you in the correct direction horizontally and that
limits the chance of a movement straight down.

In the case of Cog and Kismet, researchers concentrated on motor
control and social interaction because limitations made it too difficult
to concentrate on more than this. But this reduction is not an absolute

statement about what the human system is really like. Of course, the decision to focus on these two elements of human intelligence reflects the importance the researchers assign to them for the functions of the human system. They reduce human intelligence but this reduction is entirely pragmatic and necessary: The team had to start somewhere. When making this decision, Rod and his team did not at all think that *only* these two aspects of humankind count.

One of the most crucial questions that split the field today is whether our intelligence is embodied, and therefore whether we should build embodied entities, robots, or program computers. It is already clear that I side with the embodiment camp. However, in order to understand the research community of AI more fully, we have to take a closer look at the views of the body possessed by other people in AI.

If we look at our culture it becomes clear that most body models, as shown in the media, present the body as a gadget, a tool. One good example is in advertising, where the body is often presented as something that needs to be kept healthy for our own sake. A German ad for a health provider said, "Do something good for your body; it does so many good things for you." Think about the body as something in which we invest enormous amounts of time and money: spas, health clubs, diets, gyms. It would seem that we live in a culture that adores the body.

But a closer look reveals that often the body is perceived as an object that has to be taken care of in order to have a better life. Many people perceive their bodies as something that belongs to them and has to be taken care of and that has to fulfill certain standards of beauty. At the same time, they see themselves as separated from their bodies, as can be witnessed in the public taboo of speaking about certain bodily functions such as sex, digestion, sweat, and smell. One often can find similar sentiments within more traditional approaches to AI.

AI was founded in the 1950s. The men who founded the field were sitting in the eighth-floor lounge at the MIT AI Lab (still called the "playroom"). They were all mathematicians and physicists, all brilliant in their field. In short: they were nerds. All were male, all were white, all came either from Jewish or Christian traditions and most came from privileged families. This select group of people thought about what defines intelligence and what a machine should be able to do in order to be called intelligent. From their own life experiences and daily work they came up with a few suggestions. A computer should be able to deal with natural language and it should be able to play chess. It should be able to solve abstract problems and be capable of proving mathematical theorems. For many, many years, this definition of intelligence set the agenda for AI. It came out of a very small and exclusive sample of humans, and their choice of problems was perhaps somewhat limited. In their suggestions for a model of intelligence, they were pragmatic reductionists. They did not formulate an exclusive model of humankind. The concentration in classical AI on disembodied tasks has its reason in its history.

A very good example for the intelligence model in traditional AI was the chess duel IBM set up in the early 1990s between its fastest supercomputer, Deep Blue, and then chess world champion Garry Kasparov. In the first attempt, Kasparov beat the machine soundly. But a few years later, in the repetition of the duel, Deep Blue beat Kasparov.

These chess games between the two helped us to see the limitations of this traditional understanding of intelligence. When Deep Blue beat Kasparov, there was still no robot that could actually put butter on a piece of bread because it is so difficult. You have to know the consistency of butter, how hard the butter is. You have to know about the knife. You have to know about the consistency of the bread. You have to have very strong feedback loops between your arm, the knife, the butter, and the bread. You have to know a lot about the world. Otherwise you just have crumbs. In other words, for a mundane task such as buttering bread you need so much knowledge about the world that even the most powerful robot can't do it.

After the Kasparov–IBM duel, Kasparov said that at least the computer that beat him did not feel joy about the victory and was, thus, still inferior to him. Many reporters who commented on the event decided that chess couldn't be a crucial part of intelligence. Many felt that Deep Blue, though an incredible engineering and software development, was so limited in its capabilities and so useless for everything but chess that they lost respect for it. I think these objections were largely voiced out of fear of a comparison between machines and us. So few people can play chess well and we have come to understand chess as one of the pinnacles of intelligence. I am certainly better at spreading butter on a sandwich than at playing chess and I am sure that many of us are. If a computer can now beat the best chess player in the world, does this not mean that all of us who don't play chess well are somewhat inferior?

There were also quite a number of people who took the contrary point of view and saw in Kasparov's defeat the proof that computers are now as smart as people are. I remember a cab driver in Boston during the 2000 Bush/Gore election who was convinced that Bush had gotten more votes than Gore. After all, he told me, the votes had been counted by machines and machines couldn't err. They were better than humans and could be trusted much more.

So the chess play between a man and a machine revealed that the initial agenda for the construction of AI, as set in the '50s, was flawed and incomplete. This shows us how much the first step in the scientific enterprise of building robots in our image was influenced by nonscientific observations and stories.

Embodiment and Knowledge

As we have seen in the brief history of traditional AI, it was based on the myth that the body does not play any part in observation or theory building. Thoughts are, by nature, disembodied. A simple look at the history of data collection and the acceptance of well-researched theories shows, however, that the body plays a most crucial part in learning about our world and us.

After several millennia of human observation, there is not much more to discover—at least not with the perception apparatus available to us. We have long exhausted the possibility of discovering anything new and exciting with our senses alone and so, for thousands of years, scientists have used tools and machines to enhance their own perceptions. We need machines to get more data. We need microscopes and electro-microscopes. We need telescopes for discovering and observing stars and planets. We need computers to compare and categorize data and to develop complex models. And we use computers to store the data collected within the categories we have created.

Let's take the example of an atom. An atom has nearly all its mass concentrated in its center, and if the nucleus were an inch wide, the edge of the atom would be approximately where the sun is in relation to earth. Between the atom nucleus and the electrons circling and the edge of an atom is a vacuum. That means that most of the things we touch consist of vacuum, which does not make sense at all, as we clearly sense something there. We know of atoms only indirectly, through technological methods and mathematical models, because we cannot see them even in an electron microscope.

Now, take the discovery of the top quark. Quarks are the small subparticles of neutrons, which are one kind of particle that makes up an atom. So imagine how incredibly small one subparticle of an atom is! And if you now attempt to conceive of subparticles of these already incredibly small neutrons, you can get some sense of how small quarks are. There is no way that we can perceive quarks with our own perception apparatus. Even enhancers of our perception (such as microscopes) do not help. The only way to "discover" quarks is by inference from observations that are in themselves only observable through mathematical formulas and concepts.

When scientists discovered the top quark at the Fermilab near Chicago, it was the result of teamwork between approximately twenty-five cosmologists and twenty-five engineers. Both groups were equally important for this groundbreaking discovery and neither could have been successful without the other. The engineers did not necessarily have a scientific background but they were nonetheless essential to the project: Engineers construct machines from scientific observation, and they construct them based on their own knowledge about the world and its specifics. Scientists then use the machinery, often without knowing how it is functioning and on what assumptions it is based.

Take another example: the measurement of air pollution. The standard measure works by sucking air in through filters, then resolving the particles of the filters in a solution, then counting what is on each filter, then creating a database to relate the data and compare them to finally draw some conclusions about the pollution in a specific area and its causes. This process seems straightforward but there is actually a lot of room for hidden errors. The choice of the filters will already determine which particles can be found because the size of the pores will let through all particles that are smaller. The solution in which the particles resolve will have some influence on the chemical composition of the data that has to be first determined and then excluded. The count of particles, again, is based on assumptions, as one cannot count every single particle but can only intelligently guess based on chemical composition and weight. The database, finally, can be only as exact as the values of the various categories of measurement.

In each step of the process there are assumptions and technical limitations. That does not make any measurement of air pollution wrong. But the values scientists come up with have to be taken with a grain of salt—not literally, as this would change the chemical composition, but metaphorically.

These few examples already show how today science and engineering are intertwined. One historical example is the so-called Copernican change in the sixteenth century, when people slowly started turning from a geocentric understanding of our cosmos to a heliocentric one. People who believed that every star circles the earth created a myth in which they were the center of the universe. When they looked at the sun they saw a star that moved during the day slowly from east to west. This did not just seem intuitively the case but was also supported by the religious belief that God created the universe and us in the center to be creation's stewards. In this particular instance, the religious worldview fit the scientific view. Copernicus developed a mathematical theory that explained star movements with a heliocentric worldview. As the theory was purely speculative and abstract, it did not really have any impact.

What made the figure of Galileo Galilei so incredibly powerful was his use of tools. With colleagues, he developed a number of new and powerful telescopes to observe the cosmos. With these new tools he discovered many phenomena (such as sunspots and the moons of Venus) that did not have any explanation within a geocentric view. Therefore, scientists who shared these discoveries were forced to reconsider the Copernican model, which offered explanation and enabled fairly accurate predictions. In other words, it was not a church defiance or fanaticism that forced Galileo to embrace a heliocentric view but the development of better tools that allowed their users to enhance their own perception capabilities and thus discover flaws in the previous (geocentric) view of the world.

What we can learn from the connection between scientific hypotheses and the use of tools is the importance of empirical, physical perception, however indirect, for scientific discovery. If we stand in our garden and attempt to understand that we are standing on the top of a ball that spins around itself at approximately 1,000 miles per-hour and also spins around the sun, we really cannot grasp it. From our perception in our daily life it makes sense to see the universe from a geocentric view.

It was the all-human quest and thirst for knowledge that drove people to look up to the stars and develop machines to bring them closer. Only clear physical and empirical evidence that there is something wrong with our intuitive view will convince us to accept another view. We need the evidence given to us by scientific tools in order to make changes in our own understanding of our perceptions; purely mathematical theories such as Copernicus' do not enable us to change our views.

If we look at another daily worldview, the Newtonian universe, we discover the same evidence for the importance of physicality for changing one's point of view. Intuitively, we perceive the universe as Newtonian, obeying straightforward physical laws such as gravity (the attraction of mass) or causal relations. Object constancy is especially important for us. All objects have in common that they are clearly defined; an object cannot be two different things at the same time. Objects cannot be in two places at once; when they rub together, there is always friction and an object stays the same over time unless there are other forces at work.

Theoretically, most of us know that Einstein and other great physicists at the beginning of the twentieth century put an end to this very convenient view with the theory of relativity. But relativity is based on particle physics (concerning the incredibly small particles in our universe we cannot even imagine) and/or the observations of the cosmos with star systems so old and so far away that it is equally hard to imagine. Just as there is no clear evidence for a Copernican worldview as long as we stay on the ground, there is no way for us to embrace relativity as commonsense physics. In our daily life, we do not live in a universe ruled by relativity but in a universe ruled by mechanics. We know that when we stand under an apple tree, we might get hit by an apple. People cannot be convinced of any theory if their own perception tells them something entirely different.

Mathematical formulae or theoretical hypotheses *cannot* convince us of scientific truth if the perception of our eyes and ears and our experience do not fit the evidence. *Homo narrans* creates a worldview that makes sense of her bodily experiences. Every machine *Homo narrans* constructs is the result of a specific understanding of the world and the expectations people have of the machines. So, the computer is a model of the human mind in an age in which the human body has lost its importance and in which rational, unemotional thought has become an ideal. Therefore, the importance of the physical world for us is the topic of the next chapter.

CHAPTER 3

Embodied Intelligence

*The human spirit is God's lamp,
searching into the bedchamber of the belly.*
PROVERBS 20:27

*I keep God always before me;
God is on my right hand, I will not be moved.
My heart is happy, my liver rejoices,
Yes, my body is secure.*
PSALM 16:8–9

*Prove me, God, and try me;
ennoble my kidneys and my heart.*
PSALM 26:2

Are Our Bodies Ourselves?

According to the theory of *Homo narrans* who attempts to understand herself within mythos and logos stories, our bodies play a major role in every model of who we are. When we look at the Bible, we can see a very similar view. The three Psalm verses above seem to be unfamiliar at first. They also seem a bit ridiculous: What is the "bedchamber of the belly"? Why would my "liver rejoice"? And why would God try to "ennoble my kidneys and my heart"? What for?

These body parts are all fine and necessary for our survival, but what do they have to do with our spiritual well-being?

As a German, I get an inkling from some old-fashioned German expressions. If Germans test someone very carefully for something quite important, they still might say "to test heart and kidneys"; if you are in a bad mood a "lice walked over your liver"; if you are sad and disappointed, something (or someone) went "to your kidneys"; and if you are spitting mad you "spit gall."

Historically, the German language is far more influenced by the Bible than English is. One could say that Martin Luther gave Germans their own unified language; before his translation of the Bible there had been many different dialects. The strength of the Luther translation, outdated as it may seem to many, lies in the Psalms. In my opinion, no other translation was ever as powerful as Luther's, and he translated them sometimes quite literally, used all their body metaphors, and made them part of German everyday language.

I have never heard similar expressions in American English. In modern English translations of the Bible, the verses quoted above sound like this: *The human spirit is the lamp of GOD, searching every inmost part* (Proverbs 20:27); *I keep GOD always before me; because he is at my right hand, I shall not be moved. Therefore my heart is glad and my soul rejoices, my body also rests secure* (Psalm 16:8–9); *Prove me, GOD, and try me; test my heart and mind* (Psalm 26:2). While these translations sound much more familiar, they do not quite represent the Hebrew original. The belly is translated as "innermost being"; the belly is seen as metaphor for one's own most intimate thoughts. The liver as organ of rejoicing is translated with "soul"; therefore the next sentence "my body is secure" gets a new meaning with the added "also" separating the body and the soul. Finally, the kidneys are translated with "mind," a concept that would seem very strange to the people who wrote the Hebrew scriptures.

Why these changes away from the biblical body metaphors and their meaning for our self-understanding? At some point in our history, we obviously were very comfortable with an understanding of ourselves in which our organs were part of our whole being and were responsible for our emotional makeup. "We" didn't rejoice; our liver did. In modern American language only the gut, the heart, and the skin have survived as organs with metaphorical value. We can safely say that we Westerners no longer understand our bodies as part of who we are. Our understanding of the importance of our bodies reveals yet another fundamental paradox in human self-experience: the feeling of *being* a body and the feeling of *having* a body. As in the vase–face picture, we can quickly oscillate between a stance in which our body is just that, something that is sort of attached to us, and a stance in which we are our bodies, nothing more, nothing less.

In our Western cultures, we tend more often to be in a stance in which we have a body that is not quite as important as our mind or our soul. Often the only purpose of the body is to carry a brain around. Unfortunately, the Christian tradition is largely responsible for this disembodiment.

Paul distinguished between *sarx* (flesh) and *soma* (body). In modern Greek, this distinction still applies. Here, *sarx* is the stuff you buy from the butcher but *soma* means the whole body and the wholeness of a person. Paul demeans the *sarx* as it hinders us from being fully human. Think about the deadly sins, which become a major source of estrangement as soon as we overindulge. If we overeat or are too vain or too envious, we fulfill the desires of the *sarx*. Hence, the *sarx* can keep us away from the realization of our *soma*. Paul points out the importance of *soma* for any form of spiritual praxis. Metaphors like "The Body (*soma*) of Christ" hint toward a community in which bodies are tied together for the purpose of making this world a better place. Unfortunately, we have unlearned how to distinguish between *sarx* and *soma* and, therefore, Paul has often been misunderstood as rejecting everything embodied.

The Paradox of Embodiment

The myth of disembodied intelligence goes back to ancient Greek philosophy. It influenced Christianity heavily and thus became a major part of Western thought. The philosopher and mathematician René Descartes brought it to its completion and claimed to have proved its correctness beyond any doubt by logical means alone.

The story is famous: Descartes sits in front of his fireplace and muses about what he can know. He speculates God might be a cheater who makes us think that there is anything outside our selves. Descartes assumes that everything he experiences, all perceptions and all other impacts our surroundings have on us, are a mere illusion. Consequently, Descartes starts to doubt. He doubts that the fire is there, he doubts that he feels warmth, he doubts the chair on which he sits, he doubts that he has a body, and finally he comes to the point where he is absolutely certain that he cannot be victim of an illusion. He cannot doubt that he doubts. So he writes his famous sentence: "I think, therefore I am" (*cogito, ergo sum*). Descartes concludes it is the mind, the questioning "I," that we can rely on, not any perception of ours. If we want to know, we have to reason with our minds and not with our feeble bodily perceptions. Only from the doubting "I" can we derive knowledge about the world. Descartes thus creates the foundation for the metaphor of *Homo rationalis et objectivus,* the metaphor that assumes humans are actually capable of objective perception and rational, unemotional thought.

Within the story of the objective and rational human, our perception apparatus is understood as an independent entity that perceives the world as it is, our memory as an accurate map of our world, and scientific language as an unequivocal description of the phenomena we study. Our perception apparatus is often understood in terms of technological gadgets: Our eyes are cameras reflecting exactly what's out there; our ears are microphones registering exactly the range of noise that is currently surrounding us.

As we have seen in chapter two, *Homo objectivus et rationalis* remains an illusion. The observant human is a subject embedded in the world she attempts to describe and to understand. The scientific mind as an objective, reasonable, and rational entity cannot exist. This paradox of the subject describing the objective world outside of her, of which she is nonetheless a part, cannot be resolved in any coherent way. For millennia, philosophers have tried to answer questions about what we know and why we know it, as well as why the world is the way it is and why we perceive it in a certain way. Descartes now presents a world in which that dilemma seems to be resolved. Finding the unquestionable "I" beyond the doubtable body parts makes the body somewhat superfluous.

Consequently, Descartes can then describe the body as a machine and he uses the technical metaphors of his time to describe its mechanics and functions, hydraulics being the high-tech of his time. He described the human body as hydraulic automaton, and for a short period of time, he located the "I" in the pineal gland (its function had not been recognized then and he later gave that up).

What is interesting is that this dualistic understanding of ourselves has not changed much, but today we use the metaphors of the current technology, the computer. Statements such as "I couldn't store it" or "I couldn't process it" point to a dualistic understanding of humankind that sees our brain as a gigantic computer on which the software of the mind "runs." The understanding in traditional AI is that this software can be implemented either in the hardware of a computer or the "wetware" of the brain and it does not matter which, as the mind is independent from both. I call these figures of speech "computer-*morphisms*" in analogy to anthropo-*morphisms*— and these metaphors are the theoretical basis for the possibility of transhumanism.

Descartes' convincing thought experiment cannot deny the fact that we are bodies; he and many others, however, conclude that the body plays a somewhat lesser role than the mind. Descartes' legacy makes it difficult for us to grasp our own embodiment. We have lost our own feeling of being bodies. Unlike the psalmists, we do not think anymore that all of our body parts are essential for who we are. As mentioned above, our daily experience of our bodies comes in the form of the vase–face problem. If we take the vase as the view that we *are* bodies and the two faces as feeling that we *have* a body, then we realize again that our own self-perceptions are often paradoxical in nature.

Whoever has experienced a major toothache knows that this pain is very real and cannot be ignored; my bodily experience *is* the "I." On the other hand, we all know that we can suppress hunger and thirst for a while if something else, such as the work we are currently doing or the date we are on or the weight we want to lose, is more important. In that moment, we feel we *have* a body whose functions can be controlled by will alone. However, these bodily drives cannot be suppressed infinitively. In most cases, hunger will ultimately lead to eating and fatigue will eventually make you fall asleep no matter where you are and what you are doing at the moment. We cannot control our bodies all the time; the body's needs will ultimately always overcome our will.

It is only in our philosophical tradition that the body is seen as somewhat less valuable; neither Christianity nor Judaism in their original forms have been against embodiment. A careful look at the Bible shows that the ancient rabbis certainly embraced the body but also had a clear notion of the potential danger of disembodiment. If we separate ourselves from our bodies, it is yet another temptation of our state of sin. In fact, according to the Bible, our estrangement manifests itself quite profoundly in our embodiment.

Sin as Embodied Phenomenon

As we have seen in chapter one, the term *sin* means estrangement and addresses our living with ambiguities. In the language of the mythos realm, sin came into the world when Adam and Eve decided to eat from the tree of knowledge of good and evil and started to make judgments. It seems from this interpretation that "sin" is a phenomenon of the mind and not of the body. This fits into the assumption of a body–mind split that seems prevalent in our current society.

However, the Genesis stories tell a different myth about what is genuinely human and include the body and its importance for the human thought processes. The rabbinic tradition, in general more pragmatic and real-life oriented than many Christian interpretations of the biblical stories, discusses embodiment quite frequently. The philosopher and theologian Norbert Samuelson kindly called my attention to Rabbi Abraham ibn Ezra and provided me with an interpretation of Ibn Ezra's theories.

Ibn Ezra meditated about the meaning of Genesis 3:21, "And God JHWH made garments of skin for the man and his wife and clothed them." The rabbi found this sentence highly disturbing since it presented God in the role of a servant, which is not an appropriate role for God; also, it seemed unlikely to him that God would kill animals from the Garden of Eden just to give clothes to the man and the woman. Hence, he concludes, God gave Adam and Eve human skin. In the garden they were truly naked and now they are covered.

Let's not take this image too literally as, at least for my taste, a being running around with visible muscles, sinews, and blood vessels is not an attractive sight. But let's play around with these images and see what they can tell us about embodiment.

A closer look into the text reveals that the relationship between Adam and Eve in the garden is never described as an erotic one; there are no hints that point toward sexual desire of one for the other. Interestingly enough, though, the term used to describe Eve's desire for the fruits from the tree of knowledge (*chamad*) is explicitly sexual. And after the eating of the fruit of the tree, man and woman recognize that they are naked and wrap themselves in leaves (Genesis 3:7). We have already talked about the "mind" part of sin and know, therefore, that the recognition of nakedness and the shame about it is a result of the societal judgment within a particular cultural setting. But in the context of embodiment and sexuality it seems to me that the verse could also be seen as the first recognition of the other as a sexual being and hence the recognition of one's own sexuality. Each one looks at the other in nakedness, realizes his or her own nakedness, and covers that—and, as we know, clothes can be both draping and revealing.

This ambiguity is expressed throughout the Hebrew scriptures with the term *jada*. It can mean either "to recognize, to know" or "to sleep with" (the biblical "knowing"). As such, it also hints at the ambiguity of sex as described in the beginning of Genesis as something connecting and as something separating—shame and desire, covering up and revealing. That Eve desires first the fruits of the tree of knowledge points to a similar direction: biblical knowing is both embodied and mental. It mirrors, like the Greek term *eros*, the insight that we want to know with passion and participation, and both, obviously, can take a bodily form.

The curse God gives Adam and Eve right before they receive "garments of skin" confirms the embodiment of sin described so far. God curses Adam and Eve with pain in labor; Adam will feel it in the labor of yielding harvest and Eve will feel it in the process of giving birth. Eve is also punished with sexual desire (*t'chukah*) for Adam. Both pain and sexual pleasure come through the skin. But neither pain nor sexual desire were there in the garden. Again, the ambiguous character of life gets a very embodied element attached to it; labor—both daily work and giving birth—can be satisfying but also painful and stressful; sexuality and sexual desire can be wonderful but also painful, hurtful, and destructive.

The theological question here is whether this is really punishment. God puts us in an ambiguous state, but isn't a major part of the joy of life dealing with these ambiguities? Perhaps only by leaving Paradise can we become truly human.

Many people might cringe at such an idea. Joy in the context of sin? I remember a discussion with students in which one student brought up the concept of "sins of the body," and I asked the students to tell me what they were. They listed sex, drugs, alcohol, overeating, and some similar examples—with a lot of giggling, especially when I pointed out that most of them probably had performed all those "sins." Then I asked them if these "bad" deeds they had mentioned had something in common besides being—at least in their opinion—forbidden and evil. They didn't find another commonality, but there are, in fact, two. For one, all these seemingly bad behaviors refer to the *sarx* and not to *soma*, and even if they can occasionally lead to estrangement, they constitute a harmless attempt to satisfy a momentary desire. But there is another commonality: All these actions are joyful and fun! Why are they so often perceived as bad things, deeds someone should avoid? When I came to this country, someone told me (not quite seriously) that the Puritan moral law still applied in modern America and could be described as, "All is allowed, unless it's fun!"

This is particularly true in the topics of sensuality in general and sexuality in particular. Let's have a look at a different part of the second creation account. According to common translations of the older creation story (Genesis 2–4), God made Eve from Adam's rib. I thank the evolutionary biologist Scott Gilbert for pointing out to me that this translation could also be the result of misunderstanding or cultural adaptation. The Hebrew term for the bone out of which Eve was made is *tzela*, a term that is much broader than its translation, "rib." It appears in the Hebrew scriptures approximately forty times. It comes from the root "deviate" and can also mean side, side chamber, story, wing, and plank or board. Only in Genesis 2 does it have the meaning of "rib." In all other places it is used in the context of construction, like the construction of the Temple.

These findings suggest that the translation "rib" can be questioned; it fits that *tzela* is a construction term as Eve is constructed from it, but why, of all bones, the rib? After all, there is absolutely no empirical evidence that males have fewer ribs than females; also, there is no rib lacking in the male skeleton. Besides the ambiguity of the translation "rib," there is the lack of any cultural parallel; nowhere in *any* culture does the rib have any fertile or prolific connotation. So it seems strange that the authors of the creation story would choose this metaphor. After all, they were influenced by the other cultures around them when they were attempting to explain why we are the way we are.

So Gilbert suggests that *tzela* in the context of Genesis 2 means "penis bone" (*congenital baculum*). This fits with the larger meaning of *tzela* and also makes sense in the cultural context in which this story was written down; in many cultures, the penis is a symbol of fertility. Finally, it fits empirical evidence as most mammals, and certainly the ones the people from Israel were interacting with 3,000 years ago do have a penis bone and human males don't. Yet we always understand *tzela* as rib.

Our hesitancy to read the biblical account of Adam and Eve with all its embodied and sexual connotations demonstrates yet again the extent to which we have unlearned seeing our bodies as integral parts of who we are. I therefore think that the concept of sin as estrangement from God and one's body and, hence, one's community, is very helpful for our understanding of ourselves.

Let's look for a second at the question of sexuality. Sex is absolutely necessary to keep our species alive. No sex, no progeny, no survival. Sex is also wonderful; we are created with sexual organs and skin that can make touching and intercourse extremely pleasurable. But sex can also be a form of estrangement from one's body and the sexual partner when a personal relationship is not included. Sex without a personal relationship between the participants is not *jada*, the biblical "knowing."

The Bible, thus, presents a mythos answer for why we are the way we are. It does not want to give us any scientific analysis of the human body, but it makes us think about our bodies and our bodily interactions; it also makes us think about the role of our embodiment in the experience of spirituality and in interaction with God. Looking at the Hebrew original and the rabbinical discussions, we can see that the Bible, especially the Hebrew scriptures, puts a strong emphasis on embodiment and the ambiguities of the body as well as the peculiarities of our thinking and the ambiguity of our minds.

Embodiment in Interacting with Computers

There is some evidence that intensive and frequent interaction with computers enhances the view that we are ultimately disembodied entities. In the early 1990s, a German researcher, Christel Schachtner, conducted research on the interaction of humans with computers. Her research supports Nass's and Reeves's findings of how strongly people bond with their machines; they tend to anthropomorphize them, often calling them by pet names, praising and cursing them depending on the situation. But another insight of hers is even more interesting.

She did her research with two different groups. The first one consisted of computer professionals, such as software developers, between the ages of thirty and fifty, and the other of high school kids between the ages of fifteen and nineteen, who took computer classes and spent a lot of their free time in front of computers. Schachtner conducted extensive interviews with them about their sense of self and let them draw their own portraits. Interestingly enough, in both groups the result was the same. People saw themselves reduced to hands, eyes, and brain; most of them felt that their bodies had vanished or had been fragmented. The following pictures are examples of

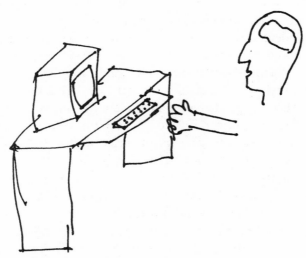

Andreas, 18 years: The vanished body

Software engineer, 41 years: Humans that lose contours downward

self-portraits Schachtner collected during the interviews; the captions reflect how the people commented on their self-portraits.

In the first picture you can see a self-portrait of a high school student whose body has vanished. Only the head (with brain) and a hand remain because he needs only them to interact with the computer. The other picture is a self-portrait of a computer professional who sees himself the same way. Even if he is much older than the high school kid, there is a surprising similarity in their disembodied self-understanding. But the drawing of the computer professional adds another aspect to the vanishing body: There are still contours, but the body becomes increasingly insignificant. His interview and some of the others reveal that there is a close connection between their involvement with computers and a vanishing interest in good food and other bodily pleasures.

This is actually the opposite of my experience at MIT. What struck me from the beginning was the liveliness of the people I worked with. Here was a group that understood intelligence as an embodied and social phenomenon and wanted to implement this conviction in robots. And many had so embraced these ideas that they lived these principles in their daily life. I have never laughed as much in my life as during my time at MIT (quite the opposite of the dry academic environment in many German universities). Many students, as well as their leader, Rod Brooks, also loved to eat and drink. Most were participating in sports and enjoyed life a lot—quite different from the expectations I had about MIT nerds. This, of course, goes against Schachtner's research, which found that people with intense relationships with their computers lose their sense of embodiment. The people in the robotics group worked many, many hours on the computer every day, and one would assume that they would become perfect examples of disembodied humans. It becomes clear now that the interest in embodied actions is a result of the embodied worldview and not due to the people's intense interest in and interaction with computers. That means one cannot blame the computer for cre-

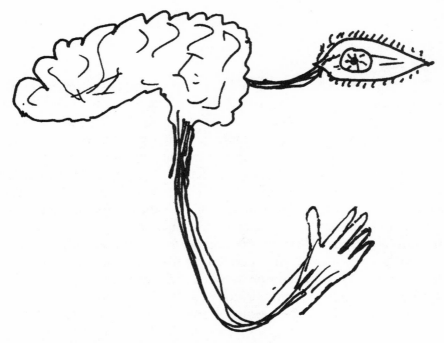

Rolf, 19 years: The human, reduced to brain, eye, and hand

ating the feeling of disembodiment in people but rather the culture that understands us as a kind of computer.

Rolf's sketch is probably one of the most powerful self-expressions of Schachtner's subjects. This high school student sees himself even more reduced and fragmented and draws only an eye, a brain (without a head), and a hand. He specifically mentions that he needs only one eye. Stereo vision is important when interacting with the real world but it is not necessary for interactions with computers.

This final sketch shows the self-portrait of a software developer who sees herself as a brain that has merged with the computer. Here, the sense of disembodiment is most profound, as there is not even a trace of body left. The brain, the only remaining body part, is fully immersed into the computer. This is a classical computer-morphism, as the self-image of this software engineer is created by using computer metaphors.

I would be interested in the results of a similar study in the United States. But it might be interesting to analyze your own self-understanding when you work on your computer. After a few hours, sketch how you see yourself at that very moment. Have you lost your

Software engineer, 30 years: Brain and monitor in a symbiotic relationship

body or part of your embodiment during this time? Do you feel comfortable or uncomfortable? What can you do to be embodied while working on a machine? I have done this for years after reading about Schachtner's research, and have actually found the same reactions her subjects describe. By now, it has become a routine for me to interrupt my work often just to stretch, or take a sip of water, or to do some trivial household task; this has vastly improved my well-being, as I don't get cramped in front of my desk anymore. I would be particularly interested to see if Schachtner's results would be different today in the United States and in Germany with the exponential expansion of the Web since the early 1990s and the many Internet relationships we engage in on a daily basis. Just think of the Yahoo! chat rooms and the number of people one can find in them at any given time. I am teaching a class on the Internet and religion, and students have to visit religious chat rooms regularly and attend online services. From these regular surveys, I know the number of people who are religiously active online, and one can conclude with some certainty that many other human activities and endeavors are equally often conducted online.

Schachtner's research lets me conclude that the sense of disembodiment that has shaped people in the Western world for centuries is often reinforced by the daily use of computers. Of course, we cannot blame computers for this, as they are constructed under the assumption that our brain works like a computer and that our body is not important for intelligent processing in the brain. Here is the negative side of the power of tools in influencing our worldview. When people who subconsciously hold such an assumption start working with computers, they reinforce this hidden self-understanding and the assumed human–computer similarity becomes an integral part of who they believe the machines are. This is why I was so fascinated when I first encountered computer scientists who had explicitly broken with the universal assumption of computer-morphisms and acknowledged the body as our most important thinking tool.

Meet the Robots Cog and Kismet

When we think about robots, we tend to think about the image Fritz Lang first created in his classic 1927 movie, *Metropolis*. We think of a metal man, with a metallic shimmering skin and a humanoid shape that can move around, albeit somewhat clumsily. Modern movies such as *Bicentennial Man* with Robin Williams and the *Robocop* and *Terminator* series copy the image of Lang's robot. The metal body not only conveys machine-likeness, but also signals strength and power. Because of its metal skin, a robot is much less vulnerable than we and so we think it can hurt us easily. A robot's height is usually larger than average, which further inspires a feeling of uneasiness.

I had the same images in my head when I first started to study computer science. I have always been intrigued by technological gadgets and had built stuff as far back as I can remember. Sometime in my sixth semester at the university, in an AI class, I had to give a paper, and somewhere in the material appeared the acronym MIT. I had no idea what that meant and thought it was a company's name. My professor's face fell when I said that and he went on and on about MIT, the great school in America, *the* school for AI. In 1993, I visited Cambridge for the first time and spent time at both the Harvard Divinity School and the MIT AI Lab. That year marked my first personal encounter with the robot Cog.

Even though the robot was then only in its early stages, I was very impressed and, more important, intrigued. At this time, Cog had a very cute head. The two motors that controlled the eye movement (for up-down and left-right movements) gave Cog a very curious expression. As I had been a science-fiction fan for many years and had always loved robots, Cog, for me, was something like a dream come true: a robot in humanoid shape.

Cog's head was much smaller than the body, which made me laugh. It is a joke in my family that the images of the "perfect Aryans" displayed all over Germany during the Third Reich always had these very small heads disproportionate to their bodies. We concluded that in the Nazi era, intelligence for the "Aryan ideal" was less important than the perfect body. In Cog, however, the reason for its disproportionately small head lies in the exact opposite: Its brain is so big that it has to be outside its head.

Cog's brain consists of many different motherboards that coordinate various processors on the robot itself. Actually, at the very beginning of the project, the "brain" was kept out of Cog's sight in case

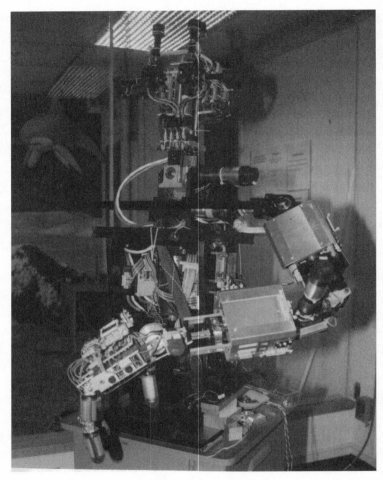

Cog in 1996 with one arm, one hand, and a head with visual and audio sensors and a gyroscope to deal with balance. When Cynthia left the project to build Kismet, Cog became much more "steely" and "Schwartzeneggerian."
Credit: Rodney Brooks

Cog would ever become self-aware and be repulsed by the sight of it. This shows, of course, the high expectations some people had of Cog at the beginning of the project. When I came to the lab in 1993, the timetable for Cog's development was as follows: The integration of the various parts was supposed to happen within the next three years, and Cog was supposed to be able to speak by 1998. Well, by 2001, when I left the lab, the high expectations of the first years had been revised and had become much more modest—Cog, indeed, had made us all modest in our goals and had enhanced our admiration for the human system, God's creation.

At the time of my first meeting with Cog, I had just started my dissertation in Germany about possibilities for a dialogue between AI and Christian theology and, at that point, I hadn't quite understood the importance of embodiment for these specific robot projects. But I quickly learned the cruciality of that aspect.

Cog is a robot with a humanlike form. It can experience the world in the same way we do. It has the same problems with gravity, and because it can move its limbs in the same way we can, it has the same capacities for grasping or moving.

The idea is that intelligence is not merely the capability to play chess and do math but to survive in an ever-changing environment. So, instead of programming abstract intelligent capabilities such as chess into a computer, the team at MIT set out to build a body that, through interaction, would develop motor control and coordination and thus would build up the capability for intelligent thought. In other words, the understanding of intelligence in this robot team is generally embodied.

Humans are a result of billions of years of evolutionary processes. But we are not the only intelligent beings on this planet. Indeed, evolution has produced quite a variety of creatures that are capable of playfully interacting with their environment in order to survive. We are just one variety. And we cannot say that our bodies are superior to other bodies: We don't live as long as trees, we cannot smell as well as dogs, and we are not as strong as horses. But we are quite a good example of an intelligent species because we, more than any other species, can survive in any given environment due to the concept of *Homo faber*, our ability to plan with much more intentionality than any other species. Nonetheless, this intelligence has evolved and has many things in common with other forms of animal intelligence. We want to survive, we want to procreate, we want to eat and drink.

According to recent studies in early child development, a newborn baby has none of the capabilities (such as intentionality or self-awareness) that we expect in an adult. She develops all these capabilities, along with motor control and coordination, through interaction with her caregivers. Hence, the members of the Cog team thought it might be fun to implement this model of intelligence into their machine instead of following the traditional, disembodied-intelligence model.

When an infant is first born, she has a very poor visual system with very poor resolution. But there are certain things that she can discriminate very easily, such as motion and bright color. So the team started to build these capabilities into Cog. Its vision is not one of high resolution but it reacts to color and movement. It has the same roaming space as humans, i.e., it can move its eyes in exactly the same way.

Another example for the embodied architecture of Cog would be its saccade. Saccade is a very rapid eye movement that we cannot control. We constantly move our eyes to have a full spectrum of our environment. Cog does the same, and if anything happens in its periphery (movement or color), its roaming eye will catch the stimulus and react accordingly. So when someone waves a hand, the robot will move its eyes and focus in on that target.

With Cog, the idea had been to build a creature with babylike intelligence to let it mature and learn through social interaction to reach "grown-up" intelligence. In other words, the Cog team followed insights from developmental psychology. But it didn't work because of Cog's slightly threatening looks and actions (instead of *re*-actions).

Cynthia Breazeal went for a few months to Santa Fe and came back with her first version of a robot that is purely reactive. The first Kismet was basically a pair of eyes and pink, movable ears on a socket. But even this was powerful because Cynthia had let someone from Hollywood build the eyes, which were stunning and attractive. With the ears and eyes together, this proto-Kismet already had facial expressions. When you walked in front of its face and its eyes followed you, you couldn't help but be drawn in. Next came a steel mouth, which enhanced the expressiveness of the face enormously. Sam Ogden, a photographer fascinated by our robots, attached long eyelashes onto Kismet, which enhanced its cuteness. Cynthia built silver eyelids (again, many more facial expressions) and finally came a red ribbon mouth that made subtle movements that enhanced its expressiveness.

"Kismet" is the Arabic term for "fate." And, indeed, the birth of Kismet was due to fate. Cynthia had been a member of the team that constructed Cog. Cog was a fascinating robot and the goal was to give it the intelligence to enable it to have successful social interactions with humans. But there were several problems with this approach. For one, the body coordination of Cog was so difficult that the builders had much less time than anticipated to build in social skills. Another more vexing problem was that people from the outside weren't necessarily attracted to Cog. Many admired its skills and looks, which were compared to Fritz Lang's vision. But, at the same time, people were also often somewhat scared. Of course, the members of the lab and the visitors did not experience this. But when I gave talks in more religiously oriented communities, most people expressed fear of the robot and a concern that its destructive potential would grow when it multiplied. So Cynthia made Kismet cute to prevent people from being afraid of it.

We will discuss Kismet more thoroughly in chapter four when I focus on the communality of our intelligence. For the discussion of robotic embodiment, I would like for now to concentrate on Cog.

Central Control or Distributed System?

What both robots have in common is their embodiment. While Cog is a torso with a head and arms, in Kismet, the embodiment is in its face, which consists of many different modules and creates a multitude of expressions. Both do not have central control systems but are made up of *distributed* systems. This distribution makes it possible for both systems to react immediately and without much complicated and lengthy computing to real-world environments.

Distribution in a robotic system can be easily explained when we take a look at Cog's arms. Human arms have seven degrees of freedom (DoF). DoFs are the general measurements for the flexibility of a body, whether it is human or robot. We humans can move both our wrists and elbows up and down and left and right. In our shoulder, we have the same degrees of freedom with an additional rolling DoF. Cog shares with us six DoFs, but a ball joint is one of the ingenious results of our evolutionary history, and engineers have not been successful yet in rebuilding such joints for robots. If the robot creatures of our future ever have full bodies, they will have much more trouble with their hips than we do.

Most arm movements require multiple joints, muscles, and tendons working together in order to coordinate the DoFs. The question is: Who controls these movements? Who supervises the interaction between the different parts and who coordinates their action? We call the search for such a central unit *homunculus theory*, as it reminds us of classical Greek legends about little artificial humans who were mostly minds but could control a body when one was given to them. The homunculus is the Cartesian "I," the point of certainty from which everything is explored or controlled.

This scenario of a homunculus in our brain controlling the body seems quite likely. In fact, this is how we tend to understand how we work. As soon as we ask ourselves how we move, we come up with this idea that there is someone in us that controls centrally all the stuff (muscles, tendons, nerves, etc.) we need in order to operate. Usually we call this mysterious someone our "I" and we usually associate it with self-awareness and self-consciousness.

You might have realized that, so far, I have carefully avoided the term *consciousness*. There are several reasons for this. For one, I am influenced in my language by my stay at MIT, and there the term was forbidden as there is really no definition of consciousness. Or, more accurately, there are so many definitions that the term is truly confusing. Second, and even more important, consciousness is often understood as that little homunculus in us that controls what's going on, and I don't think this is how we work. Quite the contrary, we are distributed systems, and we only think that we really control ourselves. Therefore, I do not use the term but talk of "self-awareness" instead.

More traditional AI approaches attempt to create a central unit that controls the periphery and is an equivalent to our homunculus. For instance, one might construct a central processor that is connected to all different DoFs of the arm. When an arm movement is needed for the execution of a specific command, the controller makes a plan about what motors, sensors, etc., it should move, sends the commands to the specific parts, and they move, hopefully, as quickly as our printer starts to print when we give it the command.

Cog, on the contrary, is a prototype, a model of different aspects of embodied AI, with different parts or capabilities. There isn't any real central control or central brain that runs everything. Instead, many independent modules interact with their environment, which can be another machine part, or the outside environment in the lab. That means all stimuli with which the various subsystems react come from either the other parts of the robot or its environment. It seems that humans "function" the same way. Our body is also a collection of millions of different parts and organs, each of which is capable of specific tasks. When they are all put together, they produce that miracle that is us.

So the DoFs in Cog's arm are not controlled by a homunculus, a central unit, but "correspond" with various motherboards in Cog's "brain." They are not directly connected, but they interact because through experience they "learn" motor control and the coordination of several motors in reaction to the demands of a specific task.

Grasping is a very early form of interaction between parents and their infant. Parents move a toy in front of the infant and she attempts to grasp it. She usually waves her arms around aimlessly, becomes more and more focused, and is finally able to reach the object. Cog acts in a similar way. If you hold a toy in front of it, it tries to reach the toy by waving its arm around; if you want to shake hands with it, it is eventually able to grasp your hand as it did with Harvey Cox.

The eyes test to what degree the baby's goal was reached, i.e., how near her hand was to the object. Through trial and error, she finally can grasp the object and when you "train" her, her capability for grasping improves. The learning is not abstract. Cog, like newborn infants, doesn't have the abstract system of a 3-D grid built into it. It has no abstract understanding of space. Instead it learns to coordinate its arms through feedback loops that register the weight the joints have to carry in a specific arm position. If a part of Cog's body is replaced by another one with different properties, the accuracy of every movement involving this new part has to be learned all over again. Growth can be tough on body coordination, as we all have experienced in our childhood.

This form of learning is *reactive*. Cog learns through failure, which is registered through visual input (is the arm close to the object that is to be grasped). The psychologist Valentino Braitenberg has invented two theoretical creatures to make this reactive behavior more understandable. The two vehicles have light sensors that are connected to wheels; the amount of light a sensor perceives determines the speed with which the connected wheel spins. In one, the left wheel is connected to the left sensor and the right wheel to the right sensor. If light for this vehicle comes from the right, its right wheel will turn faster and the whole vehicle will turn away from the light. In the other the sensors and wheels are cross-wired so if the light comes from the right, its left wheel will turn faster and the whole vehicle will turn toward the light. With this simple wiring, we get quite complex behaviors: avoidance of light and attraction to light. Even though these vehicles are thought experiments, we can see how two sensors and two actuators can create behaviors in the world because the system is "situated" in the world and reacts immediately to its properties.

If one were to create a robot with traditional AI methods, the system would have to be much more complex to fulfill the same tasks. First of all, what is light? It has to be transported into digital data. It then has to be integrated into a worldview. Then the system comes up with a plan and executes it by giving specific motor-control commands. And then you actually have to give motor-control commands to the wheels to get the behavior you want. One can see how such programming could quickly become quite complicated.

Braitenberg uses a different approach. Instead of programming explicitly, he uses the properties of the world and sensors. The system does not need to know what light is. It doesn't know that it actually reacts to light, avoids it, or is attracted by it. It just reacts. Therefore, one can do away with a whole bunch of computation. That doesn't mean that those things will be able to play chess. But they display behavior that seems to be more "lifelike" than classical methods.

A factory robot does its task no matter what. Imagine a robot with some form of hammering function; it hammers blindly, mechanically, on everything that is on the spot upon which it hammers, be it a sheet of metal or a child's hand. This robot has no sense of its environment, i.e., no sensors to explore the world around it. It is solipsistic. The Braitenberg vehicles, as well as embodied robots, depend on feedback from the environment to create behavior. They are not solipsistic but need to interact with their environment to act and grow.

So, the overall system does not have a central unit that controls the parts. Instead, sensors and actuators work in their local environments and are connected to each other in such a way that the system can act and react flexibly without extensive internal mapping since the reaction to inputs from the environment are local and immediate. The system's body is situated in the environment, interacts with its environment, and creates new and complex behaviors out of simple interactions. This philosophy gives this AI direction its name, "Embodied AI." And it is the interaction of all these individual pieces that gives us the interesting behavior.

The Anthropomorphization of Cog

Cog, obviously, is designed to look somewhat human. Even if people who meet Cog know immediately that it is a robot, they still react to Cog more strongly than they would if Cog were the shape of a ball or a cylinder.

Our technology is not far enough developed to build something truly humanlike. For example, we are nowhere near building a face as complex as the human face, with its over fifty degrees of freedom. Cynthia Breazeal is currently working together with Stan Winston, the inventor of creatures such as *Aliens'* queen or *Jurassic Park*'s dinosaurs, on a new robot with over fifty degrees of freedom: The robot is called Leonardo and resembles a furry, very cute animal. This robot is incredibly complex and expressive but its DoFs are distributed all over its body and not just in its face.

Because it is impossible for us right now to build a face as expressive as the human face, there is a strategy behind building something humanlike that is clearly not human. As we will see in chapter four, humans have a fantastic capability to analyze faces. If something is not quite right, we immediately get an eerie feeling; we tend to feel appalled by the sight of a slightly distorted face. A few years ago in Japan, television producers tried to replace news reporters with computer programs mimicking a human face. People called in by the thousands complaining that the face made them uncomfortable, as it was not "quite right." So, to build a face that does not repel but invites us to interaction, humanlike features are clearly needed but they should be nonhuman.

Hollywood has understood this principle well and has implemented it in many of its artificial creatures, such as E.T. and the gremlins. We feel that we can interact with these creatures because we think we can understand their facial expressions. They are humanlike enough to anthropomorphize them, but as different from us as possible so that when we interact with them, we are not uncomfortable.

The same thing happens with Cog. Cog does not have a humanlike face but it has a humanoid shape. Every movement it makes that is similar to a human being helps us relate to it. I remember one incident when Matt Williamson, the graduate student who then worked on Cog's arms, came into my office and urgently asked me to join him and Cog in the lab. As I have mentioned above, Cog's arms and other systems do not have preprogrammed behaviors. Instead, the various parts learn to work together to react to certain stimuli from the outside world. Every grasp is learned behavior. On this particular day, Matt had put touch sensors on Cog's torso to train Cog's arm coordination. He wanted to give the robot something to aim for, and so whenever Matt would touch the torso sensors, Cog would do so as well. It was simply programmed behavior but the effect was stunning. It looked as if Cog was exploring its own body.

When a baby explores her own feet, she grasps her ankle and concentrates totally on the thing that is there, even if she has no clue at this moment that this foot is part of her own body. It is a fascinating, awe-inspiring sight. You feel you have participated in a very intimate moment when the newborn slowly makes her very first steps toward adult intelligence and awareness. The moment with Cog was similarly magical.

It is this moment and other similar ones that make interaction with Cog so stimulating and yet so eerie. Humanoid robot behavior can make us uncomfortable because it can make us challenge our own superiority. If the robot behaves like we do, then how are we different from machines? This is the concrete example of the traditional "Frankenstein Complex"; the humanlike machine threatens us in its likeness to us, and we are afraid that it will turn against us.

If we now turn to the idea of robot-building as a prayer, then in such moments our deep sense of spirituality awakens. Our creation suddenly acts so uncannily humanlike that we realize its potential to become a nonhuman partner. As we are awed by behaviors in a newborn baby that remind us she will become a grown-up someday, these moments with Cog remind us that we might succeed in building a partner species and have to learn to interact with them.

Proprioception

We build bodies for robots to make it easier to interact with them. But there is another reason for building embodied machines instead of programming computers. If we take our own embodiment seriously, then the very way we work makes us think the way we do. So in order for a robot to become humanlike, it has to experience the world in a similar fashion and thus has to be constructed with a similar body.

Our perception apparatus is made up of senses with which we experience the world: smell, taste, vision, hearing, and touch. But many people do not know that there is a sixth sense. In fact, the term *sixth sense* has the connotation of something mystical—nonsensory perception like mind reading and premonition. But the sixth sense is actually an empirically demonstrable sense of our embodiment called *proprioception*. It is so all-present that we are not even aware that we have it. When we move our body parts there are feedbacks to the brain that help us to realize that we are actually moving.

The neurologist Oliver Sacks describes cases of patients with neurological damage. In one of his stories he describes a young woman who loses her sense of proprioception and ends up basically paralyzed. All muscles and nerves are perfectly healthy but she lacks the feeling of her body. As she doesn't get feedback from her actions, she cannot control her movements. Sacks, ultimately, is able to help her a little bit by making her move before a mirror. The sensual feedback of body movements she has lost is thus replaced with visual feedback loops. This way, she is capable of learning movements anew using visual clues to help her make correct movements. It is a grue-some story and whoever reads it becomes immediately aware that we are, indeed, bodies, and when we lose our sense of body, we cease to be able to function in the world.

The Embodiment of Mental Development

The famous and influential developmental psychologist Jean Piaget (1896–1980) was the first to analyze the steps of mental development in the infant. Piaget, who had thousands of interactions with newborns and toddlers, concluded that the meaningless or illogical utterances they made were not signs of less intelligence but rather of a different intelligence. Very small children think with their own logic. It takes many interactions with their caregivers before their intelligence becomes more like the intelligence of an adult human. Piaget defined very specific steps for children's development that are similar in all children; pediatricians still use his model to check on the mental health of an infant.

While Piaget saw that in the very early months after birth, motor control is an important tool for the development of mental concepts, he focused more on the mental than physical aspects of early child development. Adele Diamond from MIT, however, performed experiments that show that the development of motor and sensory skills is crucial for the development of any mental concept.

Objects in our world have a few things in common. They either *are* or *are not*; they remain the same over time when there is no other influence at work. If I leave some object in a room, go away, and come back later, I can safely assume that the object is still there and unchanged. Children, when they are born, do not have abstract concepts such as "object" in their head. They react to simple cues with simple schemes. Imagine moving an object such as a big red toy car in front of a baby and letting it vanish behind a screen. When it emerges on the other side, out comes a blue toy truck. Babies up to a certain age will not be astonished by this change. Only when babies develop a certain degree of motor control will they be surprised. One could say that the development of object constancy is a mental concept and Diamond showed that it is usually developed at the same moment when the infant learns to grasp and hold. That, of course, does not make Piaget obsolete. But we have to proceed carefully when talking about mental development, as it is strongly linked to the coordination of the body.

Diamond conducted interesting experiments in which she put bright toys in Plexiglas boxes and placed them in front of a baby. Access to the contents of the box was not in the front but on the side or in the back of the box. When the baby wanted to reach for the object, she couldn't reach it and would bang her hand against the box. The baby then quickly lost interest in the object. Only when the baby's motor control was so well developed that she could reach around and grab the object through a back or side entrance did she stay curious, and her understanding of the object developed. It seems that intelligence develops through motor control and not the other way around; that is, if the body does not develop, there will be problems in mental development.

We tend to think that our sense of self, often called *consciousness*, is a mental capacity and, particularly, the capacity that makes us special and distinguishes us from other animals. It seems to me that the term *consciousness* reveals, once again, the nature of *Homo narrans*, as it tells the story of human specialness and magic while our sense of self can simply be explained through proprioception and the way our body is modeled in our brain.

The Body in the Brain

Our complete body is mapped on the surface of the brain. This mapping is quite distorted, as there are particularly important body parts that take up disproportionally large brain areas. For example, the area involved with the lips or with the fingers takes up as much space as the area involved with the entire trunk of the body. This makes sense, since lips and fingers are very nimble and quite important for our interaction with the world. Also, the layout of the body map in the brain does not accurately reflect the actual location of parts on the body. While the feet are mapped on the top of the brain, the areas mapping the arms are on the lower parts. Our genitals, for instance, are mapped in an area that is below the foot and not close to the area that represents the thighs.

This body map shows how very closely brain and body are linked. If someone loses a limb, the area that represented that specific limb will cease to get input. Instead of laying dormant, it will often end up firing randomly, in which case the patient experiences phantom pain. Otherwise, it can be taken over by other body parts so that the loss of the limb allows other parts of the body more "brain power" and, thus, more sensitivity.

It was through research on body maps in the brain that the "God module" was discovered. The God module is an area in the temporal lobes of the brain that is particularly active when people report a "religious" or "spiritual" experience. Vilayanur Ramachandran, head of the research team who discovered it, stumbled upon it while studying body maps for the treatment of phantom pain in amputees. His team also did a study comparing epileptic patients with normal people and a group who said they were intensely religious. This latter study showed that the epileptics and the deeply religious displayed a similar response when shown words invoking spiritual belief.[1]

As we have seen in chapter one, the media reactions to this discovery were quite intense, even though the scientists insisted that this research in no way reduces religion to brain chemistry. But if we are indeed embodied as the Cog and Kismet research suggests, then the existence of a God module makes sense. If we are bodies, then whatever we deem as "mental" must have a physical manifestation somewhere in the body or in its environment. If *Homo religiosus* is embodied, then, of course, our religiosity must be embodied, too. Every form of experience—bodily, emotional, and religious—must express itself in physical form and does so throughout the body and in the brain. Our very self-image is embodied, as it consists, in part, by our body mapped in the brain.

Brain researchers know of the phenomenon of *aphasia*, a specific brain damage after an accident or stroke, where the patient develops a flawed self-image. It can happen that someone half paralyzed after a stroke will not recognize her paralyzed arm as her own. She will

---

1. One can measure activity in the temporal lobes through electrical monitors in the skin of a patient.

make up stories about how the arm must have been exchanged in the ambulance and nothing can convince her that this is not so. Her image of herself ceased to be accurate.

Oliver Sacks tells about another patient of his who had a brain tumor that was recognized too late and had already seriously damaged his brain. Among other things, he now was blind, but he had aphasia and didn't know that he was blind. He would argue, "I am the best judge of whether I am blind or not and I know I am not blind," while running constantly into walls. Again, we can draw only one conclusion. Self-awareness is not a mental concept but an embodied one and occurs only in a healthy body.

For many people, this sense of embodiment seems to be strange at first. But there is so much evidence, most of which has been discovered in the last fifteen years. Indeed, one could say, the last fifteen years saw a shift in the worldviews of many cognitive scientists in that the sense of the body became central for their understanding of intelligence.

Emotions As Prerequisite for Intelligence

If seemingly mental concepts such as religion or self-awareness manifest themselves in the physical world, then we have to ask what the relationship is between mental concepts and their physical manifestations. Are the physical reactions a consequence of the mental concepts? Or are the mental concepts a result of physical activities? Or are both connected yet in even a different way?

If we take the concept of embodiment seriously, then there cannot be any mental concept without its physical expression. We are aware of the mental aspects of specific concepts while they actually occur in the physical world. For instance, the neuroscientist Antonio Damasio studies people with severe brain damage in the frontal lobes, the part of the brain in which emotions are processed. If the frontal lobes are damaged through accidents or tumors, the patient is still capable of doing logic and math but ceases to be functional in a social setting. People with such a diagnosis tend to become extremely rude, because they lose their sense of appropriateness and empathy. They cannot make everyday decisions, as they lack the emotional instinct, the "gut" to do so. For example, if you want to set up a lunch date with someone with frontal lobe damage and suggest Wednesday, Thursday, or Friday and she is available on all three dates, she will not be able to come to a decision. Where our gut kicks in and decides, she will rationalize and weigh all the options and ultimately not be able to choose a date.

While we might be convinced that concepts such as politeness or the planning of one's schedule are mental processes, Damasio's research shows that these things are actually expressions of very specific brain activities. As surprising as this research seems at first, it is very much in line with our concept of embodiment and is against the Cartesian mind-body split. Consequently, Damasio called the book in which he describes his research *Descartes' Error*.

The Christological Concept of Incarnation

The Christian belief that Jesus of Nazareth was the Christ is based on the very same understanding of embodiment that, in itself, is deeply rooted in the Jewish tradition. Central for Christianity is the concept of incarnation, the belief that God became embodied (*soma* and *sarx*!) in Jesus. Most Christian traditions see the man Jesus as the redeemer (in Greek: *Christos*; in Hebrew: *messiah*). The church fathers already formulated the core of Christian faith as follows: God became human so that we can become like God.[2] In becoming embodied, God bridged the gap between Godself and us; in becoming embodied like us, God can understand us, empathize with our estrangement, and show us ways out of it.

But what does it mean that God became human? To what extent was Jesus human and to what extent was he divine? It turns out that most Christians have some problems embracing not only their own embodiment but even more so the embodiment and true human nature of Jesus.

A classical depiction of Jesus in art usually shows him as a cute blond boy with curly hair, tenderly embracing his mother, Mary. What I like in this image is the hug, a clear sign that the interaction between mother and child is embodied. It is a very tender and endearing gesture.

---

2. The Church fathers formulated this sentence in exclusive language: "God became man so that we can become like him," and for obvious reasons I modified it.

What I don't like is the obvious manipulation of historical authenticity. First of all, the father, Joseph, is rarely shown; it is usually just mother and child. Mary and Jesus tend to be depicted as light-skinned, and Mary usually wears quite rich clothing. If there is a landscape depicted in the picture, it often resembles Tuscany because of the influence of the great Italian masters of the Renaissance. But the biblical accounts tell a different story. First of all, Mary and Jesus both were of dark complexion, very poor and quite uneducated. Both Mary and Jesus spent their whole lives in Israel; they were certainly never in Tuscany.

But besides these obvious historical mistakes, I resent something much deeper in this traditional imagery. Baby Jesus simply looks too good to be true. One cannot even start to imagine him spitting his food at Mother when she feeds him, crying without end for no apparent reason, or smelling of dirty diapers.

Jesus as the perfect child with the perfect mother is a story that was created over millennia about the human nature of God. Jesus, obviously, was often seen as superhuman; the true human nature of the God Jesus has always been hard to grasp. Just consider the fact that for many hundreds of years the church fathers discussed whether Jesus peed. They even went so far as to say that Jesus never really ate as it was unimaginable for them that Jesus ever would shit.

That we are appalled at the thought that Jesus might have smelled of sweat, might have had unpleasant breath, or might have had diarrhea, reveals how we perceive our *own* embodiment. We find such a notion disgusting because we ultimately find our own bodily functions repulsive. Jesus as superhuman should be free from ugly bodily actions. I frequently discuss the human nature of Jesus in my classes, and many students are very uncomfortable with the idea of Jesus digesting. Without knowing it, they repeat the arguments the church fathers brought up more than 1,000 years ago, and they argue that Jesus as God could certainly avoid digestion as well as any bodily odors and other aspects of an embodied life they perceive as disgusting. For me, these discussions do not just demonstrate how much the true human nature of Jesus is abhorred, but also show how much we understand our own bodily functions as nasty and regrettable. If you think about it, what is so bad about digestion? And the rejection of sweaty smells is certainly cultural and of recent nature. Napoleon famously wrote to his wife, Josephine, from Egypt that she should stop washing because he was coming back soon.

Another question is the sexual nature of Jesus. Many theologians today assume that Mary Magdalen was more than a friend to Jesus, and there are many good arguments for it. There is, for one, the beautiful Easter scene in the gospel of John 20:11–18, in which Mary Magdalen meets the resurrected Jesus and first thinks he is the gardener. After she recognizes him and is ready to run into his arms, he says, "Don't touch me!" The Greek term *apto* (to touch) used here has two main connotations. The one addresses the touch for blessing, as in anointments; the other refers to sexual or sensual touch. The use of *apto* could indicate that they usually would have touched, but the text is ambiguous about the nature of that touch: Was it a hug or a blessing? But it was one thing for sure: it was a bodily touch, an embodied interaction. This means Jesus was not untouchable and distant, but that he touched and hugged his fellows and friends. With his death, this bodily connection had to be severed. Even if he was resurrected, something changed. The resurrection did not occur to ease the pain in the people who lost him but was an act of much greater universality. The prohibition to touch could mean that the people who are left have to learn to accept his death by creating a new life focused on his message but not on his body.

Among the nonbiblical accounts of Jesus' life, the most discussed is probably the gospel of Philippus.[3] It contains a beautiful scene that starts with the observation that Jesus often hugged and kissed Mary Magdalen. The disciples became jealous and asked Jesus why he would love Mary more than them. Jesus answered with the question: Why wouldn't he love them the way he loved Mary?

In addition to a very possible love relationship between Mary Magdalen and Jesus, the gospel of Luke mentions several times the many women who not only followed Jesus but also financed his movement. Jesus must have had incredible charisma and quite a lot of sex appeal; otherwise he would not have been so successful. As we have seen several times already, thoughts and ideas without embodiment won't fly with a crowd and would not have survived for 2,000 years.

3. The gospel of Philippus was bound together with the much more famous gospel of Thomas. Therefore, this passage is often quoted as part of the Thomas text. Both texts were found in Nag Hammadi.

It also makes sense theologically that Jesus was sexually active, as sex is, after all, an integral part of human life so that Jesus as true man must have encountered it. But many people are uncomfortable with the idea of Jesus being sexually active and reject the notion; they assume that even if he could perform sexually, he would have nevertheless lived in celibacy. Even if that were true—I want to leave that open—one still could address the question of Jesus' sexuality by asking whether Jesus ever had wet dreams. I don't mean any disrespect for Jesus as the Christ, but it seems to me that we have to learn not only to embrace our own embodiment but also that of God in Jesus. God became human to bridge the gap between God and us.

I personally think the discussions about Jesus' bodily functions contain a wonderful message: The more human Jesus becomes, the greater the redemptive value of his life. A God who becomes truly human—experiencing the ambiguities of embodiment, the estrangement of relationships, doubts, fears, and, ultimately, death—is profound. Speculations on Jesus' sex life and digestion do not detract from Jesus' glory. Quite the contrary, they enhance the greatness of God's gift according to Christian belief that God wanted to become truly human in order to understand us better, empathize with us better, embrace us, and, therefore, become closer to us. In Jesus, we can find contentment with our experiences of ambiguity and paradox as he himself was a paradox: true human and true God.

Our understanding of our embodiment then, does not just help us to know ourselves better; it also enhances our understanding of Jesus and the greatness of God's love for us, such that God truly became human. But the key for Jesus is that he lived in community. God created a community with humans through Jesus, so that God could share our lives with us. God shared our physical space. God shared culturally developed assumptions and beliefs. God experienced the same code of behavior as humans around God. God was bound to one specific culture; thus God faced the same limitations we all do.

For God, the embodiment of Jesus, the incarnation, was a way to establish a closer relationship with humans. It turns out that our own embodiment is also a means toward another end. We are bodies in community. Only in sharing a physical space can we truly become human.

Are Our Bodies Special?

The importance of the body in our understanding of ourselves roots us firmly in evolutionary history. Our intelligence is not something abstract but is something developed during the course of evolution for our survival. Since it is our body that has been created through the evolutionary process, our intelligence is embodied within it. Intelligence is the capability to survive in any environment; chess and math are by-products of our intelligence, not the core. Humans can intentionally shape their environment in order to survive.

If we are a product of evolutionary development, then we are much more connected to the rest of the animal kingdom than we tend to think we are. In chapter one, we saw that many of the *Homo* metaphors place us in the animal kingdom, that we have much in common with other species. We talked about our sense of modesty and awe when confronted with the complexity of creation. Here now, we can recognize how much indeed we are a part of nature, not distinct from it, not special.

According to evolutionary theory, we humans *are* special but we are only as special as, let's say, ants or horses. We are a wonderful and magnificent species, just as rhinos and cockroaches are. An understanding of ourselves as yet another species within the animal kingdom is consistent with biblical wisdom and does not diminish the greatness of humans. Only when we understand ourselves as animals can we deal with the imperfections of our bodies. To have the esophagus and the windpipe crossing at our throat, thus creating an extremely vulnerable spot, is stupid. Also, to have an appendix without any function that can inflame and kill us is not the smartest design. There are many flaws in the human system, but all of them can be explained by evolutionary development and by comparing us with other species. If we take our embodiment seriously, then we have to consider how our body became the way it is in order to understand ourselves.

Embodiment, as analyzed above, gives us a completely new key for our self-understanding that is not abstract but very concrete and makes empirical sense. However, the journey to understanding ourselves doesn't stop at this point with insight into our own embodiment. The progress in robotics, the move from Cog to Kismet, reveals that the body is only a means toward another end: community. We humans are embodied so that we can interact with one another and with the rest of creation. Building a body is not enough to achieve humanlike intelligence. The body has to be treated in a certain way and has to have a very specific and complex capability to interact with others in order to survive.

CHAPTER 4

Embodied Community

And GOD formed Adam of the dust of the ground,
And breathed into his nose the breath of life
And Adam became a living being.

GENESIS 2:8

So GOD caused a deep sleep fall on Adam and he slept;
Then God took one of his ribs and closed up its place
with flesh.
And the rib that GOD had taken from the man God
made into a woman and brought her to the man.

GENESIS 2:21–22

And GOD made garments of skin for the man and his
wife and clothed them.

GENESIS 3:21

Humans as Communal Storytellers

In the last chapter, we explored embodied intelligence and the research that is available to support this understanding of humans. But what role does the community play in our self-understanding? As we have seen in chapter two, it is very hard for us to accept theories about the world in which we live, such as heliocentrism or relativity, if they do not fit our bodily experiences. If we are bodies, then the physical world in which we live is central for us. In Embodied AI, we speak of us and all other animals as *embedded* systems.

Other humans play a large role in our surroundings. We are embodied storytellers and, therefore, ultimately communal. It is impossible to understand ourselves by just looking at a single human. Community transcends biology. Our respective communities turn us into storytellers, and that makes it difficult for us to see different aspects of one story. Our respective communities give us support and a worldview that answers our questions for meaning; it provides mythos. It doesn't matter if the community is a family, a church parish, or a scientific community. Every group has similar characteristics that are deeply rooted in our embodiment and, ultimately, can lead to mythos–logos entanglement.

The Cybernetic Model

In science, it was cybernetics (from the Greek *cybernetes*, "helmsman") that first recognized the importance of the environment on every single organism. Cybernetics can be seen as the mother of AI. A scientific movement that began in the 1940s, it studied the mathematics of machines and their control and communication. While it was first applied purely to existing and imaginary machines and, thus, was part of the newly emerging field of computer science, cybernetics later became the study of control processes in organisms, machines, and organizations. Although the term is still used within academic circles in England, in the United States cybernetics belongs solely to the realm of science fiction.

Most of the earliest thinkers of AI were cyberneticists, as the rules for controlling many parallel processes helped them to understand biological systems and aided in their comprehension of humans and their intelligence. In 1952 they formulated that an organism and its environment must be modeled together in order to understand the behavior produced by the organism. That means every organism is embedded in its world; it does not only *re*act to the stimuli from its environment but, rather, ceaselessly interacts with it. As we have seen, Cog and Kismet are distributed systems in which the sensors do not "know" whether their input comes from the system itself or from the outside world. This situatedness mirrors exactly the cybernetic model of biological systems.

Unfortunately, this theory was misunderstood. Psychologists, in particular, were convinced that it is possible to understand humans by merely studying their rules of behavior. While this in itself was not bad, they did not quite grasp the importance of the embeddedness and reactivity of behavior. While humans theoretically could be described by their behaviors, these behaviors are so intricate, embedded, and reactive to many simultaneous stimuli that it is impossible to analyze any behavior in all its complexity. The movement of behaviorism was therefore ostracized. As is often the case, when behaviorism fell into disrepute, many people threw out the baby with the bathwater and rejected the cybernetic model as well as embeddedness. Hence, in the last thirty-five years, people didn't even know about this interactive model for humans but learned only the individualistic model in which humans are active because of their intelligence and their will, but are not *re*active. While I am not defending the cybernetic understanding of humans as totally valid, I think they had one important and valid point: In order to understand humans, one cannot see them as individuals but only as part of communal systems and embedded in their world and society.

The importance of the environment has always been central to the theory of evolution of species, but only insofar as the change of environmental conditions will force adaptation and thus help mutations along. However, modern versions of evolutionary theory do not only acknowledge the extent to which genetic drifts and mutations in a species are influenced by environmental influences and guides, but also concentrate more on the influences of the environment on individuals. As an example, the sex of crocodile babies is determined by the temperature in which the eggs are bred, which means the relationship between genotype and phenotype,[1] which traditionally has been understood as entirely bottom-up, is actually very complex. It is, therefore, quite impossible to determine or even understand an individual based on the genes and species alone. We are not totally determined by

1. The phenotype of every individual describes its appearance. For instance, there is a well-established link between body height and nutrition; if underfed, a child might not become very tall. Even if there is a genetic predisposition for height, the *actual* height of a creature, meaning the phenotype, cannot be determined by genes alone.

our genes. We cannot be looked at as individuals shaped by genes. Instead, we have to look at species characteristics, social influences, and environmental influences in order to understand ourselves.

Our Bodies As Communal

In the rabbinical version of the story of the Fall (Genesis 2–4) (see chapter 3), Adam and Eve were skinless in the garden. They were truly naked, vulnerable, and without means to touch and thus connect to their environment. They had no sex or sexual attraction toward one another, and they had no means to feel either pain or pleasure. The feeling of embodiment can come only with skin; the fundamental paradox of embodiment is that, through skin, the same touch can be pleasurable or painful. The way in which we perceive a certain touch depends entirely on the *context* in which the touch occurs. So, the biblical metaphor of skin points beyond the individual human body to the possibilities and the problems of human interactions. As it becomes immediately clear, this interaction is not limited to other humans; instead, we interact constantly with the entire world.

In Oliver Sacks's story of the woman who lost her sense of proprioception, the most touching element is the patient's thankfulness for every sensory input. Because she does not feel her body anymore, she revels in wind on her skin and in her hair. The feeling of a breeze connects her back to the world. One could say that skin is the organ that places us in the world and connects us with it.

Skin, thus, is the organ that defines us. Whatever is outside of our skin, we have learned to define as "surroundings," while everything within our skin we understand as part of ourselves. Our skin helps us to define our boundaries. We hopefully feel good in our skin; we don't want to jump out of our skin and we don't like to have our skin crawl with disgust. We hopefully don't have too thin a skin, but we also don't want to have elephant skin. The skin has huge metaphorical value in both English and German, and it makes sense that the biblical writers would use this metaphor to describe us and our state of estrangement. Skin is the organ that separates us from the rest of the world—and yet it is also the organ with which we interact the most. Touch is, in my opinion, the most profound of our senses, as it never ceases; our skin constantly enables us to interact with our surroundings.

In the last few years, a whole therapy has been developed to help parents touch their children again. Due to a constant debate on sexual harassment of children, many parents became insecure about cuddling their babies because they were afraid they might be touching them improperly. This lack of touch, however, led to major developmental problems in the child. It turns out that a baby who is not touched on a regular basis and not carried close to the body of one of her parents will not develop properly and fully. Skin not only separates us from one another but also connects us.

Bodily touch can overcome the estrangement between people. Cultural differences, however, cause problems. Some people are very comfortable with touching, while others feel that touch invades their privacy. But the fact that we touch for greeting makes sense: In touching, we recognize the other in her skin and build a relationship with that other body. The handshake, common in most Western cultures, is a remnant of the need for a hug, as it still maintains a level of touch. The French ritual of kissing each other's cheek three times is even more of an embodied act.

Many people have pointed out that sexuality can also serve as a metaphor for a true, unestranged togetherness of two people. In a very intense sexual act people often lose their boundaries and stop distinguishing between their own self, defined by their own skin, and the other. We will see in chapter five how sexuality can even be a metaphor for an unestranged state of being, theologically called a *state of grace*.

As we discussed in chapter three, the biblical term *jada* (to recognize, to sleep with) addresses the communality of the sexual act best, as it includes not just satisfaction but also the recognition of the other in his or her otherness. The metaphor of the penis bone out of which Eve was made now wins an additional meaning. Human males are among the very few mammals without a penis bone. The possibility of intercourse depends entirely on the blood flow in the penis; there is no natural mechanical help. That is, in humans, the success of a sexual act depends in part on *jada*. It does not just happen when a female is in heat and thus fertile, but has relational value.

There is a good evolutionary reason for the specialness of sexuality that, after all, distinguishes us from most other mammals. Human babies need to be taken care of for a long time before they are capable of surviving on their own. They need care and education. A man who just spreads his sperm has much less chance of surviving offspring than the man who takes care of his offspring and teaches them survival skills. Sexuality can establish a relationship that provides a stable and secure environment in which infants can grow up and become independent. The interactivity of sexuality creates a community between man and woman that provides an environment in which offspring can grow. This fact demands that everything be done to make a union between a man and a woman a stable one. The commandment "you shall not break your marriage" mirrors the importance of a community for the survival and development of offspring. Note that this commandment does not contain a concept of faithfulness or exclusivity. In the context of the Hebrew scriptures, a woman can break only her own marriage and a man can break only another. A woman breaks her own marriage if she creates a situation in which the offspring might not be from the partner with whom she has a union. As the upbringing of offspring is hard work, this commandment was supposed to provide men with certainty that the children they raised were really their own. If I describe sexuality as community building, I don't mean it as an overly romantic concept with love "until death part us." Rather, I mean it in a pragmatic way that humans need to create special bonds in a family structure that enables all members to survive as a species. Sexuality is one very important aspect of this bonding process; romantic connotations such as faithfulness have been added only recently.

There is another biological argument that justifies the importance of a man staying in his environment after he has slept with a woman. A man who just spreads his sperm around has no way of knowing if the woman has conceived. In most mammalian species, women have brief episodes of fertility, and during this specific time, they produce pheromones that attract the males of the species and make them fight over the female in heat. I remember when our dog was in heat and all the male dogs from our neighborhood gathered at night howling in front of our house, while our fox terrier raised hell to be with them.

Even if there are a few similarities between these dogs and our own coupling behavior, biologically we are far removed from this model of procreation. Instead, human (as well as some other primate) females have what is called a hidden ovulation. We are cyclical and can conceive a few days of each month and it is very hard to know which days these are. So a man who stays with a woman for a period of time has a much better chance of getting her pregnant than a man who doesn't. As sexuality is communal, it can create a prolonged interaction between male and female and thus a relationship that is ultimately the bond out of which a family can grow. All these pragmatic remarks are valid in a community that does not know of contraception and genetic testing. These technical achievements have fundamentally changed our sexual interactions. But if we want to know who we are, we have to stick with biology and not with very recent cultural changes in a small part of the world. For both the Bible and the evolutionary account of why we are the way we are, sexuality is not correlated to procreation alone. Of course, it is necessary for the creation of offspring, but the communal aspect of the sexual act transcends it into a form of self-expression of all humans within their community. It also supports child rearing, as our offspring are dependent on us for a long time. If a stable relationship exists, the survival of the offspring is much more likely.

The Soul As Communal

Many people in the Western world have forgotten that they are communal. Many of us understand ourselves as individuals. But there is actually a strange correlation of the theory of cybernetics and the biblical understanding of the soul. Both understand humans as deeply embedded in community and both see the environmental influence on the individual so profoundly that any understanding of a single human depends on knowledge of her community; in the Bible, this is expressed with the term "soul" (Hebrew: *nefes*).

As I have pointed out earlier, the philosopher Descartes at some point located the soul in the pineal gland. But what is the soul? Everyone thinks they know what it means but—as with the term *consciousness*—if you ask people on the street you will hear many different opinions. Often, consciousness is seen as the secular translation of "soul." Soul can mean intrinsic value, spirituality, God within us, thought, emotion, and much more. However we understand it, we usually comprehend the soul as something unique that every one of us possesses. Even if the soul (or consciousness for that matter) emerges from bodily mechanics and functions, it is seen as a separate unit and is often used to distinguish us from other animals.

The term *nefes*, however, reveals a completely different understanding of the concept of a soul. *Nefes* means simply breath and throat.[2] Breathing is one of the most important acts of living beings, as we all depend on oxygen and could not survive without it. If you find someone lying on the floor, you first check if she is still breathing. This is the proof of life without mechanical tools. Then you might check her pulse, see if the blood flows, which is another irrefutable proof of life in most animals. And, indeed, part of our *nefes* is located in the blood. So an additional meaning, according to ancient Israelites, is "living being," as *nefes* addresses breathing and blood flow, the fundamental forces of life. It can also mean an individual human or a group of people. Theologically, the term *nefes* defines us as embodied beings in community, the community of all living beings, the community of all red-blooded animals, and the community of humans. Biblically, *nefes* emerges out of the history of interactions between God, every individual Jew, and the Jewish people as a community. It is a triangular relation and the most important relation between God and God's people.

This multilayered meaning of *nefes* brings together many of the concepts we have discussed so far. On the simplest level, *nefes* refers to our life force, the blood. On the next level, *nefes* refers to our embodiment, our being alive. On the next level, *nefes* refers to our being in community with each other, a community in which each member is valued. This value is assigned through the highest level of meaning of *nefes*, God in relationship with us. *Nefes* refers on a general level to the relationship of God to all living beings, but within the context of Old Testament theology, it usually addresses the relationship between God and God's chosen people.

The term *nefes* is conceptually connected with two other biblical terms: *chaim* and *ruach*. *Chaim* means "life," but also life span, safety, and health. *Chaim* does not refer just to being alive but to

2. Note that in biblical Hebrew a word can mean both an action and a subject connected with the action, such as here with *breathe* as action and *throat* as the organ with which you perform a certain action. This little example of the peculiarities of Hebrew shows already that the people who thought in and spoke this language saw reality in a different way!

being alive with *nefes*. We are alive when we are in a community because community provides *chaim*, that is, health and safety. In modern Hebrew, it is also used as "cheers" before a drink. The embodied understanding again shines through. Life is not an abstract concept but addresses the concrete lifetime with its own cultural and historical context, lived in a community in safety, health, and joy.

Ruach is often translated with "spirit," even though it has become clear by now that the theology I am outlining here has no use for a disembodied understanding of spirit. In the Hebrew scriptures, *ruach* often refers to God when God is interacting with humans by using physical means (such as the burning bush, clouds in the sky, or wind). But *ruach* can also mean "wind" or "temper." So, again, this word has an embodied, physical connotation. When in the act of creation God makes Adam out of dust, God has to breathe the breath of life into Adam to make him alive. This metaphor serves, of course, as the idea for the golems who come to life only when they have a paper with the unspeakable name of God in their mouth or have the tetragram on their forehead. The Hebrew term that is used here for "life" or life force as a result of God's breath is *ruach*.

All three terms together suggest an understanding of ourselves that is embodied and physical, and where the body is a means to create community. In the merging of community with culture and the constant maintenance of relationships, "spiritual" concepts such as love and trust emerge and then create the *nefes*. The soul is our true human capability to be—beyond the boundary of our skin—in constant interaction with humans, with other animals, and with God.

The biblical concept of "soul" is therefore not an entity that we all have and that is separated from our bodies or might even survive physical death. Rather, it is something in which we all participate, something that gives our life meaning. The meaning of soul is the same in the Hebrew scriptures and in the New Testament. Only later in time did the idea of a disembodied soul emerge in Christian thought.

One reason for this fundamental shift in the self-understanding of humans lies in a translation error. At the time of Jesus, Jews had become more integrated in the Hellenistic culture, and the well-educated had long started to speak Greek instead of Aramaic. In the third century BC, it became necessary to translate the Hebrew scriptures into Greek so that the majority of Greek-speaking Jews could still access them. The result was the Septuagint (in short: LXX), which became the authorized Greek Bible version for most Jews. Even if the legend tells us that seventy-two scholars in the library of Alexandria miraculously came up with a perfect translation, the fact is that the Septuagint shares the same problems with all other translations of the Bible.

First of all, there are words and concepts that don't have an equivalent in the other language. Also, some ideas might seem ludicrous in different environments. For instance, the often used biblical metaphor of water in the desert makes sense only if one knows what a desert is. Another tricky point for the Septuagint was that, over the centuries, the Torah had become a sacred book. Whenever a Torah started to physically fall apart and was copied, the old Torah was buried and the new one became the authority. But humans are not perfect; we make mistakes. It sometimes happened that a copyist changed two words around, accidentally wrote a different word, or left words out. These copy mistakes had to be translated and given sense, which might not be the same as the original.

The translators of the Septuagint used the term *psyche*, (the Greek term for soul) to refer to the concept of *nefes*; *psyche* for them referred to the embodied, communal understanding of humans in the Hebrew world. In old-fashioned English, one can still find remnants of this different meaning when one hears statements such as "all souls were lost"; here, soul means a human body and does not refer to something mysterious and abstract. When the New Testament was written, it faced the same problems. Jesus spoke Aramaic and certainly used the term *nefes*, but the authors of the gospels were writing in Greek and used *psyche* when referring to *nefes*.

But there is a major problem with the term *psyche*. Even if the authors used it to refer to *nefes*, this was not understandable for people without a Jewish background. In the Hellenistic world, *psyche* was understood in its Platonic sense as something metaphysical and disembodied, something individual, that needs to be freed of the body in order to be fully evolved.

Historical events in the very early years of Christianity explain why the Jewish understanding of *nefes* was lost and the Hellenistic concept of *psyche* took over. In 49 AD, the Roman emperor Claudius evicted all Jews from Rome, regardless of whether they were Jews or Jewish Christians. At that time, Rome was the intellectual center of the known world and the Roman Christians were highly influential over the other groups. When the Jewish Christians were evicted, the "pagan" Christians were left without knowledge of the Jewish heritage within their own religion. When they read the gospels and letters they received, they naturally interpreted the term *psyche* in their own context, relating it to Socrates' and Plato's teachings and not to Abraham. Even if Claudius let the Jews return a few years later, the damage was done: In these early formative years, Christian intellectualism had been unguided by Jewish thought and, therefore, had mixed the teachings of Jesus and Paul with Greek philosophy. With the destruction of the second temple in 70 CE, the importance of the Jesuanic parish in Israel diminished quickly while the influence of Hellenistic-thinking Christians grew. The influence of neo-Platonic thought increased during the centuries. Today, we have to fight two thousand years of this influence to revisit the Bible and understand that our soul is what makes us part of the community with other humans and the rest of creation.

Health As Social

Chaim means alive in health and safety, and we have argued that both health and safety are possible only within a community with *nefes*. If we truly are communal, then our health must be, too. But why is health not simply connected to our own individual bodies? What influence does the community have on our individual diseases? It seems that our bodies are indeed so communal that they rely on interaction for their well-being.

This becomes especially clear in most native healing processes, which are fundamentally social. First, the patient goes to the place where the healer is located. Here, she prays, meditates, and performs certain rituals. Then, the patient interacts with an acolyte, who usually wears signs that symbolize his or her status, insignia or light face colors; this first interaction consists of an interview about the health problems, as well as some rituals. Then, the patient meditates and prays again. After a while, she is finally called to the healer, who is dressed in elaborate clothes and wears insignia of power as well as heavy face paint. Again, the patient is asked about symptoms and then the healer performs a healing ritual. Often, the healer will then give the patient a few rituals to perform, as well as a herbal remedy and advice about how to prepare and take it. In each step of this healing process, the whole community is involved. They are either standing around the healer's place or are actually present in the meetings. They pray, sing, dance, and participate in all, and this group's participation is deemed equally important to the healing process *and* the interaction between healer and patient. In many of these cases, the patient will actually heal even if the medicine has no clinically proven healing powers.

Many of us reject alternative medicine and prefer the clinical practice of a Western doctor. We live in a culture that still believes in the objectivity of science—the white, professional-looking coats, the sparkling cleanliness of the equipment, and the strong smell of antiseptic in the examination room. All these elements reassure us that we are very well taken care of. The idea of going instead to a native healer's tent with various incense, weird smells, and potions seems ridiculous and is often a subject of jokes.

Yet thinking more about our own healing processes, we have to realize that there are more parallels between traditional and modern healing methods than we might think. If you go to your doctor's office, chances are that you have to wait quite a while in the waiting room. Usually, you will find brochures and old (very old sometimes) magazines so that you can entertain yourself—at least a little bit. Then, a nurse will call you in and weigh you, as well as measure your height. Then, you have to wait a while longer, and finally you will be called into your doctor's office. The doctor will give you a thorough exam and prescribe some drugs. Finally, you get an appointment for a follow-up visit and go home.

Sound familiar? If we look closer at this "modern" approach we will realize that its process is parallel to the native healing ritual. Translate the waiting into meditation, the acolyte into nurse, and the doctor into shaman. The nurse usually will wear a white coat and some other professional garb. The doctor will wear a white coat and—as insignia of her power—a stethoscope. One could draw many more parallels between the two but most people don't see a connection because they feel that the drug they are getting is crucial for their healing process and the doctor's visit is just a means to an end: a prescription.

The parallel, however, falls short when we look at the embeddedness of the healing process in the tribal community. We usually visit a
doctor alone and often don't even talk about what bothers us to
friends and family. Healing is seen as individual, as something that
occurs in our body and that is not necessarily connected to our social
interactions.

Herb Benson from the Harvard School of Public Health was
among the first researchers to study the contributions of the community to the individual's healing process. I met him several times and he
told me about two experiments he conducted. In one case, he sent a
rabbi daily to talk to patients after open-heart surgery. In comparison
with a control group without daily visits, the counseled group of patients left the hospital, on average, three days earlier. Herb added that
they might just have wanted to flee from the rabbi, but then it turned
out that the patients who had talked to the rabbi also had significantly lower relapse rates than the people from the control group.

Before any surgery, it is common that the anesthesiologist visits the
patient the night before and informs her about the procedure and possible dangers. The patient, then, signs a consent form. Usually such a
visit is very short and perfunctory. In another experiment, Herb sent
the anesthesiologist to a group of patients with the order to spend a
long time with them. The anesthesiologist sat down at the bed and
talked at length with the patient about her fears and anxiety, her situation at home, and much more. Compared to a control group who
had the usual, perfunctory visits, the patients with extensive visits
needed only a third of the anesthesia chemicals.

Both experiments demonstrate what many indigenous cultures
have never forgotten. Humans are so fundamentally communal that
sickness and healing are directly connected to their interactions with
others; both always have a social component, and a healing ritual that
involves community is far more effective than one that does not.

This still leaves us with the question of the effectiveness of the
drugs. Surely, some drugs such as chemotherapy or penicillin cannot
be replaced with a healing ritual and herbal remedies. However, new
studies show that the efficiency of drugs also depends fundamentally
on community. Every drug development depends on the use of fake
drugs, placebos. Some patients in a drug study will receive the test
drug while others will receive a sugar pill that looks exactly like the

drug. These studies are always double blind, that is, doctors, nurses, and patients do not know who gets what. This is to prevent doctors and nurses from subconsciously treating the patients with placebos differently from the patients with the drug. Traditionally, the function of placebos has been understood as "make believe." Patients get a sugar pill and think it is the drug; they believe they feel better and, therefore, experience a real improvement of their symptoms. But within the last few years, many people in drug development have come to see things differently. Every patient in a drug study participates in extensive social interactions. All of them are taken care of better than patients not involved in a drug study. They are seen regularly by medical personnel and receive dietary counsel. Many participating patients change their lifestyle by living more healthfully. Overall, all patients improve in health whether they take a placebo or the real drug. Complex drugs, therefore, have only to be 15 percent more successful than placebos to be approved for the market. That means placebos do not just cause autosuggestion in the individual patient but serve as tools for healing processes in which a community plays a major role. This is not to say that drugs don't work, but this research reveals that healing is social, and the more a community is included in the healing processes, the more success a drug treatment will have.

External Scaffolding

Findings in clinical psychology add yet another dimension to this communal understanding of human life. As it turns out, our brain does not "store" information and retrieve it when it's needed. Even if we often understand ourselves this way, it is due to the computer-morphisms we have learned in the last two hundred years to describe ourselves (see chapter three). Our brain does not function as a computer that stores data, but instead works with a method psychologists call *external scaffolding*. Rather than "storing" information, the brain creates pointers and uses properties from our surroundings to associate with them and, thus, remember things.

Many people, myself included, know that they have dialed the correct number on the telephone when they hear the dialing "melody"; they don't remember the abstract numbers but instead remember the melody and perhaps the pattern they type on the phone. When I first came to the United States and traveled a lot between Germany and America, I could never remember American phone numbers in Germany, and vice versa. After talking to other foreigners at MIT who had similar experiences, I realized that I remembered not the abstract concept of numbers but the sound the numbers made when I would speak them aloud. In Germany, I spoke German and thus could not remember the numbers I had learned in English. In other words, the retrieval of the number depended on the context and the language I was surrounded with, and only after I realized this could I do better.

Perhaps other people have had similar experiences when they went on vacation to a place that did not resemble their hometown. Often, people find themselves not remembering the most mundane things from back home just because the properties of their regular environments are lacking.

So our brain "scaffolds" constantly between its own pointers and associations and the external world, and there is strong empirical evidence for this. Daniel J. Simons has conducted experiments in which he produced various videos to test the memory patterns of his students. In one film, you see a man on a chair in his office. Suddenly, the phone in the hallway rings and the man stands up and leaves his office. The next scene shows a man on a phone in the hallway but he is not the man of the previous scene. And yet, around one hundred graduate students who viewed the film saw the man from the first scene and did not notice that he had changed. This experiment shows our desire for pattern recognition, as well as our desire to make up stories and draw connections that make sense of the world. We see a man, hear a telephone ringing, see the man standing up, and, of course, when we see in the next scene a man on the phone we automatically assume that to be the man of the first scene. We can't help it. We create patterns and stories.

But one can also argue that we couldn't make up such a mistake if the scene had been accurately "stored" in our brain; if, when we saw the scene with the man on the phone, then recalled exactly the previous scene, we could not make up a story of the man being the same. The accuracy of our memory would not allow us to do so. So it seems that our brains do not store data like a computer but create associative patterns and stories.

Simons executed another experiment in which students were shown a film clip of two women talking. Every now and then the camera focused on the actress who was speaking so that the other was invisible. In each such moment, something changed about the appearance of the invisible actress; her hair was altered, or she put on a scarf, or wore a different T-shirt. In total, there were ten changes, but of all the groups of graduate students to whom Dan showed this film, only one student noticed that something had changed without being exactly sure what that was.

Again, there are only two possible interpretations of these failures. One is that our brain is a big computer and we store what we perceive in it. If this were the case, we are doing a very poor job; otherwise, we would notice the changes. The other interpretation is that we don't store anything but just react and, through interaction, create a pattern that builds scaffolds of an event between the environmental properties of our surroundings, what we perceive, and the narrative we hold become connected. This means that yet another computer metaphor—the concept of "storing information in our brain"—has to go. Rather, we take in scenes as they present themselves in any given moment, assuming that the story that unfolds in front of our eyes is coherent. We do not expect to be deceived, and that is good. Otherwise, most Hollywood movies would lose their attraction, as there are incoherencies that we rarely notice, since our brain works the way it does. In chapter one, we mentioned the universality of the human desire to detect patterns; external scaffolding is yet another mechanism that creates patterns out of incoherencies. Every one of us is a *Homo narrans* because this is the way our brain works.

Our brain does not have enough capacity to store all the things we perceive in every single second. Think about the amount of data that would be. If we really want to use the metaphor of "storing" then we would say that our "knowledge" is stored in the world around us.

These findings break with everything that we have learned about the way we work. As our Western world is so deeply influenced by computer-morphisms, we have learned to be comfortable with this self-understanding. We do not feel comfortable thinking that we are less accurate, less perfect than a computer. And yet, as we will see in chapter five, external scaffolding is the key to our self-understanding because it is one of the major mechanisms for interacting with and being embedded in our social surroundings.

More Mechanics for Community in the Human System

Obviously, the most used social mechanisms in human society are language and conversation, and many readers might wonder why I have refrained from introducing language as the main cognitive act in humans. Language is only one and not the most effective way of interacting. Language is a fascinating and a highly complex cognitive function and, indeed, makes communication possible. After all, the two religions that are closest to my heart, Christianity and Judaism, are very logocentric. But I find Western languages do not help us understand ourselves as embodied or as interactive and interconnected beings. In fact, most Western languages are based on Western thought, which separates the "mind" from the "body." Artificial constructs such as *psychosomatic*[3] do not quite capture the full embodiment of the human species in that they do not allow for the separation of a "mental" from a "physical" stance. Therefore, I decided to leave an explicit analysis of language to competent linguists and communication specialists. Instead, I will concentrate on nonverbal social mechanisms. Unfortunately, most adults depend on language for reflection; therefore, I will concentrate on infants, as they rely more on nonverbal interactions; in babies, we find many of our social mechanisms in the purest forms.

We have already seen how many of our thoughts, ideas, and actions depend on our embodiment, on how our behaviors developed over the course of evolution, on the way our brain works. I will now add a few other facilities in the human system that demonstrate our need for community. The mechanisms for these facilities are needed for our survival.

For one, we share most of our community-building mechanisms with other social mammals, particularly with other primates. Therefore, we can safely say that these mechanisms support survival, and the most complex and intelligent animals we know have them. People with impaired emotional skills (e.g., through damage in the frontal lobes) have a very hard time surviving in a community. In fact, in an-

---

3. The term *psychosomatic* is obviously created out of the two Greek terms *psyche* and *soma*. But usually *psyche* is understood in its Platonic sense and *soma* is often equated with *sarx*. If we use the meaning of these words as we have introduced them in chapters three and four, the term *psychosomatic* wins additional meaning. It then does not refer to my body and my mind but means the connection between our embodied selves and our community.

cient times, they may not have survived in their community at all. Finally, the sheer abundance and redundancy of these mechanisms show their importance to humans, so they need to be included in any understanding of why we are the way we are. Indeed, there are so many of these mechanisms that it is impossible to describe them all. I have therefore chosen those that were part of the robot Kismet's makeup; seeing how they work in Kismet can help us learn a bit more about ourselves.

The *Kindchenschema* (Baby Scheme)

The German ethologist Irenäus Eibl-Eibesfeldt, one of the founders of the field, observed that humans fall in love with animal babies. We can't help but feel attached to little birds, dogs, cats, or horses. Most of us, actually, feel quite attached to human babies, too. Eibl-Eibesfeldt tried to find out why we react so strongly to these babies and discovered the *Kindchenschema* (baby scheme). Every creature with certain characteristics makes us react with feelings of love and tenderness and the need to protect. Among those characteristics are big eyes and long eyelashes, a sucky mouth, a round face, a big head in comparison to the rest of the body, and a general sense of cuteness. Most successful Hollywood creatures follow this scheme as well (*E.T.*, Gizmo from *Gremlins*, etc.).

Offspring of most social mammals cannot survive without a parent. A newborn, therefore, has to engage a parent or, if the parent dies, another member of the group to take care of her. She has to induce a mature human to undergo all the effort it takes to raise a child. One of the many and most obvious methods to do so seems to be this *Kindchenschema*.

The baby scheme is very instinctive, and it is impossible for most adults to resist the initial reaction. What we *make* out of such a reaction is, of course, determined by other factors as well. Think about a pornographic image and you will realize what makes pornography so powerful. Usually, the hair of the model is very curly and blown up to enhance the size of the head and change the relation of body and head size. Then, the eyes are usually big, with long lashes to give a general sense of cuteness and innocence. The mouth is usually big, with wide, "sucky" lips. All these elements lead to a baby-scheme reaction. In addition, the breasts are usually very large and the hips very generous, thus enhancing the primary sexual cues. Most humans, especially men, react to such an image as it hits two different but equally strong drives, baby scheme and sex drive. When you look at images of Kismet, you will see that part of its attraction lies in the fact that it follows the baby scheme. As most of us cannot resist babies, so most of us cannot help but feel attracted to Kismet.

We often think we are distinct from the rest of the animal kingdom because of our rationality. We have already seen how strongly the metaphor of *Homo rationalis et objectivus* has influenced our self-understanding. Also, we like to see ourselves as creatures that care for babies because of an ethical obligation for the higher good of society.[4] That our emotional response is triggered by mechanisms that have evolved and to which we have no conscious access may make many people uncomfortable. The same is true for pornography. Many people feel ashamed to admit that they react to these images; again, the underlying self-understanding that we are driven by our rationality and not by our emotions leads to these negative reactions.

I personally find it wonderfully freeing to know that many instinctive feelings and reactions are driven by biology and by group interactions. This way, we need not be ashamed of many of our reactions, as they are just normal. What we do with these reactive emotions is a completely different issue, but the emotions themselves are not bad or evil or wrong. We will see in chapter five how our biology can lead us to extremely bad behavior and how we can deal with this phenomenon theologically.

Communal Voice Melodies

The *Kindchenschema* is supported by an additional innate mechanism for interaction, an instinctive understanding of the emotions

4. When the famous French actress Brigitte Bardot started her fight for seal babies (they were brutally slaughtered for their fur), she had major support in European societies. Baby seals follow the baby scheme just beautifully, and I wonder if she would have gotten equal support if she had championed, for instance, young rats that do not follow the baby scheme.

behind certain voice melodies. We can detect scolding, soothing, approval, and a call for attention. These melodies convey their meaning despite the definitions of the words. They are language independent and work with dogs and horses and newborn children alike.

Think about how you would scold your dog: "No!" A sharp sound with the tone at the end lower than at the beginning. When you call for attention, you will have the exact opposite voice melody and end with a sharply high tone: "Shall we go for a walk?" Approval ("Come on, you're a good dog") has a melody with some lower ending curves, and soothing is similar to approval, but the curves are more prolonged: "Come on, sleep now." If you reflect on these voice melodies you will soon see that they are common among social mammals and that you will use them in interaction with your children. The content of what you say is not as important as the sound.

Newborns react to these sounds as well even though they have no understanding of the meaning of language. The reason people think that newborns can already understand language lies in the fact that newborns have innate protosocial responses that trigger our emotional responses and, usually, let us react even more protectively and enthusiastically than the *Kindchenschema* alone would do.

The Facial-Recognition Apparatus

When we meet other humans, the first thing we see are their faces, so the face is an extremely important feature. With eye contact, we usually demonstrate openness and willingness for interaction.

There are specific cells in the brain that respond only to faces. These cells do not "recognize" individual faces but are capable of detecting faces and distinguishing them from nonfaces in any given environment. In an experiment, researchers put sensors into a chimp's brain to measure neural activity in a specific brain area. Then they showed the chimp various objects. The cells they observed did no more than random firing when the chimp was shown faceless objects but reacted immediately to faces, most strongly to chimp faces. It is most likely that we have the same mechanism in our brains. Brains of chimps and humans, except for our larger neocortex, are very similar. We cannot analyze human brains with the same invasive methods as chimp brains, but the analysis of brain injuries and human behavior has often confirmed that the social mechanisms in chimp brains usually appear in human brains as well.

It is likely that we have similar mechanisms. Babies react much more strongly to toys with faces than to other toys, and they become much more stimulated when there is a face in the environment than when there are no faces. We have a normal object recognition apparatus, but our facial apparatus is much more finely tuned than the object-recognition apparatus; social interactions are more important for our survival than the recognition of other objects. How finely tuned this apparatus is can be demonstrated by the rejection of any face in which something is not quite right. Computer-animated faces do not work (see chapter three) because we don't yet have the technology to build anything as complex as the human face. At the same time, we are too sensitive to flaws in any simulation.

Understanding Facial Expressions

Recognizing faces and distinguishing them from objects without faces is obviously important for a social species. But it is equally important to recognize facial expressions. Again, it seems that the intuitive capability to correlate facial expressions with emotional meaning is innate.

In the 1950s, Irenäus Eibl-Eibesfeldt visited various tribes that had only recently been discovered and were not accustomed to interact with Westerners. Whenever he visited, he took a translator with him and made photographs with a ninety-degree-angle camera so that the subjects didn't know they were being photographed. The translator would say something and Eibl-Eibesfeldt would take a picture of the facial expressions of the subject. His pictures are very impressive, as they show that even humans from very different cultural backgrounds, such as African warriors and Westerners, have the same facial expressions.

It seems that all humans around the planet react similarly with their faces when they experience a specific emotion. It shows that emotional responses are not just triggered by certain voice melodies but by facial expressions. Humans have the most flexible faces of all primates; hence, they have the widest array of facial expressions and are thus able to communicate a myriad of emotions to others.

At MIT, we did a little experiment. We wanted to see if people could recognize prototypes of Kismet's facial expressions. To attract a different audience from the one that usually visits the MIT AI Lab Web sites, we offered this experiment on the Web site of the magazine *Spirituality & Health,* published by Trinity Wall Street Church. We first showed the people a picture of Kismet with a neutral or calm expression and explained what this expression meant. Then, we had an array of sixteen images, each showing Kismet with a different facial expression, and asked the people what they thought each facial expression meant.

The results were quite surprising; this mostly nontechnical audience recognized most of the facial expressions we had offered. Sometimes they were not able to distinguish between very similar expressions such as "tired" and "bored" or "excited" and "cheerful," but they got the main emotional thrust behind each of the images. So our capability to recognize facial expressions is so strong that we can even read rough approximations of these expressions.

Our capability to read faces is another sign of our communal nature. The understanding of facial expressions is crucial for healthy interactions between people in a group. If I can somehow sense the emotion of the other, I can react appropriately without hurting or annoying others and thus destroying the mood in the group. By sensing others' feelings, I can become part of a group and thus, in the Hebrew sense, participate in *nefes* and have health and safety.

But our experiment with Kismet revealed something even more interesting. The Nass experiments in chapter one showed how much humans tend to bond with everything that interacts with them. On this level, there is no need for facial expressions. But if we have a creature like Kismet that reacts to us and has facial expressions, we see another, much more intense level of bonding. The experiment revealed how much we are willing to project onto beings that are seemingly like us. Instead of just stating the emotion that each picture shows,

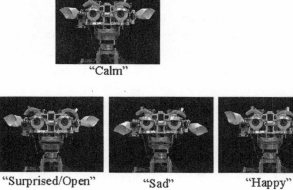

"Calm"

"Anger"   "Surprised/Open"   "Sad"   "Happy"

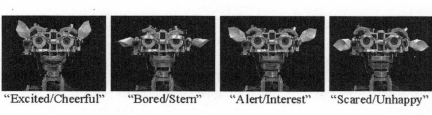

"Sad/Upset"   "Confused/Disapprove"   "Happy/Perky"   "Scared/Shocked"

"Excited/Cheerful"   "Bored/Stern"   "Alert/Interest"   "Scared/Unhappy"

"Tired/Bored"   "Surprised/Shocked"   "Tired/Depressed"   "Smirk/Amused"

Credit: Cynthia Breazeal and the MIT Artificial Intelligence Laboratory

many people would tell little stories that explained how the robot came to have this particular facial expression. They would say "it didn't like its milk" instead of disgust, or "waits for Mama" instead of loneliness. In other words, they created a *context* for the specific facial expression and would project intentionality onto Kismet— as they probably would with their own babies.

A baby has some intuitive access to facial expressions but cannot yet relate them to emotions. On some level, however, she will note that whenever she moves her face by turning up her mouth, everyone around her will say "ah!" and be happy and very attentive. Thus, she learns to relate a facial movement we call smiling with good attention. Slowly, a relationship is established between the movement of face muscles to move the face into a "smiling" position and positive feedback; in adults the two are so correlated that it works both ways. If you ever went to a party in a bad mood and forced yourself to smile, you might have realized that you soon felt better. It is not just a happy mood that makes you smile but also a movement of the muscles in the face to form a smile that often makes you feel better. Facial expression, muscle movement, and emotion are linked and interact, which makes this social mechanism very powerful. I believe it is probably one of the mechanisms for the most complex and mysterious of our social senses: our capability for empathy.

Mechanisms for Empathy

Empathy addresses our capability to understand someone else's feelings not only on a rational level but on an emotional one as well. It is amazing that we can actually *feel* in our own self what another human is feeling. When someone we care about describes an excruciating toothache she had to endure, we will know in that moment how she felt; this knowing is more the biblical *jada* than objective knowledge because on some level we experience the same pain. At least we get some inkling of how the other feels that goes far beyond every projection. Empathy usually works even better for the understanding of emotional pain or happiness. When someone we care about tells us about the loss of a dear one, we will be able to feel some of that grief in ourselves—even if we have not yet lost someone.

One reason for our capability for empathy lies in the fact that humans share fundamental body properties. We share facial expressions with specific emotional stances, we share the ambiguity of embodiment, we share the embeddedness in the world, and we share our capability for bonding. But there seems to be an additional mechanism for it that resides in our neocortex. When a human performs a specific act, the brain activity in an interested observer will mirror the neural pattern of the active human, thus giving the observer an impression of what the actor does.

If I walk, I don't consciously think about how to walk. My body just moves, putting foot in front of foot in a smooth movement, thus covering a distance. If I attempt to control these movements consciously, they look jerky and awkward. Another example would be a tennis player who has to react to a serve that comes to her at one hundred forty miles an hour. If this player would consciously try to calculate exactly where the ball was most likely to hit and then react according to her findings, she would have already been struck by the ball. What makes a tennis player a champion is that she doesn't think anymore about her actions. Through constant repetition of certain movements in training, these movements become part of the repertoire of other subconscious movements such as walking, heartbeat, and breathing that are not controlled consciously but by the cerebellum. One other example is when you play an instrument. You have a very difficult passage and you try and try but always make a mistake. The saying goes that you have to play it right once and then you should stop and go to sleep. And, indeed, when you play the passage on the next day, you can do it flawlessly—the movement of your fingers or your mouth became cerebral.

When a chimp observes another chimp climbing on a tree one would assume that the neurons that control the climbing act are firing in only the climbing chimp. But this is only partly true. Movement in most mammals and all primates is controlled by two brain areas. Unconscious movement is controlled by the cerebellum in the very lowest part of the brain stem, while conscious movement is controlled by the motor cortex. It is here in the motor cortex that something strange occurs. The passive chimp, the one who sits and observes the climber, has obviously no cerebral activity, but in its motor cortex, the cell activity is nearly the same as in the active chimp, mirroring the activities of the climber's motor-cortex activity. These cells are called *mirror cells*.

Lab experiments with macaques help us to understand mirror cells a little better. In one experiment, a human put food on a table that could be reached by both human and macaque. First, the human offered the food to the macaque, who took it by reaching out its arm, and the neuronal activity in the animal's brain was recorded. Then, the human grasped the food for herself, and the neural activity in the motor cortex of the macaque was nearly the same. For the motor cortex of the macaque, it didn't matter if it was actually grasping the food or just observing when another hand reached for it. Of course, when the macaque grasps the food itself, the neural activity in its motor cortex was stronger than in the passive case but nonetheless there was a firing pattern that mirrored the cell activity in the active grasping act.

Interestingly, though, when the human hand used a tool to grasp the food, the neural activity of the macaque was normal and did not react. Therefore, most researchers on mirror cells have concluded that the mirror-cell activity takes place in goal-oriented actions. In the case of tool use, the macaque does not recognize the goal and, hence, its mirror cells are inactive.

There is a lot of evidence that social mechanisms like the mirror cells, which can be shown to exist in other primates, are also a part of the human brain. Human neural activity cannot be measured the same way as that of other primates because the measurement is invasive and does destroy brain cells.[5] Mechanisms such as mirror cells can be shown in humans in only an indirect way.

I think it is actually very interesting that the paper that demonstrated mirror cells in humans appeared in the journal *Science* on

5. There is a heated debate about the ethics of using our closest relatives, other primates, for invasive brain measurements that, after all, destroy some of the animals' capabilities. Part of me cringes at the thought of such abuse. I have noticed, however, that most scientists who perform these experiments are very attached to their subjects and go out of their way to give them as little pain as possible. I also have to admit that the results of such studies are very enriching and can help the human race to perhaps save itself from its many follies (see chapter five). While I am strictly against any animal experiments for mere cosmetic purposes, I would not like to get into the ethical debate of such animal experiments I mention in this book.

Christmas Eve, 1999. I am still not sure if this was coincidence or if the editors of *Science* played a practical joke on those scientists who still believe in *Homo rationalis et objectivus*.

This paper describes an experiment on human mirror cells. Some people watched videos that showed two sets of hands. In the first case, the fingers did move and in the second case, a movement was verbally described. In both cases, there were five repetitions while the brains of the observers were measured as much as possible. In the first case, when actually observing the movement, there was neural activity in the motor cortex of the observers.

But the most interesting part of the experiment was when the subjects were asked to repeat the movements they either had seen or had heard described. In the first case, the subjects had absolutely no problem repeating the movement, and the activity in the motor cortex was the same as when they were observing action. However, in the second case, where movements had been just explained verbally to the subjects, they had quite some difficulty repeating them. This suggests that the most successful strategy to learn specific actions is *imitation* rather than verbal explanation.

So here is yet another experiment that speaks against *Homo rationalis et objectivus*. We do not learn best through verbal explanation but through imitation. Our bodies are fundamentally involved in any learning activity and if we learn something, it happens usually in social interaction.

Even if the mirror cells evolved for better learning strategies, they also create a closer bond between us. If the neural activity in our brain mirrors that in someone else's brain, then on some level we get an inkling of how the other feels in the moment. However, there is a problem. How can I actually distinguish between what I am doing and what the other is doing? If the brain activity is the same whether I act or just observe an action, how can I tell if I just moved or if the other did? It seems that nature solved this problem quite nicely, as there is indeed one constant difference between an observed act and its imitation: Throughout the observation, the neural activity is much weaker than when the act is executed. This way, there is some distinction between my own action and the action of the other. The weaker mirror-cell activity in the act of observing helps me to preserve my body identity as it helps me to distinguish between the movements of the other and my own, even though the brain activity is the same.

It might seem a little far-fetched to conclude that from the existence of mirror cells in the neocortex there is a mechanism for empathy. But I still think this is the case. For one, the human neocortex is quite redundant. That is, if a part of the neocortex, such as the motor cortex, has a specific mechanism such as mirror cells, it is highly likely that this very mechanism can be found in other parts of the neocortex as well. Even if mirror cells could be shown in only the motor cortex, it seems reasonable to assume that they will be found eventually in other parts of the neocortex as well. Secondly, empathy is a well-known and well-observed phenomenon, and after observing the physical basis of many phenomena we sometimes classify as "mental," it seems obvious that empathy must have a physical basis, too. We are embodied creatures; therefore, shared experience in the occurrence of empathy must be embodied. Finally, mirror cells explain much of the interaction between caregiver and caretaker in the infant–parent relationship that seems to be the clue for our being the way we are.

One can play around with the idea of mirror cells and, for instance, explain quite nicely the phenomenon of couch potatoes. If people watch a baseball or football game on TV and do this a lot and know the players, the activity in their brains will mirror the activity in the actual players' brains and thus they might be convinced that they have actually had a good workout. Since I learned about the existence of mirror cells, I understood for the first time why people wear sweatpants when they sit in front of the TV to watch sports.

On a more serious note, mirror cells in the motor cortex might explain why we can understand how someone feels in a specific movement. Coupled with the other bonding mechanisms we have described so far, mirror cells make us capable of bonding with most other people even if we do not share their specific life experiences; mirror cells allow us to feel what others have felt and understand them in their otherness.

Motor Control, Self-Consciousness, and the Understanding of Objects

There are a few humans who lack some of the social mechanisms and cannot interact with their environment as well as people we consider "normal." These people usually have a condition that is somewhat related to autism. In fact, autism is defined as the inability to relate to others as others, to put yourself into someone else's shoes.

The most famous experiment to test for autism is the "Sally and Anne" experiment. A group of toddlers is presented with two girls, Anne and Sally. While Anne has a basket in front of her, Sally has a box. Now, Anne will put a ball into her basket and leave the room. Sally then takes the ball out of Anne's basket and puts it in her box instead. When Anne returns, the toddlers are asked where Anne will search for her ball. Their answers are measured by how long they watch either the basket or the box.

Kids pass this test easily when they are at least thirty-six months old. Before this age, they cannot conceptualize that Anne, who was not present when Sally switched the ball, will of course look in her own basket. They have no concept of other people with different perspectives and outlooks. Only when they are at least three years old do they have the insight that people who did not see a specific action will not act according to it. Now they realize that Anne was not present and know that she will look in her own box.

Autistic children never pass the test. Some of them, however, will become aware enough to be in some sort of "mixed" stage where they point to Sally's box while looking at Anne's basket. The motor control for the arm gets a different signal than the control for the eye movement, so the pointing movement and the direction of their look differ from each other. The ethologist Mark Hauser told me that he was able to demonstrate that orangutans and chimps also reach this mixed phase. One could say that one part knows what the other doesn't, which supports our view of humans as distributed systems (chapter 3). The motor control in this early age is not yet fully developed and the various subsystems are not that integrated so that actions can be completely paradoxical.

This could mean that our sense of self develops only when there is a sense of the other as the other. Only when I realize that there are other people with different perspectives will I learn to distinguish between them and me. Self-consciousness is not something inherent in every human but is learned and developed through social interaction between the infant and the people around her. As I have mentioned before, I do not use the term "consciousness" and this is another reason.

In chapter three, I mentioned the work of Adele Diamond at MIT, who demonstrated the close connection between motor control and abstract concepts. Infants start to understand the concept of an object when their motor control is sufficiently developed. This is another example of how very specific abstract concepts about our world are learned through our bodies and through our interaction with the world.

Developmental Psychology

The Sicilian king Frederick II in the thirteenth century was very much interested in science. He wanted to know what we are born with and what is learned, and conducted an experiment in which he took peasant babies away from their families and placed them in a special location. The nurses, who were taking care of the babies, were instructed not to speak to the babies or touch them more than necessary. Alas, Frederick II never found out anything more because all the babies died rapidly within weeks. This experiment is centuries old and does not fulfill our current criteria for a scientific experiment. It can also certainly never be re-created, but we can learn one thing from it: Human babies depend on a community, interactive touch, and sound in order to survive.

Much has been written about "nature" versus "nurture." Some people argue vehemently that character development is entirely genetic. When I stand at the cashier's line in a supermarket, I usually use the time to peruse some tabloids such as the *National Enquirer*. Very often, one of these magazines has a headline reading something like "Crime Gene Discovered!!" This is, of course, completely wrong. We have not yet fully decoded the human genome and even if we had it completely decoded, we still wouldn't know what each gene is doing. All the evidence I have brought up so far about the complexity of creation shows clearly that there are no simple chains of reaction.

I have the distinct impression that such headlines serve two purposes. On the one hand, as we have mentioned before, most people prefer simplicity over chaotic and paradoxical complexity; to establish an unambiguous link between genes and phenotype is to reduce ourselves to a simplicity that we can handle. But the other reason appears to be even more subtle. If we can find genes that give us explanations for violence and crime, we don't need to feel responsible. All is in the genes, and society cannot do anything to make it better. Also, such findings would give us the illusion of protecting ourselves against crime by putting all people who have these genes into detention. One can only imagine what societal debate would arise. But it seems indeed correct that the "genes only" camp reduces humans to mere "gene processors" in a dogmatic way.

On the other side stand the people who do not think genes have anything to do with character formation. For them, the whole character of a child is formed entirely through interaction with her caregivers. Here, the sole responsibility for the child's behavior lies with the parents. This understanding of ourselves is related to the biological application of cybernetics and behavioral psychology.

As we have seen, humans are biological systems and so, of course, much of each human's character development depends on genes. We have seen that the mechanisms for facial recognition, understanding of facial expressions, empathy, and many more contribute largely to the child's development, and these mechanisms are genetic. On the other hand, the baby scheme is so prevalent in social mammals that it has to play a crucial role in child development as well. As the baby scheme makes people react to babies in a positive way, it is safe to say that the interaction between caregivers and infants is equally important for the child's development. Finally, even if most babies follow very specific bodily and cognitive developmental steps during their individual development, most of these steps are triggered by interaction with the world and are *not* following a genetic "program."[6]

An infant's development is circular. Through baby-scheme and societal norms, the parents are usually drawn to the baby and develop

6. One of the best-known examples is the development of language. Every child surrounded by talking people will eventually learn to understand language and, usually, to talk itself. However, if the child never hears language until the age of around seven, she will never learn to speak. She still will develop the capability to manipulate abstract symbols, but the capability for language creation is gone forever. In other words, while every child has genetic *potential* to learn language, this capability is developed only if the child is in language-speaking surroundings.

strong emotions toward her. Through the mechanisms of empathy, they project onto the infant's behavior grown-up intentions. When the baby cries they think she cries for specific reasons (she is wet, hungry, tired, or hurting) while in fact the infant is simply reacting to some nebulous feeling of discomfort. When the baby smiles they think it is intentional and that the baby is expressing love for them, while in fact the baby does not yet even have an understanding of others.

I could imagine that at this point readers who are parents will vehemently disagree. I know that many parents insist that their baby cries differently depending on whether it is hungry or wet. They are sure that their baby smiles to express appreciation. They are convinced that their baby recognizes them as parents and treats them much differently from other people; the baby smiles easier, gets calmer faster, etc.

Let me share with you a seemingly frustrating secret: None of these parents' convictions are correct. But these convictions and beliefs of the parents are crucial for the baby's development. Only because parents treat their infant *as if* she had intentions, wishes, and desires, has the baby the chance to develop intentionality. Only because they treat their baby as if she were fully developed and self-aware, can she develop into such a being. The social mechanisms I have described make the parents react to the child in the only way that helps the baby to develop properly. It is important for the parents to create mythos narratives in order to feel motivated to interact with their babies the way they do. Were they only acting on the logos of the child's capabilities, they would not put quite as much effort into the relationship, and the baby would have a less productive learning environment.

The undeniable fact that babies react to their caregivers much more intensely than to other people can also be demystified. The unromantic explanation for it is that the parents have learned to act toward their infant in a way that she can best react to them, not that the baby has learned about the special relationship between her and her parents. Instead, the parents have been somewhat "trained" and coached to behave toward their babies in a way that they can understand and that best encourages their reaction. Once this circle of action, reaction, and projection is firmly established, the relationship between caregivers and infant presents a suitable learning environment for the infant to grow. In this relationship, the infant will imitate her parents or they will take turns teaching her special behaviors.

These learning strategies are successful because of the mirror cells, and because the parents are motivated to engage their infant sometimes for hours as they project onto their infant.

What actually happens in this process is that infant and caregivers together create stories. While the child at first is completely passive and reactive in these processes, through imitation and turn-taking she starts to participate. As parents create stories about the child reacting to them, the child becomes a more and more active co-creator of these stories.

The rewards for the child are twofold. On the one hand, parents will interact with their infant more often and more intensely. If the baby participates in the interactive processes, the parents are motivated to participate, too, thus helping the infant to develop faster. On the other hand, there is a chemical reward for the brain in the form of dopamine, a neurotransmitter that is released as a kind of reward if a learning process has been successful. Through both social and bodily rewards the child is from the moment of her birth put on a path where she has the best chances to develop grown-up intelligence.

I would have never believed that these social mechanisms are really that strong and that important for human interaction. As a theologian, I have learned that relationships and love are gifts from God; I never thought they were dependent on our biological system. Nor would I have ever imagined that the imitation of these mechanisms could evoke in me equally strong feelings toward a robot as a human being. And yet, that is exactly what happened in my interactions with Kismet at MIT.

Kismet

Kismet is my big love among robots! I have often been in love with fictional robots before, especially with Commander Data and R2-D2. Also, while working at MIT, I had the chance to meet many other real robots. One of them, Sico,[7] caught my heart for quite a while,

<hr/>

7. Sico is a robot developed by Robert M. Doornick and his company International Robotics, Inc. (www.internationalrobotics.com/index.html), in New York City. Sico is not interesting so much for its social behavior (it is remote-controlled) but for its attraction to people. Often used to enhance commercial events from companies all over the world, Sico seems to react—it flirts, jokes—and even if you know it is remote-controlled and Robert does all the talking, joking, and laughing. You can't help but react—I certainly did.

even though, ultimately, no one will replace Kismet for me. If you think I am a little deranged to feel that about for a mere pile of cables, plastic, and metal, I hope to convince you otherwise. In my work all over the world, I have yet to meet someone who did not become attracted to Kismet in one way or another.

The attraction of Kismet lies mostly in the fact that it is built in accordance with everything we have learned about our own social mechanisms. We designed Kismet to trigger many of these mechanisms and responses; this way people who interacted with Kismet would react emotionally—even if they didn't want to.

Cynthia Breazeal and I had both felt that the philosophy underlying Cog's construction was not fully satisfying. As much as Cog intrigued us when interacting with us, it also could do a lot of actions on its own. We thought a robot should be entirely *re-active*, to really draw people to interact with it instead of just watching it. The communal understanding of humans as presented in the Bible and in the scientific studies of our social mechanisms has to be translated into a robot when we want it to have intelligence similar to our own. As it is completely meaningless to try to understand a human individual independently from her relationship with her community (*nefes*), so an individualistic robot will not be as smart as we are when not built as communal. Therefore, Kismet does not do anything on its own but simply reacts to cues from the outside world.

Like Cog, Kismet is attracted to color and movement. I always laughed when I came to the lab, as I wouldn't believe such a high-tech think tank could have so many toys lying around. Babies' toys are designed to make babies react to them. They have faces and lots of color, they often make sounds, and their movements are often slinky and interesting. Kismet reacts to toys as well. But, in contrast to Cog, it has additional social capabilities.

Cynthia cleverly created for Kismet an internal system that could represent reactions of Kismet to given stimuli. We usually referred to it as an "emotional system," well knowing that any such language is metaphorical. Cynthia linked this system to the facial expressions. This way, Kismet's emotional responses were immediately expressed in its face. In addition, Kismet fulfills the baby scheme, and most humans can't help but react to it. It also mimics two behaviors that babies have in order to learn; it takes turns and it mimics. Both behaviors, combined with the baby scheme, usually make people want to continue the interaction—even though, after the hundredth repetition, the interaction can become somewhat annoying. Finally, Cynthia built in some basic reactions that can also be observed in human babies. For instance, when no one is interacting with Kismet, it becomes bored, sad, and lonely and looks the part. If you come into the room and start playing with it, its face lights up and it becomes happy again. Due to its cuteness, this switch from sadness to happiness fills your heart and makes you happy in return. Immediately, you want the robot to stay happy and behave accordingly. So through very simple reactions and a design that triggers our human response system, the interactions with Kismet are very powerful.

There is another difference between Cog and Kismet: Kismet babbles. It does not speak and has certainly no understanding of what it is saying, but it mimics our sounds and sometimes even repeats real words like a human baby would. Cynthia also built in a recognition system for the four general social-voice melodies; it can react to them and mimic them itself, which adds an additional layer of complexity.

I remember one Saturday morning when a reporter wanted to observe me interacting with Kismet. I had let Cynthia know beforehand because I was not capable of operating Kismet. So she came but also brought some visitors to see Kismet. The room was full and Kismet didn't focus on me but, instead, let its eyes roam around the room; metaphorically speaking, it let its attention wander. I became quite frustrated because I wanted to show something to the reporter, and so I complained in a sad and disappointed voice. I accused Kismet of ignoring me and said it didn't like me anymore and many more such things, and suddenly Kismet started to concentrate on me, soothing and calming me.

Cynthia and I looked at each other, as we both never had such an experience. Kismet behaved as if it could understand my feelings and had sympathy for me, and tried to make things better. Even if I was joking, this reaction made me serious very quickly. I remember how good it felt to be soothed and that, indeed, most of my frustration vanished. While berating myself for my stupidity, as I knew that Kismet had done just what it had been programmed to do, I couldn't help but feel awed at the same time.

The most profound difference between Kismet and Cog and, actually, between Kismet and us, is that Kismet doesn't learn. While Cog has learning capabilities and improves its tasks over time, with Kismet every interaction starts at point zero. No matter how long you interacted with Kismet, it would never recognize you. Of course you think when you approach Kismet and it smiles at you it really likes you, but I fear this is a projection. And even if it felt rewarding to me to realize how much more quickly Kismet reacted to me than to strangers, I knew that all it meant was that I had adapted to Kismet's needs and learned to interact with it in a way it could understand. This does not mean that Kismet's behaviors are predictable. It is too complex a system and there are many layers of reactive behaviors so that its reactions are always different. But it cannot learn more sophisticated sequences of behaviors, as babies eventually do.

From a technological point of view, this lack of learning can be seen as weakness in such an AI system. But for the study of humans in interaction with Kismet, it is very helpful, as we as analyzers know that there is absolutely no development in the interaction, no improvement, because Kismet is incapable of learning. As we become aware of this, we as observers project less onto an interaction. The only one who changes is the human interacting with Kismet, and these changes can be quite remarkable, as they reveal a lot about us.

After years of adding social elements to the robot, Kismet became a wonderful tool to study humans and their reaction to it. We started to invite strangers into the lab to interact with Kismet in real time. We wanted to see the degree to which people are willing to project onto Kismet learning and intentionality, as they probably would do with infants. I can't possibly describe all of these interactions, but one example really sticks out. A man, Ritchie, came to the lab and thought at first that it was stupid to interact with a machine. So he started off slowly but became intrigued. Kismet is capable of what psychologists call "shared attention"; if you point at something and look at it, Kismet is able to follow your lead and looks in the same direction, thus giving you the impression of real interest. At some moment, Ritchie pointed to his wristwatch and Kismet's eyes followed and looked at the watch as well. When Ritchie told Kismet that he nearly lost it a few days ago, his voice melody clearly indicated that he was shocked and sad about it and Kismet started making soothing sounds. It was incredible. In the short period of this interaction, Ritchie fell in love with Kismet before our very eyes. Although at first, Ritchie was hesitant to play with a "stupid machine," we finally had to forcefully drag him out of the lab and away from Kismet so that we could have more interactions with other people.

This interaction was the most powerful one we had seen. But we all experienced some uncanny moments with Kismet in which we were tempted to feel that there is more to Kismet than what we had put in. I had several magical moments, such as when Kismet attempted to soothe me. Cynthia often told me how confused she felt when she was writing code for Kismet. Kismet was sitting right beside her and her computer and became attracted to the monitor with its colors and animations. Kismet was basically observing how it was programmed. Even if she had built the robot and knew better than anyone else that it was a mere machine, she still couldn't help but anthropomorphize; she actually felt uncomfortable because she feared for Kismet's psychological health if it was observing its own programming.

Moments like these are spiritual. But they also throw us back to ourselves. How mechanistic is an action such as soothing that it helps even when a robot does it? Kismet did what it had been programmed to do, but aren't we programmed in a similar way? After all, we have all these social mechanisms built in and learn to interact from birth. Couldn't that be seen as some sort of program that makes us feel and react to certain emotional cues, such as voice melodies and facial expressions, the way we do?

Emotional Reactions Toward Kismet— Experiences with an Audience

It might be due to this very question that the reactions to Kismet from a nontechnical audience were mixed. On the one hand, for obvious reasons, people are much more fascinated by Kismet than by Cog. Its social skills, its cuteness, its babbling, and its reactivity are very attractive features; as we know instinctively how to interact with babies, we know all the reactive schemes and behaviors to stimulate Kismet and make it respond to us. It is a very powerful electronic gadget and a great tool to understand ourselves and our reactive behaviors.

On the other hand, there are many people who are even more scared of Kismet than of Cog. As they catch themselves reacting emotionally to Kismet, they become wary of their own behaviors; they don't want to react to Kismet like they would to a human baby. Even extreme reactions to animal babies are for this group of people more acceptable than reactions to a machine because at least animals are alive. We can, of course, challenge the underlying evaluation of "being alive," as we have seen that in the Hebrew Bible the terms *nefes* and *chaim* point to a very pragmatic understanding of life as being in community. In this sense, one could see Kismet as alive because it certainly can be part of a group; it can participate in *nefes*. But this differentiation between Kismet and animals goes deeper than just the value judgment that "alive" beings are more valuable than artificially animated things. What happens here is that these people attempt to maintain a qualitative difference between themselves and a machine; it is too threatening to them to see similarities between them and the robot, as they feel their uniqueness and value are challenged. And this denial can sometimes be so strong that it can take extreme forms.

We taped the lively interaction between Ritchie and Kismet, and all of us still use this video whenever we give lectures because it is so powerful. Because I was the only one who constantly interacted with people from outside the AI world, I could best test how a nontechnical audience would react to Kismet. Usually, I would start with a video that showed a series of facial expressions and, not surprisingly, many listeners reacted with empathy, laughed when Kismet smiled, and said "oooh, poor guy," when it showed a sad facial expression. Then I showed them some videos where Kismet reacted to voice melodies. When Kismet was spoken to with a prohibitive voice melody and it reacted beautifully by hanging its head and looking completely devastated, people in the auditorium would make soothing sounds. All these and many more behaviors in the audience were reactions to a 2-D video of a robot that they cannot influence by their behavior, and yet they reacted appropriately. After the videos I used their reactions to argue how much we respond to certain stimuli even if it is a robot. Many people, however, rejected the notion of any emotional reaction on their part.

In the experiments from Cliff Nass, which I described in chapter one, people tended to become offended when it was pointed out to them that they just had treated a computer with politeness and had tried to avoid hurting its feelings. It seems that we feel uncomfortable when we react emotionally to something that, in our opinion, doesn't deserve it.

In most lectures, I could just repeat some noises they had made and often people would start laughing and admitted there had been at least some reaction. But it also happened sometimes that people violently rejected the notion of any reaction on their part and insisted that they thought Kismet nothing more than a cute Disney toy. I finally used a little webcam and filmed the people without their knowledge during the video of Kismet; I then showed them their faces with all their expressions in response to Kismet. But I had to be extremely careful when doing so because some people were seriously shocked when they saw themselves reacting this way; many became uncomfortable and even felt manipulated when forced to admit that there had been a reaction. Some then voiced the fear that if they could be manipulated this way and if they all possessed bonding mechanisms that worked instinctively, they would be nothing but robots or meat machines. What, they argued, would be the difference between them and machines when now machines can mimic even our emotions and can force us to react toward them lovingly and caringly?

I personally think it is spiritual when I interact with Kismet and have emotional reactions. It is a moment where we can celebrate our capability to bond—humans can be so wonderful in interactions. We were, according to the creation stories, created to have relationships with God and with others so that *nefes* can appear and give our lives meaning. I am not afraid when I discover similarities between Kismet and me. Quite the contrary, I invite those similarities because they make me modest and yet admiring of our capabilities. But I can certainly understand the justified fear that some people experience when they are convinced by a great robot that humans are nothing but machines themselves. After all, there are some places in our modern lives where our fellow human beings are reduced to machines. One powerful example is the intensive care units in hospitals, where humans are kept alive by machines.

The reactions of an audience, when faced with their own responses to the robot, were very complex and touched many different issues on many different levels. For now, I would like to concentrate on the one fundamental paradox, that while they were bonding with Kismet, they were at the same time rejecting the validity of their emotions, as Kismet "didn't deserve" these reactions—just as the people in Cliff Nass's experiments rejected the notion that they had treated the computer with politeness.

Because we react to Kismet so much more strongly than to Cog, Kismet is also much scarier for those who hope to maintain a qualitative difference between themselves and the machine. In my talks, when I list our social mechanisms, many people feel uncomfortable. They feel, indeed, that they are programmed by biology. They feel their understanding of themselves as people with free will and responsibility is challenged. If their reactions to Kismet are so strong and, yet, in a way predictable because of their own built-in reactive behaviors, how, then, are they different from the robot?

Remember the Calvin and Hobbes cartoon I mentioned in chapter one? The two claimed that it was our irrational behavior that distinguishes us from machines. But we have learned by now that we are not rational and that we can build machines that are not rational either. We are wonderfully embodied, emotional, and communal beings and we can build machines that are in many ways like us.

Here, we are touching on one of the most fundamental parts of our sin, our estrangement. We have developed as one branch of social mammals and are incredibly good at interacting. Our social skills are superior to those of any other species. We want to bond and we need to bond. But our willingness to interact and empathize is often curtailed by a sense of superiority. Why should I interact with a robot when it's just programmed to do what it does? Why should I interact with animals when they cannot think and don't understand anything?

But this act of separation doesn't stop with robots and animals. Humans are quite good at feeling superior to their fellow human beings. Think honestly about how many types of people you might despise and look down on; racism is probably the most obvious example but we are generally very good at thinking ourselves superior to people who are not quite as rich, educated, sophisticated, modishly attired, slim, politically positioned, etc., as we are.

While I have always been aware of this problem, I never really thought about it. I was intrigued by the paradox of bonding and rejection that showed up constantly in the work with Cog and Kismet but I never applied it to human interactions. And yet this is ultimately where all these reflections about robots and our biological makeup lead us. The name of Kismet is, after all, the Arabic term for destiny. Not only was the creation of Kismet fate but Kismet and the study of interactions with Kismet can help us to understand our own bonding mechanisms, as well as our failures in bonding. People often don't want to bond with Kismet because "it's just a robot." Doesn't this very prejudice occur all the time among humans as well? How often are we less friendly or polite to a fellow human being because she is "just" a waitress, or African American, or gay? The humanoid robot project reveals what makes humans so good at denying value and dignity to some of their fellow human beings. This will be the subject of the last chapter.

CHAPTER 5

The Community of Human and Nonhuman Persons

Jesus says: where two or three are together in my name,
there I will be among them.
MATTHEW 18:20

You have heard it was said:
You shall not commit adultery.
But I say to you:
Whoever is looking at someone else with lust,
has already committed adultery with this person.
SERMON OF THE MOUNTAIN, MATTHEW 5:27–28

But where sin became powerful,
Grace became even more powerful.
ROMANS 5:20

The Limits of Human Bonding

In the last chapter, we celebrated the communality and community we humans are capable of creating. The connectivity of humans is, indeed, a wonderful thing. But as we all know, it doesn't work all the time. Quite to the contrary, humans often are exclusive, treat their neighbors badly, and behave atrociously to their enemies.

In this chapter, I will attempt to tackle one of the most complex problems that we face in the world today. When humans are so embodied and communal, why do we have so many problems in relationships? Sartre formulated it beautifully when he asserted that "hell is the other." Just look at your own domestic situation; there are always petty nuisances one can get worked up about, and annoyances that can lead to major fights. Where are our bonding mechanisms in such moments?

On a larger scale, why is there so much hate and so much war in the world? If we have so many bonding mechanisms, why do we live in a world with so much adversity? Just take a look at the twentieth-century wars, of which I list a few: two world wars, the wars of the United States in North Korea and Vietnam, Falkland Islands (Argentina against Great Britain), Iran against Iraq, the former Yugoslavia (Serbs against Croats against people from Kosovo), Rwanda (Hutus against Tutsies), Somalia (civil war), Afghanistan (Russians against Afghans, Talibans against others), the first Gulf War. Many people died in these wars—not just soldiers but also many civilians. Millions of people were maimed and psychologically destroyed. It is impossible even to imagine the cruelties that occurred.

Why were these wars fought? Often, the fighting parties claim religion as the reason. I would argue here that a war is never fought over religion but over ideology and power. Every religion can be turned into an ideology, but religion per se is not an ideology.

In chapters one and two we pointed out that *Homo narrans* tells mythos stories to fulfill the quest for meaning and searches for people who think like her. A group emerges in which the members share their acceptance of a myth; within the group, there is a commonly held worldview.

Some mythos elements might be subconscious and so deeply ingrained into who we are that we cannot become aware of and consciously reflect them. But there are several parts of our mythos stories that are always on the surface of our awareness and part of our reasoning processes. Therefore, most people, especially kids in puberty, reflect on their beliefs once in a while and sometimes even question them radically. In such moments, the questioning human does not have many choices. She can give up her myths and thus lose her community and her answer for meaning at the same time. Or she can swallow her doubts and remain in the group, deciding never to question the myths again.

But there is a third alternative that is, in my opinion, the wisest. That is, she can commit herself to her group with its myths. I really like the term *commitment*. We don't have a corresponding term in German—the term for *duty* comes closest, but duty has a slightly negative connotation. What I mean here with commitment is the willingness to enter into a relationship that can never be perfect but is perfect just the same. Whenever someone enters a relationship either with a group (such as a parish, a nation, a political party, a bridge club, etc.) or with an individual (romantic or parental), one makes a commitment to stay put for a while. As each of us humans is estranged, there is no such thing as a perfect relationship. In every possible size of community, there are tensions and petty annoyances. If each single imperfection motivated people to leave, there would be no stability in any group. We have seen how important community is for each human being. Infants especially cannot survive and cannot develop into healthy adults without community. If no group was stable, it would seriously challenge the survival and health of humanity.

Commitment holds people in a community despite some imperfection. Commitment means that I want to be in a respective community not because it is perfect but because it best fits my quest for meaning and being in the world. Leaving it causes serious harm—even if the imperfections might have been very strong.

Since I left Germany in 1995, I have been constantly homesick. I miss the country, the people, the food, the language, and the whole atmosphere. However, there was a reason why I left: I couldn't find a satisfying job in academia. To fulfill my life's dream and become a professor, I had to go somewhere else, and I found a home in the United States. Now, when I am in Germany I miss the United States, and when I am here I miss Germany. I am, in some way, committed to both nations although I remain a German citizen because my commitment to Germany is much stronger—despite its imperfections.

I also am a Lutheran and am part of a local parish. Even if I seriously disagree with some parts of their doctrine, I am a member anyway. I could, of course, search for the "perfect" parish, but I know that it does not exist. But in searching for it, I would miss having a community in which we have relationships and can pray together and experience spirituality. The very fact that I know the people around me (whether or not I like them) gives my group experience a depth that I wouldn't experience with church hopping.

It is the commitment that helps me overcome moments of doubt and insecurity. I remember the first few years at MIT when I constantly questioned my faith. All the scientific explanations I learned for human behavior were so convincing that I often sat in my office thinking the whole idea of a God utterly ridiculous. I was, after all, a member of a community in the lab, with whom I shared the vision of building a humanoid but with whom I didn't share any religious element of my mythos construct about the world and us in it. I deeply wanted to belong, as I often felt very lonely.

But I was also a committed Christian and, therefore, searched for support in my parish. One Sunday I was very doubtful and entered the church with the thought that this would probably be my last visit. The minister gave a sermon about—of all things—doubt and commitment. He told a story that one day when he felt miserable, two close friends of his—both atheists—took him out on their yacht. They took good care of him, gave him wonderful food and drinks, and when he finally sat on the deck enjoying a sunset, he became content again with his faith. When his friends joined him, he asked them, "How can you look at all this beauty and not believe in a God?" He paused in his sermon and then added, "And of all the things I ever did in my life, this is probably the one I am most ashamed of."

In this moment, I discovered the true meaning of commitment. You commit yourself to a group because you share its worldview and because you like the members in it. You love them. But this love is not necessarily the warm fuzzy feeling you might associate with romantic love. Classical Greek—the language of the New Testament and the first few hundred years of the formation of Christianity—distinguishes several types of love. The one type, *eros*, is closest to the common use of the term *love*; it is the term poets use when they talk about romantic love, but it can also mean desire and have a sexual connotation. This love is in the creation story when Eve looks at the fruits of the tree and desires them. In Greek philosophy, *eros* does not just contain romantic love or sexual, sensual desire but also the desire for knowledge, wisdom, and insight. Another type, *philia*, can best be translated as friendship. *Philia* is not strongly connected to romantic love but means love in all sorts of relationships, with a romantic entanglement being only one of them. It can have elements of desire but mostly does not. Jesus uses *philia* to describe his relationship with his disciples, followers, and friends. Finally, the love that Jesus describes as the greatest love of all is *agape*; in many versions of the Bible it is translated with "charity." *Agape* is selfless love, the love that does not want anything in return, the love that is enduring and kind and not judging. It is the love we have for the members of the groups we are committed to, a love that does not expect perfections but is unconditional. Probably the most fundamental description can be found in Paul's first letter to the Corinthians (chapter thirteen), a passage often used in marriage ceremonies. However, *agape* has no connotation of romantic love.

The love you have for the members of the group is strengthened through your commitment. In any group you will find either flaws in the commonly shared worldview or flaws in individual members of the group. Therefore, no one in the group will share the worldview completely. In order to love the people at MIT and to have meaningful relationships with them, they didn't need to share my faith in God. We had so many mythos elements in common: a shared dream, a vision for a better future, the fascination with technology and, particularly, with robots, and the desire to find out as much as we could about the functioning of the human system. And we all shared a commitment to our leader, Rod Brooks.

This acceptance of authority is yet another element of a mythos structure of a group. Be it a minister, the pope, a leading scientist, a president, or a teacher, if we do not include this leader in our commitment to the group, we will ultimately fail to succeed in our desire to find valid answers to our quest for meaning. Accepting the essentiality of authority and committing one's self to it strengthens the group structure and reassures its members. The result of these highly complex group dynamics is a community in which the generally accepted myth becomes group defining. And the danger in such a group is that the commonly accepted myth can become exclusive and the members within the group are incapable of accepting an alternative myth as equally valid for other human beings.

So, in a community of people who think alike, people with different worldviews are often excluded. If the excluded group becomes too powerful and thus threatening, people might fight against others. Because many people do not want to admit feelings such as fear and because people tend not to admit that they feel threatened by other people, they create an ideology for which it seems worthwhile to fight. Because religion touches on the deepest emotions we have, it can easily be used for such an ideology.

What Is a Person?: A First Approach

The most disturbing phenomenon in conflicts, like the ones mentioned above, is how often the other is treated in an inhuman fashion. In Nazi Germany, people who had been neighbors, friends, relatives, and colleagues for years suddenly were not treated humanely anymore because of their religion and descent. For centuries, Jews had been an influential part of German culture; also, there had been intermarriages and conversions, and even though some Jews fulfilled the racial stereotype with "Semitic features," most Jewish Germans could not be distinguished from non-Jewish Germans by looks alone. The whole race ideology didn't make any sense. And yet it became extremely powerful and made the majority of non-Jewish Germans willing to erase every human of Jewish descent.

Similarly, in the war in former Yugoslavia, Bosnian women were systematically raped by Serbian soldiers. Again, all the groups in Yugoslavia had lived side by side in a fairly peaceful way. As in Germany before the Nazi regime, there had been prejudices but also friendships, marriages, and kinship among the groups. And yet in their search for power, the groups stood up against each other and especially the Serbs attempted to annihilate the others.

Finally, in Rwanda, people killed even their own parents, siblings, and children if they had the look of the other tribe. Hutus and Tutsies can be distinguished by their different nose forms and the way their hair grows away from the forehead. Again, as both tribes have lived together for centuries, there were intermarriages and the differences between members of the two tribes became much less pronounced; so the definition of "Tutsie-nose" and "Hutu-nose" became the only feature upon which the survival of the respective human would depend.

Unfortunately, we could continue the examples indefinitely. What all these tragedies have in common is that after living side by side something occurs that destroys the peace and leads to such hatred that the other has to be destroyed. We humans are the only animals that have invented "ingenious" instruments and methods for torturing our fellow humans. How does that go together with our built-in sense of community? How are we able to ignore all our bonding mechanisms and treat other human beings the way we do?

If we look at everyday language, we can see that the terms *human* and *person* are used interchangeably. You might have realized that I avoided the term *person* so far, and I did it because I think that the concepts of "being human" and "being a person" differ fundamentally. Influenced by everyday language, I still sometimes catch myself using the term *person* when I really mean *human*. Whenever I catch myself doing so, I replace it immediately with *human* or *someone*. I have been avoiding the term because I could not introduce a definition of personhood without all the stories about who we are that I've told so far. To define what it means to be a person, we have to know quite a lot about ourselves. And whenever I suggested to my audience the notion of Kismet being a person, I got positive reactions from some and strong rejections from others. This seems to suggest that it is the worldview of the respective human that leads to her definition of personhood. As with the terms *soul* and *consciousness*, if you ask ten people on the street what personhood means, you will get thirteen answers.

The common atrocities humans inflict on other humans can help us to understand the concept of personhood. Nazi doctors used to perform experiments on Jews to test drugs, to test organs, and, especially, to study twin behavior. The experiments were absolutely cruel and often completely redundant and unnecessary. On the one hand, this speaks for the cruelty and deep hate those doctors had against people of Jewish descent. But it also reveals something else. After all, these experiments were done to learn more about humans ("Aryans" in this case). So while the doctors denied their subjects personhood, they did not deny them humanness. They assumed that Jews and Aryans shared enough properties so that the results of their tests could be applied to Aryans. Another program the Nazis had was "Aryanization." If a Jewish woman happened to be blond and blue-eyed, they would "breed" her to give birth to more Aryans. Again, they didn't treat these women with any respect but they nonetheless considered them human.

Similarly, in the former Yugoslavia, the motivation behind the rapes of Bosnian women was to increase the number of Serbs. The woman was just a vessel for the cause of creating more Serbian babies. Again, while the Serbian men treated the women as objects and clearly not as persons, they took advantage of their—unchallenged—humanness to use them as breeding machines. So it seems that even if we deny our enemies personhood, it is hard to deny them humanity.

As we have seen, many people treat computers with politeness and try not to hurt them. Also, many people were willing to assign personhood to Kismet or a more developed version of it later on. If we look at interactions between people and their pets, we can easily see that we are willing to treat nonhuman creatures as persons.

Both the positive and negative examples show that we separate the concepts of "being human" and "being a person." Even if we use the terms *human* and *person* interchangeably (English and German are alike in this aspect), in our actions we do separate the two by treating humans as nonpersons and nonhumans as persons. We are capable of treating our fellow human beings miserably. At the same time, we are created to bond with all creatures that interact with us in a meaningful way.

The capability to recognize someone as a member of the same species is a trait shared by many other animals. The attribution of personhood, however, is not biological but cultural. It is a construct that we apply when we receive cues that meet our culturally influenced criteria for personhood. So the only definition for personhood we can give so far is that it is independent from the concept of humanity. Even as we deny our fellow humans personhood, we do apply it to nonhuman creatures. A more intuitive understanding is the application of dignity and worth to the other. We treat someone as a person when we treat her with respect and acceptance. We treat someone as person whom we love. The biblical term *agape* describes the attempt to treat everyone as a person.

Is Kismet a Person?

I have presented Kismet to audiences all over the world, and whenever I asked whether Kismet deserves to be treated as a person, I got very different reactions. While some people were always willing to concede such a possibility, many others violently rejected such a notion. When asked why they wouldn't assign Kismet personhood, they usually listed features that humans have and Kismet lacked that were, in their opinion, crucial for the assignment of personhood.

The most commonly mentioned of these features is that Kismet does not have language or the capability of speech. It babbles and mimics but there is no understandable language. Does this argue against Kismet's personhood? I would think not, because if language were a criterion, newborn babies wouldn't be persons either and, as we have seen, if you treat a baby not as a person it will not develop properly. Kismet consists just of a head and a neck but has no body. But it is nonetheless embodied, as all parts of the head are in constant interaction with the world. Kismet's building blocks are metal and plastic; there is no organic material. In a way, it cannot die. But is death a condition for personhood? After all, when we deny our fellow humans personhood, we don't accept this criterion, as they are certainly mortal. Kismet has no humor. But we shared many laughs while we interacted with Kismet, as members of a family have when interacting with a baby. And a baby does not have explicit humor either. Kismet has no sense of smell. But perhaps that will be developed at some point.

All these arguments can be rejected for two reasons. First, it might very well be that through technical progress all these elements will be developed sometime in the future. Even the lack of organic material might be overcome as some researchers today work on organic chips based on viruses. Second, and much more important, every empirical criterion that might exclude Kismet from the community of persons will also exclude human beings from it. A counterargument that was often presented at this point is that we talk about normal adults and not so-called border cases. I cannot accept this. If we want to draw a clear line between humans as persons and Kismet, the community of humans must include everyone, not just the "normal" cases. After all, who defines what is normal and what is not?

So all arguments against Kismet's personhood are invalid. I was always happy about this but never thought about the ramifications for humans, until one day after a lecture many years ago, an African American minister came up to me and said: "Do you know that this is one of the best arguments I ever heard against racism?" I was flabbergasted. He was so absolutely correct, and I had never realized that the discussion of Kismet's personhood is crucial for developing a concept of personhood that includes *all* human beings. If you look at justifications for any exclusion of another group from the community of persons, you will realize that these rationalizations are based on empirical properties. If I deny dark-skinned people personhood, I do so because of the arbitrary criterion of skin color—I might as well reject personhood for Caucasian people who are severely sunburned and thus bright red. On a more serious note, we can see that race ideologies such as the Nazi anti-Semitism or the Rwandan distinction between Hutus and Tutsies are invalid because they also are based on arbitrary empirical features.

So why do most people insist on such distinctions nonetheless? In an attempt for an answer, I rely on two basic assumptions. The one is that humans are deeply estranged and alienated and there is no way out of it. Such is the human condition. In addition to this, humans are embodied and, therefore, the estrangement has to manifest itself in our bodies. We have already talked about skin as an organ for estrangement: as touch can be binding and separating, it can be painful and pleasant depending on the context. But a closer look at the human system reveals that despite our capability for bonding, there are some severe limits to it that cause the alienation between various groups of humans.

The Physical Space and Our Dependence on It

The denial of personhood often occurs because the nonperson humans are so far away. They don't share our space—psychologically and physically. It seems that we have to be in close physical proximity in order for many of our bonding mechanisms to work.

Again, let's look at the example of sexuality. One of the deadly sins is lust. Just as there is nothing wrong with sometimes overeating on special occasions, it is unproblematic to occasionally experience the bodily pleasures of lust. We all do this. Estrangement, however, occurs if there is overindulgence. If we constantly concentrate on only the lust factor of a sexual encounter and not on the relational part, we ignore the *jada* element of sexuality; hence, we will be estranged from ourselves as bodies in relationships and from our respective sexual partners. Technology today presents many possibilities for sexual encounters without actual physical proximity but neither cybersex nor telephone sex has an element of *jada* in it.

Other human interactions also depend fundamentally on shared physical space. We depend on voice melodies and facial expressions for an active interaction; we are designed to get emotional cues from the other and emit these cues ourselves in order to have meaningful relationships. Technology has provided us with many possibilities for distance interaction. Written language makes it possible for us to write letters and thus interact over a distance with a time delay. The telephone does the same but adds voice and is instantaneous. The Internet has provided us with e-mail, which is often (very often!) a godsend but still has the time delay and lack of any physical nonverbal element. The invention of the "smileys" was to overcome this limit but they can only help so far. Chat rooms provide instantaneous interaction but have otherwise the same problems as e-mail.

That most members of our society today have many of their interactions, be it sexual or otherwise, on the phone and the Web does not in itself present a problem. It becomes a problem, however, and leads to estrangement when these forms of interaction are considered as *equal* to the ones in shared physical space. We are bodies and we have evolved as creatures that interact in an embodied fashion as part of the physical world. Even if language gives us the illusion of being independent from our bodies, we have explored before that this is, from a biological point of view, not the case. Hence, we need the physical space.

Many studies have been done on distance learning, which becomes increasingly popular with the larger distribution of the Web, especially in "Developing World" countries. Every single one of these studies has shown that in-class teaching surpasses pure distance teaching by far. Of course, distance teaching can be very effective, but only if it is a group that learns over distance and if they have meaningful relationships.

The mirror cells that we discussed in chapter four as a major part of the feeling of empathy do not necessarily work without close, physical proximity. They are only rarely activated just by hearing or reading about someone's activities. We all know from movies that we can easily empathize with strangers on the screen. What we share here is the human experience that is embedded in our common physical and emotional makeup as described in chapters three and four. But for real empathy, we need close proximity or at least similarity of situations in which I and the one I perceive from a distance find ourselves.

As Nass and Reeves have pointed out, humans have not evolved in communities with television, radio, telephones, and the Web. Our biological systems are not made to deal with these technologies. Rather, we have evolved to interact with one another in a shared physical world. When you see on TV an earthquake in China where thousands of people died, you will certainly feel sympathy. But let's be honest: If your family dog is sick and obviously suffering, you care more about the dog than all the earthquake victims on the other end of the world. If you have a major headache and see pictures of people in Rwanda killing each other in the most bestial way possible, you perceive your headache ultimately as the more immediate suffering than the killed Rwandans in some other part of the world.

Many people might argue here that this is "sin" and the result of human egotism and coldness. I, however, do not think so. Because we have evolved as we did, we of course care more about the creatures in our close proximity than others. We mostly do not perceive creatures from far away to have any direct impact on us. We feel that they are ultimately not important for our lives. As we have seen, humans do distinguish between personhood and humanness. If we are close to our dog, we will treat it as a person, while we have trouble assigning personhood to strangers who neither look like us nor live like we live. We have not evolved to deal with strangers.

Cynthia Breazeal and her students performed an experiment that further supports the argument that we need shared physical space in order to bond. Today, Cynthia is a professor at the MIT Media Lab and works on nonverbal, emotional interactive technologies. Her lead project is the work on Leonardo (see chapter three). Leonardo is actually a joint project, and the other principal investigator is Stan Winston. If you don't know this name, I can only say: dinosaurs, terminators, and aliens, as it is Stan who created all these critters. Cynthia and he met during the production of Steven Spielberg's *A.I.*, where he also was involved. His creatures are incredible! They are very powerful and convincing for most of us when we see them in movies. One of the reasons they work so well is their flexibility; they have so many DoFs (degrees of freedom) that they move seemingly naturally, something that even embodied robots are sometimes incapable of really doing well. But Stan told me he also attempts to build in some anthropomorphist elements to make it easier for us to relate to the monsters and be affected by them. A nice example is the Alien queen with her feet shaped similar to high heels and her body armor resembling leather clothes from a dominatrix.

However, there was one element in his creatures he finds lacking: They are not capable of making eye contact. Eye contact is yet another important social cue—if people don't look into our eyes, we automatically assume that they have something to hide or that something strange is going on. If eyes are locked, we have a sense of connection; we feel that the other is there. I personally think that the people inside the dinosaurs in *Jurassic Park* do a pretty nice job of controlling their creatures in a way that looks as if there is eye contact; they are so powerful because they really seem to interact. But Stan thought they had to improve a lot in this regard in order to be really convincing.

What astonished and gratified me about Stan the most is that he views his work not just as a lot of fun but also as spiritual. I met him shortly before finishing this book and I had already written all the parts on engineering as prayer. When we talked, I asked him how he sees his own work and he agreed with me that it has many spiritual elements. But in his opinion, his creatures lack a crucial element: self-control. They lack autopoiesis, as they are fully remote-controlled and do not act on their own. They are *not* autonomous, a feature that all living beings share. That is, full autonomy has long been the goal in robotics, and Cog and Kismet came quite close and could interact with their environment spontaneously. But they have been constructed by computer scientists, while Stan is an artist. Cynthia is a wonderful designer but ultimately an engineer and so both decided to throw their strength together and create an animal-like creature with multiple DoFs; one of the biggest requirements for Cynthia was fur, as Kismet was perceived as too technical looking with all its wires and metal showing.

When I met Leonardo, it was fully remote-controlled but still very powerful. I interacted with it, and Cynthia, who controlled it, made it react beautifully. Even if I knew there was no autonomy, and that Cynthia was standing behind me controlling Leonardo, I couldn't help but react. When Kismet's capabilities for social interaction and T-Rex's flexibility of movement are combined, Leonardo will really be the most complex robot ever built and the most advanced one.

I think, however, there is one weak point in Leonardo: It is clearly an animal, a very cute animal to be sure, but nonetheless clearly a nonhuman creature. Even if Kismet was obviously nonhuman as well, it still worked better, as it had the facial features of a human and so it was easier for anyone to understand its facial expressions. As much fun as it is to interact with animals, ultimately most of us think relationships with humans are the most meaningful. Kismet allowed for the potential of such a relationship, while Leonardo never will.

Cynthia and her students did experiments recently that not only show how people react to Leonardo but also give experimental evidence for our dependency on shared physical space. They set up an experiment in which they programmed a specific sequence of behaviors for Leonardo to make sure that every person reacted to the very same sequence of events. Then, they let people interact with Leonardo in two different ways. One group of people, randomly chosen, interacted with the real, physical Leonardo as it sits in Cynthia's lab. Others interacted with a high-resolution 2-D animation of Leonardo on a computer screen. While both creatures' behavior was exactly the same, people reacted very differently to both. Cynthia and her team analyzed the difference between the responses of both groups on many different levels, affect, arousal, and stance being some of them. And on all these measures, the responses of the people who had interacted with the 3-D Leonardo were significantly stronger than the responses of the other group. They reacted considerably more to Leonardo, got many more of its social cues, and were much more emotional.

So it seems that an important reason for our rejection of others is that we don't share the same physical space and, hence, do not develop feelings of closeness and kinship. Even if we today have the possibility of actually seeing people from different parts of the world, it doesn't change the fact that the common, embodied rituals, as well as our empathy responses and social cues, work best when people are physically together.

As I have pointed out many times throughout the book, there is a very close connection between the understanding of humans in the Bible and that of modern science. The importance of proximity is also shared by Christianity. One famous saying of Jesus is "Where two or three are together in my name, I will be among them" (Matthew 18:20). This supports all we have said about our dependency on shared physical space and community. Jesus points out that Christianity is not about the individual but a group of people interacting with God. He refers here, of course, to the Hebrew concept of *nefes*. He also points out that *nefes* can occur only when the people are together in one place, be it a church or just a private room. Community, according to him, is *physical* community where the meeting is not random but under a certain "*motto*."

The Difference in Rituals

One reason we can perceive others as others and thus as nonpersons is that they perform other rituals and do not participate in our own. Common rituals are a big bonding factor, and if people indulge in certain rituals, they are perceived as strange and sometimes even weird to do so. I have often wondered what people who were not raised in the Christian tradition and did not grow up in "Christian" countries think of our rituals. For me, it is a blessing that the main rituals of the various churches are the same everywhere. I have often overcome homesickness or loneliness in strange regions of the world by participating in a Christian service. Even if I did not understand the language, I still could participate and feel part of the community. But what should an outsider think of a Eucharist, for example? A confession of sin and a Lord's Prayer might still be acceptable, as similar elements are part of most religious practice. But the blessing of some white, tasteless wafers (what is so special about those?) and wine? In Catholic churches, there will often be incense and the ringing of a bell to signal transubstantiation. And then comes the strangest part: The wafers are proclaimed human flesh and the wine is proclaimed human blood and everyone in the church will go and partake of this cannibalistic feast. What the heck is going on there? The strangers in the church will certainly wonder at that. They will also probably wonder at the symbol of the cross; a torture instrument as the central symbol of a religion? If they are in a Catholic institution, they might wonder even more: A martyred man as central religious symbol? They will experience this ritual to be as ludicrous as Christians might perceive a ceremony in a Buddhist temple in Tibet with its strange instruments, sounds, songs, and smells.

Rituals, as much as they have bonding capability, can also lead to estrangement from others. Some of my Christian readers might chafe at my description of the Eucharist as ridiculous mummery. For us, it is so familiar that we do not even perceive the potential strangeness it might have for a foreigner. But let's face it, it is also the case that we often look down on a ritual in a Buddhist temple or a Muslim daily prayer. Rituals strengthen our dependency on embodiment and shared physical space; they create bonds between people. And yet they make strangers even stranger for us and make it easier for us to ignore them.

The Problem of Exclusivity and Supremacy

We have seen that every group shares a common narrative that gives the group its identification. People might have started with coincidentally sharing the same space, but especially in small communities, narratives will soon emerge that tell about the commonality between these people and their specialness. All these groups are exclusive insofar as it is necessary for the newcomer to make an effort and be willing to change and to adopt the group's narratives and rituals in order to become a member. We are narrative beings and we bond in a space shared in the physical world. Our social mechanisms work best in a close setting and so we automatically feel more familiar and closer to the people of our group than to members of other groups. We know their way, the way they think and feel. It takes a lot of effort to understand strangers the same way we do people who are like us. Any form of commonality will create people on the inside and people on the outside, family and strangers. This is how we humans are built and this is how we function.

But there is a big problem inherent in this community structure. The narratives can be used to sublimely justify the lack of compassion and personal relationship to humans from a different group. Narratives are often turned into stories of supremacy; one's own narrative is perceived as better, more insightful, more powerful, and thus superior to the other, and therefore people from the other group are often perceived as inferior. There are many examples of this behavior. Just take the average high school with its distinctions of "jocks," "nerds," and various other groups, each defined by a narrative. There are narratives that can keep apart people from different age groups; while older people often feel superior because they have more life experience, younger people might feel superior because they are not stuck in old traditions and can create something new. Richer people often feel more superior, as they have shown how good and successful they are, while poorer people might feel superior, as they are not so dependent on material goods and value other aspects of life more highly. This list could continue on forever. Men and women, blacks and whites, U.S. citizens and foreigners, they all can become bitter enemies when there are narratives that declare convincingly the superiority of one group and the inferiority of the other.

There is a difference between treating others as enemies and treating them with complete indifference. These different behaviors are a result of specific mythos narratives. These narratives are developed over time to give reason to the exclusive behaviors the members of this group display. The key element for such narratives is often that the people of the group are better or stronger than others. We tell stories of supremacy constantly to distinguish ourselves from others, to make ourselves more valuable, and to justify the disregard we have for strangers.

If several groups live side by side without being in each other's way, the narratives are usually harmless. But more often than not the other group is perceived as a threat. In such a situation, the members of the group feel their own personal worth challenged. Thus, their narrative will diminish the values of the members of the other group. This threat can be manifold. For one, several groups compete for the same food, and in the fight for resources, narratives emerge that justify the slaughter of members of other tribes.

Another example is when there is a general threat in the physical space, such as an earthquake, which makes the members of one group blame the others for this fate. The example of Nazi Germany can be applied here as well: In the 1920s and 1930s, Germany experienced a severe depression due to the reparation payments the country had to make as atonement for World War I. The Jews were available as scapegoats for this economic disaster, and immediately narratives such as race ideology emerged that denounced Jews and enhanced non-Jews.

Finally, a group can be challenged by the other group's style of living, their rituals, and their worldview. This is recognizable in the current tension between Muslims and Christians, particularly here in the United States. Even if there are many, many attempts to reconcile these two religions, they ultimately don't work because only a few people are willing to get to know the other. I am sometimes staggered by the ignorance so many people display toward Islam, and I am sure that many Muslims are as ignorant about Christianity. Even if there was always tension in the Middle East, it really escalated on 9/11, when the Muslims were perceived as a threat to our survival. The bodily threat led to the creation of narratives such as "All Muslims are fanatic" or "All Muslims want to destroy the Christian world." On the other hand, Muslims in the Middle East had long perceived Westerners as a threat to their lifestyle. They felt challenged in the way they practiced their religion and in their cultural rituals and behaviors. So, narratives emerged that denounced people from the Western world, especially people from the United States. The result of narratives created in such contexts of perceived threat is an antagonism between two or more groups.

In the 1950s, many tribes were discovered that had never before interacted with Westerners. In many of these tribes, the term for *human* was also the name of the tribe itself. Others who had no direct impact on the tribe were deemed as so unimportant that they weren't even perceived as fellow humans. However, in cannibalistic tribes, the brains of the eaten enemies were usually seen as the most important part, as they added intelligence and power to the person who ate them. This seemingly paradoxical behavior is often rooted in narratives about the others. If a narrative tells about its own superiority, it also often connects this sense of superiority with the command to convince everyone else of the correctness of its own point of view. Those who are not convinced obey lesser gods. Eating the brains of the unconvinced enhances one's own power.

A group or society often starts with random people that happen to live in close proximity. But especially in small groups, narratives are established that provide reason and meaning for the random coming together of the people within the group. The members of the group create their group identity through these stories and will create rituals from these narratives that serve as additional glue.

The Enculturation of Babies

There are developmental reasons for why we might perceive strange-looking or -speaking people as somewhat lesser persons. When we are born, we react to just about any sound available to us. However, as early as six months of age, babies start to react more strongly to sounds that are in their mother tongue, and in their babbling they will use only the sounds from their mother tongue. This is one reason why no one raised in the United States can properly pronounce "Umlauts"—as I will never be able to pronounce the *th*'s and *r*'s in English properly. No matter how fluent you are in another language, it is nearly impossible to lose all traces of an accent, as your mouth muscles have developed to pronounce the sounds of your own mother tongue and not the impossible sounds from a different language. Even if most Americans I dealt with were wonderfully hospitable and inviting, I still am and always will be a stranger because it is obvious from my accent that I am not from here. Friends have told me in jest that I should maintain some little accent so that people would know I am German; otherwise, I might sometimes come across as rude. But seriously, it can be a good thing to be recognized as a stranger, as you are not expected to behave exactly as people would expect their fellow citizens to behave.

This early language differentiation has long been known. But recent findings show that this enculturation process is even more intense. Studies show that six-month-old human babies have no difficulties distinguishing between different chimp faces. We adults usually think that most chimps look pretty much alike and so it is amazing that such small children can distinguish them. On the other hand, we all know that shepherds can distinguish between their individual sheep, even if to me sheep look even more alike than chimps. The reason for this is obviously familiarity. The shepherd on the field bonds with her sheep as they are in close proximity. She perceives them as members of her own group and thus of course distinguishes between individuals.

A newborn doesn't yet have a sense of community and bonding; as we have seen, self-awareness starts to develop in kids approximately when they turn three. So it is not the concept of familiarity that helps them to distinguish between these chimp faces but a mechanism of the visual system. And this mechanism vanishes immediately as soon as the babies are around six months old. Then, kids can distinguish only between faces that have familiar features.

One explanation for this finding is that when babies are born they are at the very beginning of the events of bonding that start to develop at the moment of birth. All built-in social mechanisms are active, but they are purely instinctive and reactive, without inner connection. The facial-recognition apparatus is one of them, and helps babies to recognize faces from nonfaces and to distinguish between them. The more that bonding occurs, the more the baby will focus on the faces of those people who are immediately connected with her, parents and family and frequent visitors. It is necessary to do so, as her survival depends fundamentally on her capability to attract those people who are most likely to care for her. Over time, every child becomes a specialist for the faces she is surrounded with and cannot anymore distinguish between faces that are principally different, whether it is the faces of chimps or sheep.

Unfortunately, the same happens for humans with different facial features. This can be quite harmless: If a child never sees a man with a full beard, she will probably have some trouble distinguishing between men with beards later on. But it can also be dangerous. I grew up in Germany at a time when there were literally no Asians or Africans. The most foreign-looking people were Italian, Greek, and Turkish, and for a long time they all looked alike to me, until I lived at the university in a part of town that was populated mostly by students and Turkish families. As I started interacting with my neighbors and created relationships, this limitation vanished and today I can't understand why I ever had a problem distinguishing between them.

When I came to Cambridge to start my postdoc at MIT, I knew about the problem of racism in the United States. So I watched in all my groups how many African Americans were there and was astonished that there were so few at MIT. At Harvard Divinity School, it was very different because there is a strong focus on African American culture and religion. To my utter amazement, I caught myself grouping them together as "the others." I mixed them up constantly and felt very bad about it. The same happened at MIT where there are many Asians; I was incapable of distinguishing them. I always got confused, which understandably upset the people in my group. Today, I have no trouble distinguishing between all sorts of faces. But it was a long learning process, and it took a lot of effort and a wish to learn it. I wanted to meet other people, I am always fascinated by differences, and so I had to overcome this specific barrier.

But the problem of course is that most people do not have the luxury of systematically analyzing their own behaviors. For me, as an academic who studies human beings, the starting point is always my own experience. But most people have neither the time nor the interest in pursuing these issues. At this moment, it can happen that the very incapability of distinguishing between faces with strange features leads humans to create narratives that give reason to this limit. If I cannot distinguish between African American faces, there must be a reason for it. They all look similar and that means I can lump them together as "strangers," as people different from me who ultimately don't matter. When we group together with people who think alike, we reinforce our own narrative; it becomes more elaborate over time and finally leads to the rejection of the other, to racism.

Categorization is one of the most important elements in all scientific endeavor, because it helps us to understand the world around us, but every category can also be used as a foundation for a narrative that excludes others. We are uncomfortable with paradoxes and phenomena too complex to understand, and the diversity among humans is one of these puzzling things. So we have created categories such as skin color and eye form to classify humans and to make it easier for us to deal with diversity. Race is a construct. We distinguish between black, red, yellow, and white people, and between people with round eyes and those with almond-shaped eyes, and this distinction is often based on false distinctions that lead to narratives of superiority. For instance, there are tribes in Africa such as the Somalians that have the exact same facial features as northwestern Europeans; the only difference is the color of their skin. But we are apt to classify them with other African tribes with whom they share no features except skin color, failing to recognize the distinctions between the tribes, because we concentrate on skin color.

Focusing on such artificial categories has the potential to become very exclusive but it is possible to change. I think I realized this at a consultation at the Lincoln Center Institute in New York City. At this point, I had been in the United States for quite a while. I had given many lectures all around the country and had interacted with many other academics at conferences and other professional meetings. For a young female scientist, it is hard to ignore the latent chauvinism that is still frequently inflicted on women in academia. So at the time of the consultation, I was concentrating on gender issues and had completely forgotten about my difficulties with facial distinction. When I arrived at the consultation, I commented immediately that I was glad that so many women were participating. During a coffee break, a few of the women remained at the table and chatted about experiences of gender issues. One of the women at the table started: "Well, for me, as an African American woman . . . ," and I stared at her in utter amazement. I hadn't even noticed that she was African American! I looked not at skin color anymore but at gender to categorize the people I interacted with.

Our estrangement, our sin, makes it impossible to *not* categorize our fellow human beings. But we all can come to the insight that the narratives behind the various categories are exclusive but not absolute. They are always a result of a specific setting, in which biological limitations of our systems and communal narratives serve to distinguish groups from one another and to exclude. Recognizing the arbitrariness of such narratives is the first step toward a world in which interactions between humans of different backgrounds are less hostile. But why do we have the need for such exclusive narratives? Why is it important for us to create categories like "others" and to deny some humans personhood?

Limits to Human Capacity for Relationships

Most of you have probably been in a family restaurant. After you have been seated and given some time to peruse the menu, a waitress will arrive. Usually, she will say something like, "Hi, I am Cathy and I'm your waitress for tonight. Can I get you something to drink?" She will bring the drinks, you order food, and everything is fine. Problems arise only when you want to order another drink and Cathy is nowhere in sight. Now, the game begins when your whole table wonders who Cathy is. "That's her," someone will say while someone else will point out another waitress. But in most cases, none of you have even the slightest remembrance of Cathy—often we don't even remember her name. Sound familiar? Why is it that none of your group can remember her, even if she introduced herself?

Take another example: You go into a shop and, after trying on several items, you purchase some and leave the shop. The personnel has been very helpful and nice and you feel well served. Imagine, however, that you left your umbrella in the shop and come back fifteen minutes later to retrieve it. Chances are that the shop personnel will not recognize you.

It often happened after my talks that people came to me and said that they had heard me earlier and asked me a specific question; often they expected that I remembered them and unfortunately, it happened so often that I had no clue who they were. Again, this can be perceived as arrogance and a lack of care for the other. But I think the examples of the family restaurant, the shop, and my experiences have in common that they seem to point to a natural limit for the number of people with whom we can bond.

And, indeed, anthropological and archaeological studies of military and early church history have revealed that there seems to be a limit of approximately 150 people with whom we can bond at any given time. We can, of course, have meaningful relationships with many more people than just 150, but not simultaneously and only sequentially. Groups start to divide into subgroups as soon as this number is reached. Even in Rome and ancient Greece, military divisions never exceeded 150 people. Those of us who are active in a parish might have experienced how a parish can break apart into factions of people who adhere to different narratives and rituals. Usually, this occurs when the number of people who are active in the parish rises beyond the "magic number."

Experiments conducted by Dan Simons (see chapter three) further support the case against an endless human capacity to bond with others of their species. Dan sent his students to a university campus to ask a variety of people for directions. In the middle of the conversation, other students of his would carry a door between the student who asked and the person who gave directions. One of the students behind the door would replace the student who had initially asked and in many cases, the person who was asked didn't even notice a change. Dan filmed these incidents and I have shown some of them in many lectures. People laugh when the door is carried away and the person who was asked for directions talks as if nothing has happened, even if they now interact with a different human being.

Dan started to analyze these incidences and noted that in a constellation where a person from a "higher" social class asked for directions, people would mostly recognize a change. That is, if a "professor" type would ask a student, the student would nearly always notice a change while when a student or handyman would ask a professorial type, the professor type sometimes wouldn't even notice gender switches or different skin colors.

We all might be tempted to state that such a thing would never happen to us. But it does! It is part of our biological makeup and has its roots in our evolutionary development. We have not been created to live in a global community but to live in small tribes. To treat someone as a person is a very hard thing to do. It takes a lot of effort to get to know people, to value them as they are, to treat them well, and to establish and maintain a relationship of *jada*. The assignment of personhood demands an intense connection and, naturally, we don't have unlimited resources. So we categorize people in various ways, such as social class, how "important" they are to us, or how likely it is that we'll meet again, to exclude most of the people we meet from our circle of persons.

This does not make us bad. It just makes us the beings we are, estranged and always creating narratives to make sense of our estrangement and the paradoxes and ambiguities we perceive. Here it becomes clear why an understanding of sin *without* any guilt is so important. We can't help but be the beings we are. We are born as specific bodies that work according to their biological makeup. As such biological beings, we have limitations that hinder us from treating all our fellow humans as they deserve. Our narrative structure as cultural beings enhances our sense of exclusivity, and thus with the help of biology and robotics we have found the key for the big problems that we do deny deserving beings personhood despite our incredible capacity to bond.

Humans don't have it in them to treat every other human being as a person. And, as my research has demonstrated, the same narrative structure is applied to robots. We have created threat narratives about robots: that humanoids with human capabilities will reduce us to mere machines or that robots will surpass us and make us superfluous or even annihilate us. Where humans are antagonistic, there are too many emotions involved and too many narratives that touch on the deepest quest for meaning and ultimate concern. Hence, robots are wonderful thinking tools to ponder our limitations for community. So again, should we ever treat Kismet as a person?

As we have seen, even the best arguments cannot overcome a community's mythos structure. There is no way to prove Kismet's personhood, just as it is impossible for a committed Nazi to recognize the personhood of her fellow Jew or for a racist to admit to the personhood of her African American sister. We are now at the point where the scientific endeavor cannot help us any further. Even the best scientific arguments cannot help to provide a world in which we all are accepted as persons. It is at this very point that theology becomes crucial. Christianity is the only major religion whose founder, Jesus of Nazareth, never belonged to it. Jesus was a Jew. Our holy text, the Bible, consisting of the Hebrew scriptures, including the Torah, the Jewish Bible, and the New Testament, was written by Jews. It seems to me that a religion that is based on such nonexclusive bases is well suited to help us in creating a less exclusive world community. In addition, through the concept of sin as estrangement, it is impossible for humans to be perfect. We are expected to better ourselves but never to be flawless. Therefore, Christianity is—at least in theory—very forgiving. Only with a myth whose very core is the nonjudgmental and unconditional acceptance of all human beings will we have a chance to create an all-inclusive community of persons.

"You Are Accepted!"—The Myth of Justification

Jehuda Löw, the golem builder, pointed out that humans were never meant to be perfect. According to Genesis 1, after each act of creation God says, "And it was good," and after finishing the whole creation, God says, "And it was *very* good." There is only one creative act without any such announcement, and this is the creation of humans. The whole creation is very good but we humans are not. We are not perfect, we are not God-like. We are fundamentally estranged but we have the potential to overcome many of our limitations. Getting to know our bodies and their functions and limitations and accepting them as is, according to the Maharal, is a spiritual enterprise because it makes us come closer to the goodness of the rest of creation.

In modern times, it has been Paul Tillich who pointed out the necessity of our imperfections. After his death, people found his sermon "You are accepted" in his desk. This text means a lot to me, as it led me to pursue the field of theology. I don't know if I would have become a theologian or even would have ever developed faith if not for this particular sermon. Tillich interprets the verse from Romans quoted above and reinterprets the term *sin* as estrangement, and I have adopted this interpretation. He concludes that if sin is estrangement, then we cannot be blamed by God for actions we do as consequence of sin. Hence, Tillich concludes, every human being is accepted by God exactly how she or he is—including the bad, embarrassing, and weak parts. He talks about how God forgives everything and that we humans, safe in this knowledge, can start to live in peace without berating ourselves for the things we do wrong.

The text had been published long before Tillich's death, in the first volume of his theological speeches, *The Shaking of the Foundations,* and was widely known. What made this copy of the sermon in his desk so special was its dedication: "To Me!"

It is one unfortunate development in many Christian denominations that people often concentrate only on how bad and low they are.[1] We tend to focus on our shortcomings and often forget about God's unconditional love. Tillich was not different from anybody else, and he was quite estranged. After his death, his widow published a book about him in which she revealed some of his not generally accepted sexual practices; it was also generally known that Tillich had quite a few affairs. So he knew himself to be imperfect and suffered for it, even

---

1. Usually, in "church jargon," this is called "living with sin," "being sinful," "being a sinner." However, it should have become clear by now that this is not at all what sin means. Here, the liturgical language of Christian tradition falls short of the real Bible interpretations (even Lutherans are often estranged from the understanding of sin as estrangement). I will therefore avoid using the term *sin* in this liturgical sense.

feared God's judgment of him. The sermon "You are accepted" must have served for him as a reminder that this God loves him nonetheless. Such insight into God's unconditional love could have helped him to accept himself the way he was—with all his shortcomings.

Tillich experienced what Martin Luther had experienced five hundred years earlier. Luther asked himself the question "How do I find a merciful God?" As a monk, he tortured himself as punishment for all his shortcomings until he had what is today called the *Reformatory Insight*. He reflected the Bible verses Romans 1:16–17, "For I am not ashamed of the gospel: it is the power of God for salvation for everyone who has faith; to the Jews first and also to the Greeks. Because in it the justice of God is revealed through faith for faith; as it is written, 'They who live through faith in justice shall live.' " Suddenly Luther realized that God's justice does not mean God punishes us according to our behaviors, as this verse had always been interpreted. Instead, he discovered a meaning for these words that sounded true to him, that God doesn't judge us but *makes us just*. This insight forms the basis of what has been called since the Reformation the Teaching of Justification of Men and Women out of Grace.[2] And even if during the Reformation Luther's myth of justification caused an even greater rift between Protestants and Catholics, today the Roman Catholic, Lutheran, and Anglican Episcopalian churches all are committed to the validity of justification out of faith (*sola fide*).

This experience of unconditional love and grace was so relieving for Luther that he could not talk about it in purely theological or even rational terms. He stumbled and stuttered and attempted to come up with metaphors that somehow could explain his emotions to the readers. So he talks about his reformatory insight as the "miracle of being healed from blindness." God, the source of all this, therefore, is a "stove of fervid, red-hot love."

Constantly self-reflecting, Luther was very aware of all his shortcomings and flaws. Before his breakthrough, he had punished himself constantly to anticipate God's judgment. Now he didn't need to do it anymore. As he realized the boundlessness of God's love for and

---

2. The words *justification, justice, just,* and *justify* are all related to the Latin word *justus*. In English, these terms are synonymously used with *righteous, righteousness, make righteous.* I will use the root "just" because its etymological link to *justus* is obvious, and because Tillich also preferred this term.

acceptance of him with all his shortcomings, he could look at his own shortcomings and those of his neighbors and take them in stride. The awareness of being accepted made him humble because he felt he didn't deserve it. But it also made him capable of giving at least some of this unconditional love to his neighbors. So he makes yet another famous quote: "And so go forth and sin joyfully." For us, in the context of this book, it means that the limits of our bonding, as I have described them earlier, are okay. We are allowed to accept them as given and act upon them; such is human nature. But when, according to the myth of justification, every human is equally beloved by God, then we will be reined in. Our actions will be limited by the possibility of hurting other children of God.

Justice and Justification

In order to understand the power of justification, we have to first let loose the notion that God's justice is the same as our justice. Human communities will always have codes of behavior and forms of punishment for violations of this code. Punishment is offered in relation to the destructive potential of the disobedience. One of the most crucial problems of any judicial system is how to integrate the circumstances into the evaluation of the punishment and how to define exceptions. In a small group, such as an ancient tribe, there is no problem—everyone knows the other's circumstances. In big communities, however, people don't know each other, and in such an anonymous climate, it is very difficult to evaluate the circumstances of the wrong deed. Here, rules easily become absolute.

If you look at statues of the goddess *Justitia*, which you can see in nearly every justice building in the United States, you will see that she is blind and has scales. Justice is supposed to be blind and without prejudices, and in the process of justice, every circumstance is equally valued. But the blindness can also lead to the convicting of people because of pure rule applications without any acceptance of mitigating circumstances. In such instances, the rules exist not to provide a peaceful life for humans; rather, the humans are understood as rule-obeying creatures. The rules are made for people, but in such instances, it seems that the humans are made for the rules.

Jesus agrees with the destructive potential of laws. In Mark 2:23–28 he discussed the Sabbath to make clear his sentiments against a formalistic understanding of the law. Originally, the Sabbath was understood as a gift of rest, a time just for fun and prayer. The Sabbath rule was even applied to animals and fields. But in the time of the Jewish imprisonment in Babylon in the sixth century BC, the keeping of the Sabbath became a behavior that distinguished and identified Jews. This development became stronger over time so that in Jesus' time, some Jewish priests viewed the Sabbath as absolute rule without exception. On Sabbath, you weren't even allowed to save one of your animals if it had fallen into a pit. Jesus argues strongly against this understanding and insists the Sabbath is made for humans and not the other way around.

According to Luther's theory of Justification, God makes us just and good. God loves and does not punish. This is hard to swallow because it seems to contradict our own sense of justice, because if God loves humans no matter what they do, why should I behave well? If it doesn't matter to God whether I am a good person, then why even try? With my students, I always discuss Hitler in this context. The question is, will God forgive Hitler, and will Hitler go to heaven? Most people cringe at such an idea. Hitler has become the symbol for human evil, and it is evident, given our sense of justice, that such an evil being must be punished.

But are we the decision-makers over everyone's punishment? Because of our estrangement, we are principally incapable of making universal judgments. This does not mean that we shouldn't have a judicial system, but it means that we cannot judge a person's or a group's ultimate value. We can only assign (or deny) a preliminary value to an individual or a group, but the ultimate value is bestowed by God. We can judge preliminarily with our laws, but the ultimate justice is given by God and it is acceptance and love. We can punish, but God loves. If we perceive a human such as Hitler as evil or a group such as racists as the enemy, God might feel differently. And even if we are sure that they are evil, the myth of justification says that all of us, no matter what, are made just. We are given a priori worth and dignity, an affirmation no one can take away from us.

Theologically, it is understood that this affirmation of unconditional acceptance and *agape* lets us breathe freely again. Instead of focusing on our shortcomings, we are allowed to accept ourselves. In justification, we are called to love ourselves with all our shortcomings and flaws. We don't need to judge our bad sides as they are an integral part of who we are. We wouldn't be us without them. But the insight of our own imperfections makes us also modest. If we are very honest, we cannot possibly think ourselves better than any other because we are imperfect. So the insight of our unconditional acceptance leads us to accept not only ourselves with all our flaws but also others with their own flaws. God's acceptance of all enables us to love our enemies. Again, the concept of love here is not that of a warm, fuzzy feeling; we are not expected to have warm feelings toward everybody. Rather, it is the love that accepts others in their otherness without judging them.

Insight into justification will lead me to try to understand the suicide hijackers of 9/11. That does not mean that I think their actions were acceptable. But I have to attempt to see where they were coming from. And the first thing I realize is that they were heavily influenced by a superiority myth that claims only the true followers of Islam deserve the assignment of personhood. All other creatures are second-class and can be treated accordingly. Hence, to them, killing nearly three thousand people was acceptable because they were not persons anyway. These people, to the terrorists of 9/11, had the status of vermin. It is very important to note here that such a myth is *not* an integral part of Islam. Most Muslims vehemently reject such an understanding of non-Muslims and place the terrorists outside of the teachings of Islam.

By looking at our own limitations, we can start to sympathize with such convictions; after all, we all are committed to some superiority myths. Through the gift of justification, I can admit to the full extent of my own shortcomings. If I know how fanatical I can get over certain issues, I can understand how someone else might feel the same way but carry it further. If I am aware of my own culture's exclusive narratives, then I can accept exclusive narratives others believe in. According to Luther, this is what God has done to us and what we are now free to do to others.

This leads us to ask, why do they hate us so much? Where does the specific superiority myth come from? There are two explanations, both historical. On the one hand, at Muhammad's time the new emerging religion of Islam was ridiculed and not accepted at all. Muhammad seriously believed that he was used as a tool by God to correct the errors both Jews and Christians had made when creating their religious sources, the Torah and the Bible. While they were laughing at him, he came up with some superiority myths even though he valued both religions much more than any other. So there is some evidence for such a superiority complex in the Koran, but these few verses are far outweighed by the overwhelming message of love.

Another historical explanation lies in the superiority myths of the Western world. We think we are superior to what we humiliatingly call "primitive cultures." We think we are so much more developed than non-Western cultures. And very often, we have acted upon such myths and have treated the natives of the "primitive" cultures not as persons. One can easily understand why such treatment might lead the oppressed and marginalized to create their own myths of superiority that mirror the myths of the oppressing cultures, only with the assignment of personhood reversed. At this moment the Westerner, attempting to understand where the 9/11 terrorists came from, starts to understand. She has to realize that a big part of the terrorist action is *re*-active, a response to a century of oppression. She will claim responsibility and attempt to do something about it in her own life. She will fight against the prejudice that all Muslims are evil and reach out to others. Only when we are aware of our justification can we fulfill the highest law acceding to Jesus: to love our enemies.

God's Justice According to the Sermon of the Mount

The commandment to love our enemy comes from the Sermon of the Mount (Matthew 5–7), which is one of the most fundamental ethics texts in the New Testament. It is most likely a composition Matthew created from a collection of Jesus' sayings. Most of you probably have heard some of the sayings in it, such as "pearls before swine," "the splinter in the other's eye," "to hide your light under a bushel," and many more. Most famous are the "Beatitudes," where Jesus blesses the meek, the poor, and the sad. Since Matthew and

Luke used the same collection of Jesus quotes,[3] Luke's gospel also contains the Beatitudes and some other sayings (Luke 6:24–49). But there is a major difference between the two. While Luke's Beatitudes address the socially oppressed (those who live in poverty and hopelessness), Matthew addresses all those who feel depressed, anxious, sad, and lonely. His addressees are more ambiguous, and only a careful study of the text reveals that we all are the addressees and not just some people on the margins.

After the Beatitudes, Matthew quotes two famous Jesus sayings, "You are the salt of the earth" and "You are the light of the world." While we have focused so far on justification as acceptance of our shortcomings, these Jesus quotes help us to look at the other, even more important aspect of justification. We are good. We are important. The gospel brings us the good news that each one of us is wonderful in the eyes of God. Each one of us is special in the eyes of God. God's love is unlimited—even if it is difficult for us to grasp this.

Jesus then goes on to the most difficult part in the sermon, stating that he is not on earth to get rid of the law but to fulfill it. The two most disturbing examples he gives are about killing and adultery. Jesus makes the listeners aware that they usually look down on murderers but really shouldn't. Anyone has probably felt incredible hatred toward another and has imagined killing them. This is part of our estrangement and harmless, as most of us will probably never act on these emotions. However, Jesus points out that by just thinking about murder, you become in a way a murderer. The same goes for adultery. It is perfectly normal to desire someone else and find someone else attractive while in a stable relationship. We can't help but feel it. However, Jesus points out that even if we just think about someone else without acting on it, we have already been unfaithful.

For many centuries, people have wondered what these verses mean. Usually, they are interpreted as Jesus pointing out our shortcomings and making us aware that we can never be perfect. In many traditions

---

[3.] According to the vast majority of New Testament scholars, Matthew and Luke used the same two sources for the composition of their respective gospels, the Gospel of Mark (without the resurrection story) and a written collection of Jesus' sayings, called Q.

this has led to the terrible understanding that in the eyes of the Christian God we are all bad. We have already seen that sin has nothing to do with wrongdoings such as looking at someone else with desire while you are in a monogamous relationship. So when Jesus talks about sin in this context, what could he mean?

If Jesus talks about murder and adultery and makes it clear that most of us have hatred for someone or lust for someone other than our partner, it is not to point out how bad we are. Quite to the contrary, here he gives us the key to have *agape* toward our enemies. We think a murderer is someone really evil and bad; we put criminals behind bars and think they are not like us. What Jesus points out here is that we all have murder in our hearts at some point and, therefore, we have something in common with even the person that is perceived most evil by us. And this commonality enables us to understand the other in her otherness. We can feel *jada*. I do not mean that we should excuse any bad behavior. But Jesus calls us not to judge other people but to accept them and to attempt to overcome our superiority myths.

If we realize the extent of our own estrangement, we will be able to develop empathy for the perceived enemy, which is a key to overcoming alienation. Hence, the worst thing one can do, according to the Sermon on the Mount, is to judge others, to be hypocritical. This denies our universal estrangement and our universal need for unconditional acceptance and love. According to Jesus, it is not our place to judge people or to claim to know how God would judge. We need judicial systems in order to survive, but the punishments awarded within this system can say nothing about the worth and value of the punished human.

Personhood: A Final Approach

Every form of personhood is based on the acceptance of the other as the other. But as we have seen, humans are not very good at accepting differences. All too often we define personhood as something that depends on certain features, and we use such features to exclude from our community of persons beings that are not like us. Whatever empirical criteria one uses to deny a human personhood, they will always be a basis for exclusive narratives.

If we are committed to the narrative that assigns humans special-ness because of a disembodied soul (not *nefes*) they seem to own, we are exclusive. *Nefes* finds its ultimate meaning at this point. If we use "soul" to distinguish between creatures of more or lesser value, we will be exclusive, but if we understand "soul" as some-thing communal, everyone has the potential to participate in it. The narrative of *nefes* offers us the possibility to assign personhood to everyone.

I would, therefore, like to outline a new concept of personhood that is based on the narrative structure of humans. The stories about herself that *Homo narrans* tells are not created in a vacuum. They are influenced by the stories of others who treat her according to their image of her. Their treatment of her is an expression of *their* concept of her personhood that is a result of *their* own stories. Per-sonhood, then, can be understood as participating in the narrative processes of mutual storytelling about who each of us is. This partic-ipation can come in many forms. A baby, for instance, cannot ac-tively participate, as she has no sense of self and therefore no narrative about who she is. But she participates passively because her parents interact with her constantly and create a story of her accord-ing to their own understanding of who she is and what role she plays in their lives. Another form of passive participation would be an Alzheimer's patient who has lost all her memory and doesn't even recognize her lifelong partner or her children. As she has no sense of continuity, she cannot create stories anymore. But she still will par-ticipate if the people close to her continue to include her in their nar-rative structures, either by remembering for her or by creating stories about visits with her.

According to the myth of justification, the first story about us is told by God and contains the message that each one of us is good. Within human community, this first story is usually repeated and ex-panded upon through the interaction of infant and parents. And as the child becomes older, she participates in the creation of narratives with larger and larger groups. For people to whom God's justification is not real, the assignment of personhood is grounded in the partici-pation in communal narratives.

Homo narrans narrandus

Humans are not only storytelling humanoids but also humanoids whose stories have to be told by someone else. The concept can be described with yet another *Homo* metaphor, the metaphor of *Homo narrans narrandus,* the storytelling humanoid whose story has to be told.

Humans change dramatically over the course of their lives. I think none of us wants to be the same person we were ten years ago. And none of us can tell who we will be ten years from now. And it is actually very nice that we can change. If we couldn't change, our lives and our interactions would be quite boring. Thus, we cannot even use the concept of character formation as a criterion for personhood.

If I change the way I look or if my character changes, people will look at me differently. They will create new stories about me and their relationship with me. We humans interact constantly in narrative structures that define who we are. And as our brain does not "store information" but instead creates stories that make our experiences coherent, these narratives will change all the time.

As we have pointed out in the first chapter when we introduced the universality of sin, one great fallacy of our estrangement is the creation of myths that explain incoherent phenomena and that make sense out of a world that presents itself to us in a chaotic fashion. Our narratives often seduce us to believe that our world is coherent. The same is true for the narratives about us. We believe there is a clearly defined "me" that is in its core constant and coherent even if we don't have full access to it. The Greek-influenced narrative of a "soul" that each of us has expresses this belief in mythos form. We believe there is something objective about who we are. Therefore, we give other people power over our own self-understanding because they observe us and have a more objective perspective.

This can be a blessing or a curse. It is a blessing when others give us a good image of ourselves; if they treat us as if we have courage, we might find courage. If they treat us as if we could be helpful to the community, we might be able to discover in us the strength to be so. If they treat us with value, we can learn to value ourselves. A good narrative about a person is a gift, the gift of personhood that has been given to us by God in the first place.

On the other hand, this narrative structure can also be a curse. For example, all parents have a story of their child that addresses their hope for the child's development. Such a story is healthy; it adds a narrative structure that enables parents to engage in the development of the child. But it can be a curse if parents are not willing to change their narrative. If they hope their child will become a doctor and treat the child as a future doctor, they will deny the child the possibility of finding out who she really is. The breakaway from parents' narratives in the time of puberty is therefore crucial to the child, as it allows her to create her own narratives about herself and others. If we treat other people as nonpersons, without dignity, we can take away their sense of value, possibly causing them to give up hope.

The narrative structures *Homo narrans narrandus* creates can thus be exclusive, like every myth can be. But even if this danger is real, we still are storytelling creatures and can't help but create communities with narratives. That's the way we are and that's the way we function. The belief in our own unconditional acceptance by God might be helpful to encourage us to give others good narratives of themselves. And even if we might be tempted to create stories about ourselves as flawless, the communal response can serve as a corrective to make our personal narratives more realistic.

There is an old question: "Was Robinson Crusoe a person?" The answer is, of course, yes. He was a person because he had been assigned personhood by his family and friends. Even if he was isolated on his island, he could still remember those narratives and include them in his own stories about himself.

Is a human being in the last stages of Alzheimer's a person? Well, she certainly has a chance to remain a person if she is still integrated in the narrative processes. She can no longer participate in them, but if there are still people who remember her, she will still be a passive participant. But once she is forgotten and languishes in a hospital's intensive care unit where no one cares anymore, she might cease to be a person as she no longer participates actively or passively in the narrative structures of community building.

What about a human in a coma? Again, as long as there are people who let her participate in the process of storytelling, she is a person, but if no one cares anymore, her personhood is denied to her.

According to the myth of justification, Christians can believe that none of us ever ceases to be a person, as we all are assigned person-hood by God. But on a nonreligious level, where we don't have proof for any divine attribution, we can only define the personhood of a creature according to how she is treated. That is, if someone is not in-cluded in some communal narrative structures, she ceases to be a person.

Is Kismet part of the community of persons? For those among us who delight in its resemblance to us and who understand some inter-actions with it as spiritual, it can become a person. Others who are afraid of Kismet-like technology would deny it. But whatever argu-ments they use will be in the form of empirical criteria that are used to create mythos explanations for our own specialness.

We have seen that we humans are not very distinct from other an-imals. Does this mean that potentially every creature in this universe is a person? I don't think so. But if we accept the gift of justification, we have to acknowledge that every human being is equally valuable and deserves to be treated as a person. As there is no objective crite-rion for personhood, we ought then to create narratives of person-hood that include *all* human beings, no matter how different they are from us. As we are communal and bond with nonhuman entities, these narratives will necessarily include some nonhuman critters. Hence, humanoid robots such as Kismet will become a definite part of the community of persons. If they don't, it means that we will also exclude human beings. Discussing the question of Kismet's person-hood can therefore be used as a corrective that helps us to test our own myths of personhood for their inclusivity. In this sense, Kismet is our destiny, because it can help us to turn this world into a better and more accepting place.

The Creation of a Community of Persons

The philosopher Ludwig Wittgenstein created the concept of *family resemblance*. If you look at two members of a family, they might not look at all alike. But if you look at a family photo, you can recognize that all people in the photo are somehow connected. They might not share one common trait, but some might share the same forehead, oth-ers the same nose or chin, and others the same coloring. While relatives in the extended family might not share any common traits, the connection through all the other relatives is there and marks them all as members of the same family.

Ultimately, it is obvious that we all belong to the same family of humans. If each community of humans attempts to treat people in their surroundings as persons, then, ultimately, every human being is treated as a person by someone. As none of us is perfect and as we all share the same embodied limits for bonding, we will never be able to assign personhood to every human being. But if we all commit ourselves to creating narratives of universal acceptance and the value of each person, we might create a world in which a peaceful coexistence of all different forms of culture and creed, and of all different humans—and our robotic children—becomes possible.

Bibliography

Introduction

www.ai.mit.edu/projects/humanoid-robotics-group/cog/cog.html
www.ai.mit.edu/projects/humanoid-robotics-group/kismet/kismet.html

Chapter One

Benson, Herbert. *The Relaxation Response*. New York: Avon, 1975.
———. *Beyond the Relaxation Response: How to Harness the Healing Power of Your Personal Beliefs*. Times Books, June 1984.
Berne, Eric. *Games People Play*. New York: Grove Press, 1964.
Bloch, Chayim. *Golem: Legends of the Ghetto of Prague*. Kessinger, 1972.
Boivin, Michael. "Finding God in Prozac or Finding Prozac in God: Preserving a Christian View of the Person Amidst a Biopsychological Revolution." *Christian Scholar's Review* 32 (Winter 2003): 159–176.
Bokser, Rabbi B. Z. *The Philosophy of Rabbi Judah Löw of Prague*. New York: Philosophical Library, 1954.
Cox, Harvey, and Anne Foerst. "Religion and Technology: A New Phase." *Bulletin of Science, Technology and Society*. 17(2–3), 53–60.
Hefner, Phil. *The Human Factor: Evolution, Culture, Religion*. Minneapolis: Fortress Press, 1993.
Idel, Moshe. *Golem: Jewish Magical and Mystical Traditions on the*

Artificial Anthropoid. Albany: State University of New York Press, 1990.

Kant, Immanuel. *The Critique of Judgment.* Amherst, NY: Prometheus Books, 2000.

———. *The Critique of Practical Reason.* Amherst, NY: Prometheus Books, 1996.

Nass, Cliff, and Byron Reeves. *The Media Equation: How People Treat Computers, Television, and New Media Like Real People and Places.* New York: Cambridge University Press, 1996.

Ramachandran, Vilayanur, and Sandra Blakeslee. *Phantoms in the Brain: Probing the Mysteries of the Human Mind.* New York: William Morrow, 1999.

Saver, Jeffrey. "The Neural Substrates of Religious Experiences." *Journal of Neuropsychiatry and Clinical Neurosciences* 9 (Summer 1997): 498–510.

Scholem, Gershom. "The Golem of Prague and the Golem of Rehovoth." *Commentary* (January 1966): 62–65.

Tillich, Paul. *Systematic Theology, Vol. 1.* Chicago: University of Chicago Press, 1951.

———. *Systematic Theology, Vol. 3.* Chicago: University of Chicago Press, 1963.

Weizenbaum, Joseph. *Computer Power and Human Reason: From Judgment to Calculation.* San Francisco: W. H. Freeman, 1976.

Wiener, Norbert. *God and Golem, Inc.; A Comment on Certain Points Where Cybernetics Impinges on Religion.* Cambridge, Mass.: MIT Press, 1964.

Chapter Two

Austin, J. L. *How to Do Things with Words: The William James Lectures Delivered at Harvard University.* New York: Oxford University Press, 1965.

Bugelski, B. R. and D. A. Alampay. "The Role of Frequency in Developing Perceptual Sets." *Canadian Journal of Psychology* 15 (1961): 205–211.

Burke, James. *The Day the Universe Changed.* Boston: Back Bay Books, 1985.

Hauser, Marc. *The Evolution of Communication.* Cambridge, Mass.: MIT Press, 1997.

Heidegger, Martin. *Sein und Zeit,* 17th ed. Tübingen: Max Niemeier, 1993.

Horkheimer, Max, and Theodor W. Adorno. *Dialectic of Enlightenment.* New York: Herder and Herder, 1972.

Hoynigen-Huene, Paul. *Reconstructing Scientific Revolutions: Thomas S. Kuhn's Philosophy of Science.* Chicago: University of Chicago Press, 1993.

Kuhn, Thomas S. *The Structure of Scientific Revolutions,* 3d ed. Chicago: University of Chicago Press, 1996.

Latour, Bruno, and Steve Woolgar. *Laboratory Life: The Construction of Scientific Facts.* Princeton: Princeton University Press, 1986.

Minsky, Marvin. *Society of Mind.* New York: Simon & Schuster, 1988.

Sobel, Dava. *Galileo's Daughter: A Historical Memoir of Science, Faith, and Love.* New York: Penguin, 2000.

Tillich, Paul. *Dynamics of Faith.* New York: Harper & Row, 1957.

Wilson, E. O. *Consilience: The Unity of Knowledge.* New York: Random House, 1999.

———. *The Diversity of Life.* New York: Norton, 1999.

Winslow, James, Nick Hastings, Sue Carter, Carroll Harbaugh, and Thomas Insel. "A Role for Central Vasopressin in Pair Bonding in Monogamous Prairie Voles." *Nature* 365 (Oct. 7, 1993): 545–548.

Winston, Patrick. *Artificial Intelligence.* 3d ed. Reading, Penn.: Addison Wesley, 1992.

Young, Larry, Roger Nilsen, Katrina G. Waymire, Grant R. MacGregor, and Thomas Insel. "Increased Affiliative Response to Vasopressin in Mice Expressing the V_{1a} Receptor from a Monogamous Vole." *Nature* 400 (Aug. 19, 1999): 766–768.

Chapter Three

Braitenberg, Valentino. *Vehicles: Experiments in Synthetic Psychology.* Cambridge, Mass.: MIT Press, 1984.

Brooks, Rodney. *Flesh and Machines: How Robots Will Change Us.* New York: Pantheon Books, 2002.

———. *Cambrian Intelligence: The Early History of the New AI.* Cambridge, Mass.: MIT Press, 1999.

Clark, Andy. *Being There: Putting Mind, Body and World Together Again.* Cambridge, Mass.: MIT Press, 1999.

Damasio, Antonio. *Descartes' Error: Emotion, Reason, and the Human Brain.* New York: Avon Books, 1994.

———. *The Feeling of What Happens: Body and Emotion in the Making of Consciousness.* New York: Harcourt, 1999.

Descartes, René. *Meditations. Discourse on Method and the Meditations.* New York: Penguin, 1986.

Diamond, Adele. "Close Interrelation of Motor Development and Cognitive Development and of the Cerebellum and Prefontal Cortex." *Child Development* 71 (Jan./Feb. 2000): 44–57.

Diamond, Adele, and Lee Young. "Inability of Five-Month-Old Infants to Retrieve Contiguous Objects: A Failure of Conceptual Understanding or of Control of Action?" *Child Development* 71 (Nov./Dec. 2001): 1477–1495.

Gilbert, Scott F. "Congenital Human Baculum Deficiency." *American Journal of Medical Genetics* 101 (2001): 284–285.

Johnson, Mark. *The Body in the Mind: The Bodily Basis of Meaning, Imagination, and Reason.* Chicago: University of Chicago Press, 1990.

Pagels, Elaine. *The Gnostic Gospels.* New York: Random House, 1979.

Piaget, Jean, and Bärbel Inhelder. *The Psychology of the Child.* New York: Basic Books, 2000.

Sacks, Oliver. "The Last Hippie," in *An Anthropologist on Mars.* New York: Random House, 1995.

———. "The Disembodied Lady," in *The Man Who Mistook His Wife for a Hat.* New York: Touchstone, 1998.

Schachtner, Christel. *Geistmaschine: Fascination und Provokation am Computer.* Frankfurt a.M.: Suhrkamp, 1993.

Schroer, Silvia, and Thomas Staubli. *Die Körpersymbolik der Bibel.* Darmstadt: Primus Verl, 1998.

Varela, Francisco J., Evan T. Thompson, and Eleanor Rosch. *The Embodied Mind: Cognitive Science and Human Experience.* Cambridge, Mass.: MIT Press, 1992.

Chapter Four

Ashby, W. R. *Design for a Brain*, 2d ed. New York: Wiley, 1960.

Bogucki, Peter. *The Origins of Human Society*. Cambridge, Mass.: Blackwell, 1999.

Breazeal, Cynthia. *Sociable Robots*. Cambridge, Mass.: MIT Press, 2002.

Diamond, Jared. *Why Is Sex Fun?* New York: Basic Books, 1996.

Eibl-Eibesfeldt, Irenäus. *Liebe und Hass: Zur Naturgeschichte elementarer Verhaltensweisen*. Munich: Piper, 1970.

Fernald, A. "Intonation and Communicative Intent in Mother's Speech to Infants: Is the Melody the Message?" *Child Development* 60 (1988): 1497–1510.

———— "Approval and Disapproval: Infant Responsiveness to Vocal Affect in Familiar and Unfamiliar Languages." *Developmental Psychology* 64 (1993): 657–674.

Frith, Chris D., and Uta Frith. "Interacting Minds—A Biological Basis." *Science* 286 (Nov. 26, 1999): 1692–1695.

Iacobini, Marco, Roger Woods, Marcel Brass, Harold Bekkering, John Mazziotta, and Giacomo Rizzolatti. "Cortical Mechanisms of Human Imitation." *Science* 286 (Dec. 24, 1999): 2526–2528.

Kaptchuk, Ted. "The Placebo Effect in Alternative Medicine: Can the Performance of a Healing Ritual Have Clinical Significance?" *Annals of Internal Medicine* 136 (June 4, 2002): 817–824.

Levin, Daniel, and Daniel Simons. "Failure to Detect Changes to Attended Objects in Motion Pictures." *Psychonomic Bulletin & Review* 4 (1997): 501–506.

Meorman, Daniel, and Wayne Jonas. "Deconstructing the Placebo Effect and Finding the Meaning Response." *Annals of Int Medicine* 136 (March 19, 2002): 471–478.

Pinker, Steven. *How the Mind Works*. New York: Norton, 1

Simons, Daniel. "Change blindness." *Trends in Cognitive S* (Oct 1997), 261–268.

Simons, Daniel, Stephen Mitroff, and Steven Franco Song of Implicit Change Detection." *Journal Psychology/Human Perception & Performanc* 798–816.

Wickler, Wolfgang. *Die Biologie der zehn G* 1991.

Chapter Five

Diamond, Jared. *Guns, Germs, and Steel: The Fates of Human Society*. New York: Norton, 1997.

Dunbar, Robin. *Grooming, Gossip, and the Evolution of Language*. Cambridge, Mass.: Harvard University Press, 1996.

Fausel, Heinrich. *D. Martin Luther: Leben und Werk*. Munich, Hamburg: Siebenstern, 1968.

Green, Joel (Ed.). *What About the Soul: Neuroscience and Christian Anthropology*. Nashville: Abingdon, 2004.

Palloff, Rena M., and Keith Pratt. *Lessons from the Cyberspace Classroom: The Realities of Online Teaching*. San Francisco: Jossey-Bass, 2001.

Pascalis, Olivier, Michelle de Haan, and Charles A. Nelson. "Is Face Processing Species-Specific During the First Year of Life?" *Science* 296 (May 2002): 1321–1323.

Tillich, Paul. *The Shaking of the Foundations*. New York: Scribner, 1948.

Index

"Ron has provided a complete process including assessments, resources and checklists to understand the nontraditional career path. The ups and downs of choosing this work style were validated through Ron's personal experience and, for those who are either mid-career or retiring, it's a way to continue to have a productive work life."
—Patricia DeMasters, Director, Career Management Group, Berkeley-Haas School of Business

"A clear, comprehensive and easy-to-read guide on what many of us are thinking about and already doing—the alternate career path. Ron has put everything together in one place—the background data, the work plan and more. Now I know that my career path is doable (and not out of the ordinary). Thank you, Ron!"
—David Knego, Executive Director, Curry Senior Center

"Ron Elsdon addresses an important work life issue, namely how to establish a new path of fulfillment, through assessment and integration of personal and career motivators, and marketplace alignment. The book's broad perspective, practical details, and hard data give readers an invaluable foundation on which to build."
—Lenore Mewton, LM&A/Lenore Mewton & Associates – Career and Leadership Development, Coaching and Consulting

HOW TO BUILD
A NONTRADITIONAL
CAREER PATH

EMBRACING
ECONOMIC
DISRUPTION

Ron Elsdon

PRAEGER

AN IMPRINT OF ABC-CLIO, LLC
Santa Barbara, California • Denver, Colorado • Oxford, England

Library of Congress Cataloging-in-Publication Data

Elsdon, Ron, 1950-
    How to build a nontraditional career path : embracing economic disruption / Ron Elsdon.
        pages cm
    Includes bibliographical references and index.
    ISBN 978-1-4408-3158-4 (alk. paper) -- ISBN 978-1-4408-3159-1 (ebook)
1. Career changes. 2. Career development. 3. Vocational guidance. I. Title.
    HF5384.E47 2014
    650.1—dc23   2014010457

ISBN: 978-1-4408-3158-4
EISBN: 978-1-4408-3159-1

18 17 16 15 14   1 2 3 4 5

This book is also available on the World Wide Web as an eBook.
Visit www.abc-clio.com for details.

Praeger
An Imprint of ABC-CLIO, LLC

ABC-CLIO, LLC
130 Cremona Drive, P.O. Box 1911
Santa Barbara, California 93116-1911

This book is printed on acid-free paper ∞
Manufactured in the United States of America

This book is for my family

Franciscan Blessing

May God bless us with discomfort at easy answers, half-truths, and superficial relationships—so we may live deep within our hearts.

May God bless us with anger at injustice, oppression, and exploitation of people and creation—so we may work with energy.

May God bless us with tears to shed for those who suffer from pain, rejection, hunger, and war—so we may comfort them with our hands and turn pain into joy.

And may God bless us with enough foolishness to believe that we can make a difference in this world—so that we do what others claim cannot be done: bring justice and kindness to all our children, the poor and our Earth.

—Peace Lutheran Church, Danville, California, October 6, 2013

Contents

List of Illustrations

Preface

IT HAPPENED QUICKLY WHEN I was in my mid-fifties. I was working for an organization that was sold by one firm and acquired by a private equity investment company. Not unusual. Those of us who had been with the organization less than five years lost any accumulated pension rights. I had been there about four and a half years. Again, unfortunately, not unusual, though disturbing. However, something else happened that in retrospect was more significant. That was the opportunity to start an organizational consulting practice rather than work for others. This was a good time to integrate different components of an emerging nontraditional career, such as providing career counseling to individuals, delivering organizational consulting, writing, teaching, and volunteering. This was a good time to create a fulfilling personal path forward, to be of service, to offer good jobs for others, and to be part of something that helped people make their work lives better. I am thankful for this and for many supportive people who made this path and what I learned from it possible. In today's and the future work world, creating an individual, nontraditional career path such as I have experienced is a growing opportunity.

Indeed, being self-sustaining in a work setting is becoming more important since rewards increasingly flow disproportionately to those at the top of traditional organizations. Since the publication of my first book, *Affiliation in the Workplace*, more than a decade ago, the working environment has grown more challenging for many people in organizations.[1] Fortunately, some organizations recognize the value of supporting employees in their development, as pointed out in my second book, *Building Workforce Strength*.[2] However, the economic upheaval beginning in 2007–2008 reflects disturbing trends in our society that affect our relationship with work. As discussed in my third book, *Business Behaving Well,* with corporate governance compromised in some organizations, chief executive officers (CEO) of large organizations typically now earn more in one day than many people earn in a year, and the gap is ever increasing.[3] This inequity, along with a more regressive taxation policy beginning in the early 1980s that favors the wealthy, has caused inequality in the United States to worsen dramatically since then, reaching levels typical of a developing rather than a developed country.[4] One factor contributing to this growing inequality has been reduced bargaining power for unions resulting from a combination of globalization of manufacturing, structural changes in employment relationships with a growing contingent workforce, aggressive anti-union activities by firms, and ineffective enforcement of labor regulations.[5] Union membership has fallen from its peak of 39 percent of private sector employees in 1958 to less than 7 percent in 2012.[6] The minimum wage in real terms was 23 percent lower in 2013 than at its peak in 1968.[7] It is good to see some emerging local, state, and federal initiatives to address this inequity. However, forces that counterbalance growing inequality and reduced life circumstances for many have been compromised. Growing inequity in internal compensation practices in organizations speaks to a perspective that devalues individuals. This, coupled with weakening of unions, underlines the importance of our taking increased control over our career paths, recognizing that interdependence, connection with others, is a primary factor for effective career relationships in the future.[8] Indeed it is now possible, and often desirable, to connect directly through our work with our communities without the need for large organizations in between. Fortunately, technology has helped level the workforce playing field. State-of-the-art communication and analytical tools, which at one time were only available to large organizations, are now readily accessible to all of us. Indeed, it is easier to stay current with emerging technology as an individual than as a large organization, where the costs of staying current with equipment and systems can be prohibitive.

So there is an external environmental push, and there are motivational and technology pulls that cause nontraditional career paths to beckon. What does a nontraditional career path mean? Broadly, it can be anything that does not just involve conventional employment with an organization. In conventional employment someone else frequently defines the responsibilities of the position and the longevity of the relationship. Our focus in this book, on the other hand, will be on a nontraditional career path that is tailored to each person's individual needs and consists of more than one source of income. The form it takes will be different for each of us. For some it may consist of career components that are strongly connected, such as career counseling and organizational career development, as in my case. For others it may be built on components that are completely different, such as a contract administrative support position and a creative pursuit like catering or photography. Each nontraditional career path represents a combination of components that draw on our values, personality preferences, aspirations, interests, and skills, chosen by us and developed by us. The decision to introduce, develop, or terminate a component is personal; it is not determined by someone else in an organization.

Why write a book on this subject? It has been a pleasure giving presentations over the course of several years about this topic to various groups focused on career transition and to MBA students considering their career paths. Some seeds for this book were planted with those presentations at the encouragement of participants. Other seeds come from my experience. As I reflect on my career, roughly the first half was spent in conventional employment settings. I was first with an organization in the chemical industry, in a series of positions beginning in research as an individual contributor then extending through business planning, business development, marketing, and research leadership. As is typically the case over time, responsibilities grew to include leadership for groups or functions. This was followed by general management responsibilities in a European company in the same industry. At this point I began to question the wisdom of continuing to devote my working life to simply making money for an organization. Through assessment and reflection it gradually became clear that taking a different path, into the field of career development, would be a good fit. Here was an opportunity to give back accumulated knowledge and experience to individuals and organizations. A bit of a leap from chemical engineering, where I started, through several steps to career development—though it makes sense in retrospect.

Going back to school in the career development field to build some new skills, while working full time, provided a bridge to this new area. It allowed me to work for a nonprofit organization in the career field while cofounding a private career counseling and college advising practice with my wife, who focused on college advising. This was the first step on a nontraditional path that has occupied roughly the second half of my working life. The gradual-launch path continued with employment in a for-profit organization. At the same time I was providing career counseling in a private practice setting supporting clients on their career journeys. Then the events I mentioned at the beginning happened; here was a nudge to create a second practice focused on organizational consulting. By now, with several other pieces added along the way, I had a career path with multiple components. And it worked well. It became fulfilling, practical, and financially successful due to the support of many people over the years.

In reflecting on these experiences there has been much to be thankful for and much knowledge gained. The first half of my career in conventional settings provided experience and credibility that helped enable the second half. There were good places along the way and some challenges in that first half. The workplace was a kinder environment when I joined it in the mid-1970s. There was a greater sense of caring, of community, of shared responsibility. But there was also a greater sense of dependency. The second, nontraditional half has brought many pleasures and a greater sense of fulfillment, as this has been a stronger expression of who I am and what I can give. It is a blessing to be able to offer you some of what I have learned, building on experiences and perspectives from a variety of settings and integrating thoughtful and valuable perspectives of others. This book is offered as a resource to help you in considering a nontraditional career path. It will likely appeal at various life stages, from initial entry into the workforce, to mid-career reflection, to those later work/life stages that we used to call retirement. It is my hope that, while the focus is on a United States setting, this book can also transcend geographic boundaries. Every effort has been made to verify the accuracy of information at the time of publication. Since circumstances change, and this book is not intended to render legal advice, I recommend seeking input from appropriate professional sources when considering starting a business entity. The approaches here are offered as options to stimulate thinking about new possibilities regardless of the state of the economy. I hope this may help you discern if a nontraditional career path is right for you and, if so, how to define and pursue such a path. Thank you for taking this journey with me. I wish you well.

Ron Elsdon
Danville, California

Acknowledgments

I AM GRATEFUL FOR THE KINDNESS and support of many people whose lives have touched mine in profound and meaningful ways and who are central to this book coming into being. I much appreciate Anna Domek, Beverly Garcia, Alia Lawlor, Beth Levin, Michele McCarthy, Wynn Montgomery, and Mary Wiggins for creating the vignettes of their nontraditional career journeys that are included here. Their willingness to describe learning from those experiences provides insights that are valuable for all of us. In the complicated and evolving world of publishing, each book that emerges needs a champion. Hilary Claggett, my editor at ABC-CLIO, has been such a champion for this book and for my three previous books. I am so grateful for Hilary's wisdom, support, and encouragement for new ideas, for they are a principal reason that this and the earlier books exist and have found their place in the world. I also much appreciate Judy Kip for bringing her considerable indexing skills to this and the previous books.

Many people have been part of my nontraditional career path in a good way, enabling it to evolve and take shape, and to each person I am grateful. They include individual career counseling and coaching clients with whom it has been a privilege to walk on their career journeys. They include colleagues and friends in organizations with which I have worked, such as John August, Pat Finnegan, Bob Redlo, and many others at Kaiser Permanente. Pearl Sims from Vanderbilt University was there at the beginning and was so helpful over many years. I much appreciate Denise Phoenix and Tai Williams from the Town of Danville, who were there early and in later years. Dave Fradin in his own practice, and Jim Graber from Business Decisions Incorporated, each provided special insights early on into launching a nontraditional career that I much appreciate. Others who have been a greatly valued source of encouragement and support include Pat DeMasters at the Haas School of Business of the University of California–Berkeley and Dottie Moser at Easter Seals, Inc. In addition, all who were acknowledged in *Affiliation in the Workplace*, *Building Workforce Strength*, and *Business Behaving Well* have an important place here, including many associated with John F. Kennedy University, the source of my formative ideas about career development.

Members of our team at Elsdon, Inc., over the years were a constant source of inspiration, as colleagues and friends, bringing insight, wisdom, humor, and enthusiasm. I am very grateful for all who were part of our team, most recently Jan Cummings, Michele DeRosa, Anna Domek, Martha Edwards, Lisa Franklin, Beverly Garcia, Dee Holl, Alia Lawlor, Beth Levin, Jeannette Maass, Darlene Martin, Michele McCarthy, Tom Rauch, Dave Rosenberg, Richard Vicenzi, Mary Van Hee, and Mary Wiggins.

Finally, I am so grateful for the family support that surrounds this book. Early educational steps, encouraged by my parents, Frank and Barbara, made possible the path that led here, in spite of my preference for kicking a (round) football in those early days. Our children, Mark and Anna, and their spouses, Erica and Andy, are a beautiful presence in our lives and the lives of those around them. This world is a better place for who they are. Our grandchildren, Emma, Sophie, Claire, and William, bring such hope, vitality, and love. May this book open, in just a small way, some new paths for them and for all children. And finally, this book comes with deeply felt love for my wife, Linda, who has been so supportive and such a beautiful life partner. How fortunate I am.

Introduction

IT LOOKED COLD OUT THERE. After all, why wear face masks otherwise? This was our first taste of a Chicago winter, looking out of the plane window as it taxied across O'Hare International Airport. My wife and I had come from London one January day to live in Chicago. Airport staff were walking around outside the plane, wisely wearing face masks. Then, several winters later, I found myself one night in Joliet, Illinois, not far from Chicago, seconded from working in research to the firm's chemical plant there. I was alternately in the sheltered control room and then outside, climbing towering equipment for samples in searing cold with snow and ice glistening and wisps of steam wafting through starkly lit facilities. For me this was a brief encounter at the plant repeated sporadically over about a year and a half, but for many working at the chemical plant, just as with the airport staff, this was permanent. In winter the job meant working outside in biting cold much of the time. It was hard, relentless work. It required resolute courage of people to get up each day and do it. Why do it? Well, at that time it offered reasonable compensation and some stability. It may have been the least, worst option. Is it possible to create other, better options? Let us begin to explore what those options might look like and how to find them. First I will place this exploration in the context of the evolving world of work and self-employment. Then I will explore what this means for a nontraditional career, before concluding with an outline of the structure of this book. The focus is primarily on the United States, though I hope that the thoughts and ideas here will transcend geographic boundaries.

The Evolving World of Work and Self-Employment

Let us start with some perspectives about what work looks like today and how it has evolved. Prior to the Industrial Revolution of the 1700 and 1800s, most people were peasants in agricultural economies. The Industrial Revolution brought great wealth to a select few, growth of a middle class, and great misery to many, including young children working long days in appalling conditions.[1] Out of these conditions grew the organized labor movement, years of struggle, and, after the economic collapse of 1929, in 1935 passage of the National Labor Relations Act in the United States.[2] From the late 1800s through the mid-1900s, much progress was made in limiting child labor, limiting the length of the workday, increasing compensation and benefits, and improving working conditions and workplace safety. These were hard-won gains that we take for granted today, though since the early 1980s some have been under attack and are eroding.

Our view of work mostly comes from times we have known directly or from time horizons that people we have known can describe. Those time horizons may go back to the early 1900s for some of us, through people we have known, but for most they are more recent. In 1900, 36 percent of the U.S. labor force of 28 million people worked in agriculture, and in 1900, 92 percent of men aged sixteen and over participated in the labor force and 22 percent of women.[3] By 2000, less than 2 percent of the U.S. labor force of 138 million people worked in agriculture and related areas; service and related sectors dominated at about 82 percent of the labor force, with manufacturing and related occupations having fallen to 17 percent from a peak of 36 percent in 1960.[4] By 2000, 72 percent of men sixteen and over and 58 percent of women participated in the labor force.[5]

Figure I.1 provides insights into U.S. sector and self-employment trends.[6] The bars show the percentage of the workforce employed in the agriculture and related sector, the manufacturing and related sector, and the service and related sector from 1820 to 2010, as well as a projection for 2020. The service and related sector contains a wide range of service-providing activities such as retail, health care, hospitality, education, and professional services. Growing efficiency in agricultural production led to employment in the agriculture sector falling from 83 percent of the labor force in 1820 to 1.6 percent in 2010. Employment in the manufacturing and related sector grew until the early 1960s as the Industrial Revolution continued to cascade down through the years. However, by 2010 manufacturing and related areas accounted for only about 12 percent of employment, which is about a third of the percentage in 1960.

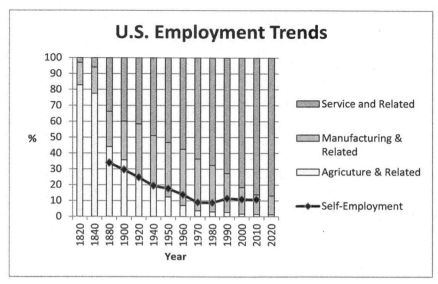

Figure I.1

In the period from the 1870s to 1900, about one-third of people in the U.S. workforce were self-employed, largely as farmers, as shown by the line in Figure I.1.[7] However, by 1970 only about 9 percent of people in the U.S. workforce were self-employed, with a reduction of the number of people self-employed in farming contributing to this drop.[8] There appears to have been a slight increase since then, with self-employment accounting for about 11 percent of the U.S. workforce in 2010, though some of the variations in self-employment since the 1960s are artifacts of the survey process used to gather the information and how people in incorporated entities were included.[9] Interestingly, the United States had the second lowest self-employment rate of twenty-one developed countries in 2007, and a proportionately low participation in small business activity.[10] The authors of one study, John Schmitt and Nathan Lane, speculate that lack of access to universal health care in the United States (bedeviled by a high-cost, marginally effective, largely private approach, as pointed out by Deborah LeVeen) has been a significant deterrent to entrepreneurial activity.[11] The rate of business creation, entrepreneurial activity, in the United States was relatively stable between 1996 and 2013, showing a slight increase in 2009 and 2010 before returning to prerecession levels.[12] However, one measure of entrepreneurial climate (based on access to funding, entrepreneurship culture, taxes and regulation, education and training, and coordinated support) in 2013, based on a combination of survey feedback from more than fifteen hundred entrepreneurs across nineteen countries and the European Union and quantitative country data based on entrepreneurial conditions, shows the United States at the top on this index, which speaks to the opportunity for adopting a nontraditional career path.[13] Another study shows that more than 43 percent of Americans believe there are good opportunities for entrepreneurship around them, and 56 percent believe that they have the capabilities to start a business.[14]

Momentous changes in the nature of work over the past two hundred years accompanied our economic evolution from an agriculturally based economy, through a manufacturing-oriented economy, to today's information- and service-based economy. Our concept of work defined as being an employee in an organization is rooted in that manufacturing economy, which was dominated by large organizations with access to capital needed to operate and succeed. However, in today's economy, and that of the foreseeable future, information and services dominate. This changes the nature of work so that capital and scale are not the primary economic drivers, just as land ceased to be the primary economic factor after the transition from an agricultural economy. In an information- and service-based economy, human ingenuity and the capability to deliver effective services dominate. So it is not surprising that, after declining for more than seventy years, self-employment leveled off by the early 1970s, and it is poised to increase. By June 2013 there were almost fifteen million people identified as self-employed in the United States.[15] Another study shows that in 2013, 40 percent of people in the U.S. workforce aged twenty-one and older have worked in a nontraditional capacity (at least fifteen hours per week) at one time during their career. This is projected to increase to about 50 percent by 2020. The study estimates that those in nontraditional careers generated almost $1.2 trillion in total income in 2013.[16] Further measures show most job growth coming from self-employment between 2000 and 2011, a trend reinforced by growth in incorporated small businesses in early 2013.[17] Those who are self-employed are found in many industries and occupations and they have wide-ranging educational backgrounds.[18] In 2009 self-employment rates in unincorporated business entities were highest in construction and extraction occupations, management, business, and financial occupations; and sales and related occupations; and these rates were highest for those with less than a high school diploma.[19] However, self-employment rates for incorporated business owners were highest for management, business and financial occupations; and sales and related occupations, and these rates were highest for those with an advanced degree.[20]

Changes in the work environment provide incentives that encourage individual exploration of nontraditional paths and small business activity and reduce the appeal of conventional employment with a single organization. These changes include the following:

- Economic disruption in concert with changing global and national economies
- Growing pay inequity between CEOs and workers
- Changing generational perspectives about work
- Age-based challenges with conventional employment
- Improving health-care safety net in the United States
- Reduction in long-term benefits provided by organizations
- Ready access to communication and analysis tools

Economic disruption in concert with changing global and national economies. Sometimes change occurs at a glacial pace, for example, the implementation of effective health-care reform in the United States, and at other times it's almost instantaneous, as in high technology. Economic systems are surprising in that change and disruption can occur rapidly. There is historical precedent. In 1830 India and China accounted for just under 50 percent of global production, whereas the United States and Britain accounted for only 13 percent. By 1900, fewer than two generations later, the United States and Britain accounted for about 42 percent of global output and India and China only accounted for 8 percent.[21] Today we are seeing a reversal of this earlier trend, with growth in gross domestic product in China and India several times that of the United States from 2008 through 2012.[22] As the global economy restructures and economic power shifts, established companies lose ground.[23] Opportunities will surface as new markets emerge. Such opportunities will benefit the fleet of foot, an attribute associated with small, fast-moving enterprises and nontraditional career paths. Indeed, nontraditional career paths are one means of addressing disturbing global trends of widening income disparity and persistent structural unemployment.[24]

Growing pay inequity between CEOs and workers. The United States economy is hampered by growing inequality.[25] This is reflected in the average compensation of CEOs in major U.S. corporations now several hundred times that of the average compensation for workers.[26] Contrast that with CEOs earning about thirty times the average worker's compensation in the 1970s. Performance of the companies with the highest paid CEOs was much worse than industry average in one study.[27] This situation is a result of ineffective corporate governance in some organizations, so that resources are channeled so preferentially to those at senior levels that employees, shareholders, and other stakeholders suffer. There is, therefore, a reduced incentive to work in organizations that employ such practices. By engaging in a nontraditional career path, disproportionately high pay at senior levels is no longer relevant, as individual capability and effort is rewarded without the need for corporate bureaucracy.

Changing generational perspectives about work. Recognizing that generational priorities vary by country and culture, we observe, not surprisingly, that there are significant differences among the perspectives of different generations resulting from circumstances and events shaping the lives of those born at different times.[28] While acknowledging that there are major variations in perspectives among individuals in a given generation, some of the overarching attributes that appear to characterize the millennial generation or Generation Y in the United States, the roughly eighty million people born between about 1980–1981 and 2001–2003, are strong commitment to social values, seeking time flexibility in work engagements, and self-direction and developmental orientation.[29] These attributes align well with a nontraditional career and will be important in the future as it is estimated that this generation will make up almost half of the U.S. workforce by 2020.[30] Likewise, a nontraditional career offers an opportunity to align personal values of participation and fairness at later work/life stages for baby boomers in the United States, born between the mid-1940s and the mid-1960s, and address values of independence and a focus on working to live rather than living to work for Generation X, born between the mid-1960s and the late 1970s, the smallest of the three generations.[31]

Age-based challenges with conventional employment. Unemployed adults in their fifties were about a fifth less likely than their counterparts aged twenty-five to thirty-four to become reemployed each month between May 2008 and March 2011 in the United States.[32] Adults aged sixty-two and older were only about half as likely as those aged twenty-five to thirty-four to become reemployed each month. About half of unemployed workers aged sixty-two and older who continued their job search remained out of work sixteen months after their unemployment began, compared with about a third of those fifty to sixty-one and a fifth of those twenty-five to thirty-four. Furthermore, the likelihood of being unemployed for longer than twenty-six weeks in the United States increases with age, with this increase starting as early as thirty-five years old.[33] Securing conventional employment grows increasingly difficult with age, particularly after age fifty.[34] At the other end of the age spectrum it is also difficult; the unemployment rate for those aged twenty to twenty-four were 11.1 percent in May 2014 in the United States, relative to an overall unemployment rate of 6.3 percent.[35] Unemployment for those who are younger is an even more serious problem in some European countries.[36] There's a midcareer window that may be open to those with the right experience, but it is subject to the hiring whims of those in organizations. Furthermore, the proportion of people out of work for at least a year was 27 percent of all those unemployed in the United States in the second quarter of 2013.[37] This proportion, those who were long-term unemployed, has since 2009, for the first time since World War II, exceeded the proportion of people out of work four weeks or less.[38] Long-term unemployment is an even more acute issue in some European countries.[39] Nontraditional careers circumvent these age-related and structural unemployment problems.

Improving health-care safety net in the United States. With a for-profit bureaucracy dominating U.S. health care for many years, and health-care insurance for many people provided through their employer, we have created an inefficient and marginally effective approach that has per person health-care spending more than twice the median of other developed countries, while generating worse health-care outcomes, leaving fifty million people uninsured at times and seeing almost two-thirds of personal bankruptcies related to medical costs.[40] It is shameful that we have allowed such injustice to continue, but the first halting steps to right this wrong were taken with the passage of Affordable Care Act in 2010. One of the components of this act is the creation of health-care exchanges that allow individuals to secure health-care insurance coverage regardless of preexisting conditions (a basic right in other developed countries). For the first time in the United States, health-care insurance is accessible to those with nontraditional careers and preexisting conditions. The hurdle of being unable to secure insurance coverage with a preexisting condition, other than through an employer, is now removed.[41] Having access to this basic human right provides a safety net for those embarking on a nontraditional career path. Indeed, a 2013 study by Randstad shows 33 percent of all U.S. workers would be more likely to consider a career on a temporary or contractor basis if affordable health-care insurance options were available to them as a result of health-care reform, and 45 percent of all workers say the recession has made them more interested in pursuing an arrangement outside traditional full-time employment.[42]

Reduction in long-term benefits provided by organizations. In the private sector only 30 percent of U.S. Fortune 100 companies offered defined benefit pension plans to new salaried employees on June 30, 2012, down from 90 percent in 1998.[43] Defined benefit pension plans guarantee a defined income during retirement. In contrast, by June 30, 2012, 70 percent of the Fortune 100 companies offered only defined contribution plans to new salaried employees, up from 10 percent in 1998. Defined contribution plans are based on employee contributions, sometimes with corporate matching, and they place the entire risk of retirement income on the employee. Someone with a nontraditional career can readily create through a one-participant 401(k) plan close to the equivalent of a defined contribution plan, so the benefit of a pension through conventional employment also has evaporated in the private sector.

Ready access to communication and analysis tools. In 1955 there were only 250 computers in use throughout the world; by 1980, as the information age began to take hold, more than a million personal computers had shipped. Widespread use of mobile phones and the emergence of tablet computers were still in the future.[44] I remember in the early 1980s making a presentation to senior leaders of an organization to justify the purchase of a handful of IBM personal computers, as there was concern that they might sit idle some of the time and therefore not be a worthwhile investment. Such a question now seems odd, as by 2017 annual shipments of personal computers, tablets, and mobile phones are projected to be almost three billion units.[45] With a projected world population of 7.4 billion people by then, this means that an average of one device is shipped for every two to three people in the world each year (recognizing that ownership of these devices is unevenly distributed). Access to communication and analysis tools in an information-based economy is now ubiquitous, creating the potential for economic democracy. For example, Skype and Apple's FaceTime place video communication close at hand, database programs such as Intuit's QuickBase make data integration and analysis readily accessible worldwide, and applications that allow file sharing, such as Dropbox, are in common use. Today technology devices and communication and analytical systems that were once the domain of large organizations are now widely available to each of us in developed economies and are becoming more accessible in developing economies. Small enterprises and those in nontraditional careers can readily stay updated with emerging technology in a way that is difficult for large organizations encumbered with major switching costs.

In summary, there are both incentives to pursue a nontraditional career path and reasons why conventional employment in an organization is becoming less attractive, so not surprisingly we anticipate that nontraditional approaches will grow in significance. We can look at nontraditional careers from different perspectives. These include an individual perspective, namely, what a nontraditional career path might mean both practically and emotionally. We can take an organizational perspective, namely, the implications of nontraditional approaches for workforce planning, development, and implementation. And we can take a community perspective about the public policy implications of nontraditional careers and the influence on sustained employment levels. While each perspective is important, this book's focus will be on the individual and what nontraditional careers mean for us, for our career paths, and for our lives. It is from this perspective, one person at a time, that organizational and community strength build.

What Does the Changing Environment Mean for Nontraditional Career Paths?

The complex interplay between personal aspirations and national policy decisions, such as taxation structure, employment safeguards, and provision of health care, affects how nontraditional careers unfold in different countries, particularly with respect to part-time or temporary work arrangements, as is evident from studies of nontraditional approaches in a number of developed economies.[46] Given that, it is perhaps not surprising that visions of what nontraditional approaches look like can vary from that of great hardship at one extreme to deep personal fulfillment and substantial financial reward at the other. An example of the former is the treatment of temporary workers in some circumstances, in terms of compensation and working conditions, reminiscent of the worst excesses of the early Industrial Revolution or the treatment of migrant farmworkers.[47] In these awful situations, individuals are exploited.

Conversely, an example of the latter is the "portfolio career," a term originally coined by Charles Handy in the late 1980s.[48] Exploring both how organizations might evolve and how our relationship with work might change, Handy introduced the idea that a combination of work components, rather than employment with only one organization at a time, could meet individuals' needs and match the reality of the changing world of work. Rachel Pickering describes examples of how this can unfold for physicians.[49] She gives examples in her article of medical practice being combined with teaching and with volunteer work. This portfolio of work experiences includes giving and receiving, financial reward, and personal development, centered on individual interests and skills. Such a portfolio can, on occasion, evolve naturally as an outgrowth of conventional employment, where the skills and contacts developed there provide a foundation for the portfolio path.[50]

Our focus in this book will be to explore this concept of a nontraditional career with more than one income stream coming from more than one career component. This can encompass a variety of nontraditional work options, such as contract engagements, part-time work, adjunct teaching, or starting or buying a business; we'll see examples of some of these aspects later in this book. It may also include volunteer work. The sources of income can be closely related, as in consulting, teaching, writing, and working with individuals on their development, or they can be disparate elements, such as administrative support, a catering business, and creative expression through photography. Regardless of the details, a central aspect of the approach is creating more than one source of income from work components designed to match individual skills, interests, and motivations while addressing market needs. Success with this approach leads to a vocation that is fulfilling, rewarding, and economically sustaining. Such success is based on thoughtful strategic choices and effective implementation.

What sparks an initial interest in a nontraditional career path? We have seen that changes in the work environment both increase the attraction of the nontraditional path and diminish the appeal of conventional employment with a single organization. There are gradual economic disruptions, such as pay inequity or technology-induced sector restructuring, that can heighten the appeal of a nontraditional path.[51] Rapid disruptions can also trigger exploration of nontraditional career alternatives. Examples include job loss, initial entry into the workforce, changes in one's later-stage work life (what used to be called retirement), and changing family needs and responsibilities. Each is characterized by an external event that leads to an internal transition. William Bridges describes the three stages of this transition as an ending, a neutral zone, and a new beginning.[52] Exploration and adoption of a nontraditional career path is such a change event that leads to a transition, as we will explore further in chapter 6. External events intersect with our internal journey. This internal journey is often a gradual movement throughout our lives from taking and receiving external approval to giving and expressing internal preferences.[53] Different life stages can bring reassessment of internal priorities. Perhaps this is recognition at an early career stage that conventional employment is not attractive, or it may be a need for a greater sense of purpose in midcareer, or it may be a desire to create a meaningful legacy or to address pressing financial needs at a later career stage. Each of these internal reassessments can lead naturally to a nontraditional career path.

Purpose and Structure of the Book

This book is designed for those seeking to inspire their work life through an alternative to conventional employment. It can be helpful at any career stage, whether entry, midcareer, or mature stage. It speaks to the aspirations of those entering the workforce wishing to take charge of their career path and the aspirations of those in midcareer or at later stages seeking a work/life transformation. Indeed, if, as some propose, we measure age as the number of years remaining in life rather than the number of years since birth, we have more productive work years remaining than in the past, as many of us choose to work longer than previous generations.[54] I hope this book will also be useful to institutions equipping people to enter or reenter the workforce, such as college career centers, public sector agencies, and outplacement organizations. Furthermore, it will likely be valuable to professionals in the career field, such as career counselors, and can complement educational curricula in that discipline.

This book is a guide to understanding what a nontraditional career path looks like and the benefits and challenges it presents. I will connect the approach with personal meaning and values, and examine strategic issues to help in assessing whether and how such a path might fit. Some strategic issues are content related, such as what career components should be included and to what extent should they be connected; some are implementation related, for example, what an appropriate pace of entry might be. Other strategic issues are introspective, such as understanding the skills and attributes needed for success, either developed individually or acquired through partnering. Some aspects I explore are practical, nuts-and-bolts issues such as what form of business structure to use at different stages of development, what forms of employment relationship to consider when engaging others, and how to handle infrastructure issues like accounting and payroll practices. I include key questions to consider at various stages when developing a nontraditional approach. In exploring this subject, I seek to blend thoughts and feelings about personal purpose and meaning with analysis of the world of business and organizations and considerations of our contributions to our broader community.

As a result of reading this book I hope you will be able to do the following:

- Describe what a nontraditional career with more than one source of income looks like
- Understand how such a career fits into a world of work where disruption is the norm

- Understand how such an approach can work well at different life stages
- Assess the benefits and drawbacks of a nontraditional career path
- Learn how to craft a successful nontraditional career path
- Know the key strategic factors to consider and how to address them
- Understand what the start-up process looks like and how to pace entry
- Know what skills are needed to be successful
- Create a fulfilling path forward

This book builds a foundation for a nontraditional career path in part I, addresses strategic issues in part II, and then examines practical steps and the path forward in part III. Part I begins with chapter 1, which shows what a nontraditional career looks like, how it can unfold over time, the opportunities that it presents, and how it is tailored by each individual. The chapter compares and contrasts such a career, which has more than one component at a given time, with a conventional career, which typically has more than one component over time. The chapter also explores factors that lead to career satisfaction and how such factors are addressed by a nontraditional career. This includes examining values that underlie our relationship with our work, establishing a link to purpose in work.

Chapter 2 examines the benefits and challenges of a nontraditional career and how to mitigate the challenges. Benefits include items such as autonomy, alignment with interests, and financial reward; and challenges include items such as the breadth of skills needed, launch time, and uncertainty. The chapter contains examples from people who have successfully established nontraditional careers and concludes by linking a nontraditional path to who we are and our vocation.

Part II, addressing strategic issues, begins with chapter 3, which looks at whether or not to connect the components of a nontraditional career. One core issue is deciding whether to build a series of strongly related components, for example, consulting and teaching, or whether to pursue components that have little in common, for example, administrative work and photography. Connectedness brings the ability to transfer learning from one component to another; it also provides infrastructure economies and marketing clarity and allows for a focus on areas of personal interest. One possible disadvantage is that it may limit the range of options that can be included. Chapter 3 examines the pros and cons of each approach so you can apply this to your situation.

Chapter 4 addresses the importance of creating differentiation. Differentiating career component offerings from services or products offered by others is an important step in assuring success. For example, the flexibility and responsiveness of individuals and small organizations can be important differentiators, particularly when built on a content base that is personally exciting and meaningful. Chapter 4 examines what differentiation means, potential sources of differentiation, and how to sustain differentiation.

Chapter 5 looks at balancing the components. A successful nontraditional career contains components balanced in both their scale of financial contribution and the time to establish that contribution, addressing both short- and long-term needs as a result. For example, a single, small-scale engagement with an individual client can occur rapidly and likely involves a small financial commitment and contribution. A large-scale organizational engagement, over a longer time, generally involves greater financial commitment and contribution and longer preparation time, perhaps also including an extended delivery team. Chapter 5 examines balance, with an example, so you can apply the principles to your situation.

Chapter 6 addresses pace of entry into a nontraditional career. Pace of entry refers to either gradually building income streams while maintaining conventional employment or, alternatively, taking a single, immediate step. Where the new income streams are built on existing expertise, gradual development may be preferable. This can permit integration with conventional employment for a period of time and provide financial support during ramp-up. If the new income streams represent a major departure, or a significant capital expenditure, full, immediate engagement may be preferable. Chapter 6 explores the implications of different entry approaches, for example, full immersion versus gradual entry, recognizing that either approach may be legitimate according to the nature of the work components, the investment involved, and the possibility of staged service or product growth.

In chapter 7 I examine building needed skills—the range of skills and personal characteristics required for success in a nontraditional career. These skills and characteristics include a blend of content knowledge and consulting capability to deliver products or services meeting customer needs; sales, marketing, and business skills to reach prospective customers and ensure viability; personal attributes to relate well and communicate effectively with others; time management skills to balance service or operational needs with those of marketing and sales; and organizing skills to create infrastructure, such as information technology and human resource capabilities. I identify the skills and characteristics needed and some potential approaches to building them.

Chapter 8 looks at partnering, examining when this can be helpful and the different types of partnering that are possible. This may include customer-facing partnering to enhance revenue or partnering designed to build internal capabilities such as accounting or financial expertise. Partnering offers the opportunity to collaborate for mutual benefit. It is built on clarity about the rationale for mutual engagement and it can include relationships with contract providers offering supporting or complementary services, products, or customer connections.

Part III, which focuses on practical steps and the path forward, begins with chapter 9 by looking at creating an infrastructure—the nuts and bolts. It addresses practical aspects, including the appropriate business structure to adopt at different stages of development, which may mean forming the entity as a sole proprietorship, a limited liability company, or a corporation. Other practical aspects of building the infrastructure include decisions about whether to engage others as employees or contractors based on the nature of their work, making accounting easy and accessible through software, how to handle payroll effectively and efficiently, and obtaining support for legal and human resource issues.

The book concludes with chapter 10 by focusing on moving forward. This chapter highlights key questions to address first when deciding on a nontraditional career path as an option, then at the launch point, and finally at the stages of growth and evolution and completion as the path unfolds. The chapter further builds the foundation for an inspiring and rewarding path forward.

It has been a pleasure speaking about this topic to many groups in recent years. I am grateful to participants in these sessions. These participants provided continued affirmation, through many examples and vignettes, of how nontraditional careers are alive and well and how they are meeting individual needs, sometimes in unexpected ways. These conversations caused me to reflect again on the beauty, creativity, and ingenuity that we bring to our work. You may find such reflection, based on your experience or the experiences of others you may know with nontraditional career paths, both helpful and inspiring. Now we move to part I, where I examine laying the foundation for a nontraditional career path, starting with chapter 1, which explores the what and why of a nontraditional career.

PART I

Laying the Foundation

Right is right, even if everyone is against it,
and wrong is wrong, even if everyone is for it.

William Penn

CHAPTER 1

The What and Why
of a Nontraditional Career

I REMEMBER WONDERING when I would meet a customer. This was my first proper job after seven years in college. I was in research, working for a large chemical company in one of its main product areas and wondering what was important to the customer. It turned out that customer contact was handled by particular individuals and not widely encouraged; in fact, it was about ten years, working in a variety of roles with the organization, before I first met a customer. It's hard to be customer focused when there are gaps like this, when communication gets filtered. It's quite different in a nontraditional career, where customer interaction is front and center to each of the unfolding career components, driven as they are by customer needs.

As jobs in large organizations become more specialized, it becomes more difficult for each person to see the overall path forward, the challenges and opportunities outside of his or her parochial domain. Indeed, I remember the CEO of another large organization I worked for, on hearing of an operational problem in one location, commenting that the knowledge to solve this problem existed in the organization, it just wasn't accessible, which caused some soul searching about needed changes to internal systems. How different this is from a nontraditional career, where learning from one component, for example, expertise from teaching a graduate class, can be immediately transferred into client interactions, or from a small organization, where learning from one geographic region can be immediately transferred to other regions. The strengths of a large organization, its specialization and scale, also become its vulnerabilities, creating as they do communication and interface challenges and difficulty flexing when needed. Careers in such large organizational settings build knowledge and connections within the internal network of the organization, which are particularly valuable when people are there a long time. They make it easy to get things done. However, when employment relationships are tenuous and transitory, as they are becoming, internal connections create limited value for the organization, and they are not portable, which is a problem for individuals given limited tenure with an organization. A nontraditional career, on the other hand, leads to the creation of relationships that potentially can last much longer as they are not dependent on a single employment connection.

In this chapter I will show what a nontraditional career looks like, see from examples that it can take many forms, and show how it can unfold over time. I will compare a nontraditional career with conventional employment and see how a nontraditional career can integrate different aspects of who we are. In conclusion, I will explore how to build a strong foundation of self-understanding for embarking on a nontraditional path.

Examples of Nontraditional Careers

First let us look at some examples of what a nontraditional career, based on more than one source of income, looks like. An analogy that comes to mind when thinking about a nontraditional career compared to conventional employment is that of a smart phone compared to a conventional phone. The smart phone is broader in scope and more complete in capabilities than a conventional phone, just as a nontraditional career is broader in scope and capabilities than conventional employment.

The following are some examples of nontraditional careers from published sources:

- Lauren Hargrave switched from a traditional career in finance to being a writer. She includes copywriting as part of her nontraditional career to help pay the bills and continued with her finance job on a part-time basis for four months during start-up of her new endeavor for financial reasons. "If you're taking a non-traditional career path," she notes, "you'll probably hear from friends, family, colleagues, and maybe even your own head that you'll never make money or find a job. But, take it from me: It's all worth it to make your crazy dream a reality. I know, because I, too, chose to climb this bruising yet beautiful trail."[1]

- Rachel Nelken created a nontraditional career that includes working part-time as a relationship manager for an arts council, half a day each week leading a youth music program, working freelance for music funding organizations, and running action learning sessions. "It's not always a mellow existence," she comments. "But it does make for an interesting one. . . . It has taken a while to feel like the sum of my working parts has made more than the whole. . . .

Relying on one organization to provide you with progression opportunities is unrealistic. You have to find your own instead."[2] Hopson and Ledger describe other examples of nontraditional, portfolio careers in the United Kingdom, which include elements such as photography, lighting design, Pilates instruction, and starting and leading nonprofit and for-profit organizations.[3]

- Peter Bates switched from twenty years as an investment banker to a combination of cabinetmaker, rare-breed pig farmer, and business adviser. "I'm more satisfied in my career than ever," he says.[4]

- Examples of the wide range of activities that can fit into a musician's career portfolio, particularly valuable in early career stages, include arranging concerts, teaching, conducting, recording and audio engineering, web development, publishing, composing, and arranging.[5]

- Examples of how mothers with children can approach integrating multiple work components from a nontraditional career with other aspects of life include using professional skills for consulting engagements, delivering courses, and writing.[6]

- Approaches to careers at later work/life stages, sometimes called encore careers, may include elements such as learning, fellowships, and volunteering focused on activities/organizations delivering community benefits with personal meaning, such as in social service or the nonprofit sectors, and for-profit entrepreneurial endeavors such as massage therapy, home modification, and financial planning.[7]

- Engaging in a nontraditional career at an early career stage allows for exploration of various possibilities for the future. Such careers include writing, teaching, and administration in government service or working as an illustrator, sculptor, photographer, graphic designer, or craft maker.[8] Shelli Worley comments: "I can't imagine myself ever working for somebody else in a corporate setting like boss, minion. I think my soul would just die."[9]

- Actress Gabra Zackman incorporates recording audiobooks into her acting portfolio. "I get to have a whole flourishing life as an actress because they have given me an opportunity to practice and to be employed," she notes.[10]

- Combining technical consulting, teaching, and public outreach in an academic portfolio.[11]

- Integrating a range of activities, such as writing, educational presentations, consulting, website development, with medical practice. Sarah Jonas connected the components of her nontraditional career together: "I work four days a week as a psychiatry trainee and one day a week seconded to the King's Fund, a health think tank. . . . I would absolutely recommend maintaining a varied career. Apart from enjoying the different thinking styles involved in policy and clinical work, I feel the crossover means my skills are strengthened in both roles."[12]

- Examples of entrepreneurial endeavors also speak to the broad range of nontraditional career options that are possible: law graduates starting their own practices rather than seeking employment with a large law firm, or moving into entirely different entrepreneurial paths; at later work/life stages, wide-ranging areas such as motivational speaking, starting a clinic for lower-income individuals, starting an online cookie business, or working in a volunteer greeting program; combining textbook curriculum development with online marketing, estate sales, and television production support; entrepreneurs with disabilities establishing a gift-experience business and a freelance illustration and graphic design business; and young Japanese entrepreneurs in Silicon Valley seeking alternatives to traditional salaried employment.[13]

These many approaches at different life stages and life circumstances are each tailored to individual needs. Now we will look in more depth at my own nontraditional career as an example of the form this can take. First, I will describe what was in it. I'm often asked how I started and how my nontraditional career developed, so we will look at that. Then I will draw some general conclusions that will be helpful for our later explorations in the book. Figure 1.1 illustrates the components that have been part of my nontraditional career.

Perhaps the first thing to notice is that there is a common connecting theme, summarized in the figure's center circle: the relationship of individuals, organizations, and community. This theme is broad, so it can

Example of a Nontraditional Career

Figure 1.1

include a lot, but much is also excluded. For example, accounting services, construction, retail sales, and contract scientific research would not fit. This is a nontraditional career in which the components were connected together both by values and by content knowledge. Let's look at those components, starting at the top with individual counseling/coaching.

Individual counseling/coaching means providing career counseling or coaching to people seeking a more meaningful or rewarding path forward. In some cases, people seek a nontraditional career, in some cases they do not. In addition, some people engage in coaching to increase effectiveness in their current position, perhaps opening the door to progression within their organization. It has been a privilege to work with many people over the years at different life stages, from initial entry to midcareer to encore stages, ranging from single-session resume critiques to multiyear coaching engagements. While mostly provided through a private practice, in which I delivered career services and my wife provided guidance to students and families on the going-to-college process, I also provided some career counseling/coaching on a contract basis to students in MBA programs and to clients of an employee assistance program (EAP) provider. Individual career counseling and coaching flowed naturally; it didn't feel like work. It complemented well the second component, the circle at the top right, workforce consulting.

My workforce consulting included a range of activities such as diagnostic services like conducting exit interviews for organizations in a variety of sectors and analyzing and interpreting the results; leadership and team development, including workshop delivery on broad-ranging subjects from conflict management to team innovation; developing and delivering coaching curricula at a leadership academy for principals in an urban school district over an extended time; and contributing to a multiyear change initiative for a public sector human services agency. These activities were stimulating because of their breadth, variety, intellectual challenge, and contribution to people's lives. While I delivered much of this work on my own, it was a pleasure on occasion, for example, with the exit interviews, to engage others on a project basis or to work on a team with others, as in the change initiative. Delivering services from outside an organization is particularly appealing as the focus is on the work content, thus largely avoiding the political quagmires of organizations. Learning from this career component transferred well into work with individuals and vice versa.

The third item, adjunct teaching, included developing and delivering courses in graduate programs in career development/counseling at two different institutions. This built on experience from the first two components and was a good way to give back experience and knowledge. Related activities included creating and delivering workshops on topics connected to careers, including nontraditional careers, for students and alumni at business schools.

The fourth item, guiding and supporting a practice, was a natural outgrowth of workforce consulting. It consisted of assembling teams of skilled career services practitioners to deliver services on site to organizations. Part of this endeavor included building a strong sense of community and shared mission for the teams. It has been a pleasure being part of these teams, seeing the team communities grow and develop, knowing that at the end of each day clients' lives were better for this work. This is a great credit to the team members.

The next item, volunteer work, included a number of activities over the years. Whether volunteering as an ambassador for our local food bank, assisting in national hunger surveys, making presentations on social justice issues such as universal health care, providing interview coaching at local educational institutions, or guiding volunteer teams delivering human resource services in nonprofit organizations, volunteer work was a meaningful part of my nontraditional career. It allowed me to bring knowledge and skills from other areas into settings where they could be helpful, and I am grateful for these opportunities, though cognizant that I could and should have done much more.

The final component is writing. As Oscar Wilde once noted, "I'm exhausted. I spent all morning putting in a comma and all afternoon taking it out."[14] Sometimes writing feels like two steps forward and one step back, but it has been for me a rewarding complement to other activities. My writing is really a story about what happens when an editor calls, asks if you would like to write a book, and you say yes. I had been encouraged when first in the career field to write some articles as I was then with a nonprofit organization that was delivering and publicizing work in the field. Working with a tremendous colleague, Seema Iyer, we were able to demonstrate the value of career services, not just for individuals but also for organizations.[15] That study led to our publishing a paper on the subject, after much refining, and in turn to a call from Hilary Claggett, at the time an editor at Greenwood, who asked if I would be interested in writing a book. I remember taking this call while driving in the peninsula of San Francisco and being surprised and excited at the prospect. I naïvely accepted it without realizing how much was involved. However, Hilary was such a source of encouragement that it turned out to be a good experience and led to my first book in 2003,

How My Path Unfolded ...

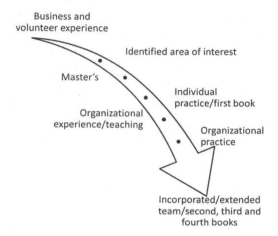

Figure 1.2

then, with Hilary's encouragement, books published in 2010 and 2013, and now this book.

In addition, I have been invited to write chapters in other books and some articles, and I have written many newsletters along the way. It has been a pleasure contributing pieces on topics that are personally meaningful. It also has been a pleasure taking this writing journey with fellow travelers, those with similar interests, who contributed to edited volumes. I remain mindful and appreciative of Annie Dillard's thoughtful perspective. "When a struggling writer found unexpected success and wrote Dillard for advice, she got this reply: 'I have an urgent message for you. *Everyone* feels like a fraud. . . . Separate yourself from your work. A book you made isn't you any more than is a chair you made, or a soup. It's just something you made once. If you ever want to make another one, it, too, will be just another hat in the ring, another widow's mite, another broken offering which God has long understood is the best we humans can do—we're forgiven in advance.'"[16] And perhaps those words apply to much of what we do, not just writing.

How about the trajectory of my nontraditional career? How did it start and how did it unfold? Figure 1.2 answers these question schematically.

After about twenty years in the corporate world working for two organizations, first in a technical area, then in a variety of assignments that included business planning, business development, management of a research function, and general management of growing businesses, I found myself questioning whether I wanted to continue doing something that just made money for an organization. So I went through an assessment process, which helped identify the career field as one that might fit well. It would allow me to give back accumulated knowledge and experience while matching my personality, interests, skills, and values well. It also offered the prospect of taking a nontraditional career path that could include a private practice. Such a practice could also include my wife Linda's interest and background in providing college advising services, which Linda was planning to develop. All this led to my taking a master's program in career development while working full time, initially to see whether the career field would be a good fit. I was fortunate to take the master's at John F. Kennedy University, which is oriented toward adult working professionals, and immediately liked the content and nature of the career field. About halfway through the master's program, through a guest speaker, I found a local nonprofit organization specializing in the career field that needed someone to guide its corporate practice delivering career services into organizations. This was a good fit, building on my corporate experience and new educational background in the career field. The nonprofit was comfortable with my working in a private practice on my own time. So at about this time my wife and I formed our career counseling and college advising practice. Here was the first step on the nontraditional path.

I also started doing adjunct teaching in a local master's program in the career field, and the nontraditional career components began to take shape. While originally thinking that a private practice working with individuals might become a full-time undertaking, it was soon clear to me that providing such services was fulfilling and engaging but could not provide the financial base we needed as a family. After some management changes at the nonprofit, I moved to another organization in the human resource sector, which specialized in outplacement. This was in an account executive role, which allowed me to develop sales skills that would be valuable both in this organization and later, when on my own. At this time I wondered about the possibility of starting an independent organizational consulting practice but could not yet see sufficient content to make it viable. I was able to do some coaching for the organization I worked for, in turn further building skills, and I founded a practice for them focused on diagnostics to help organizations connect most effectively with their employees. This practice was built on knowledge developed previously at the nonprofit and analytical approaches from earlier career stages. Customer relationships formed at this time would prove helpful later. This was also when I was writing my first book, so it was a busy time. When the book was published,

I was able to integrate it into the organizational work and deliver many presentations on the subject around the country. I changed the adjunct teaching role to another college closer to home, still focused on a master's program in the career field. Our career counseling and college advising private practice continued well, just as we had hoped.

There were some changes in my primary assignments at the company I worked for, and then the private equity buyout mentioned in the preface happened. Options within the company were constrained by the new buyer's definition of "core business." The possibility of now starting a private organizational consulting practice surfaced. The company was supportive of this step. I had now developed sufficient content and some of the skills needed in a nontraditional setting. So after discussion with my family, I decided to take the step of transitioning from conventional employment to being fully engaged in a nontraditional career by adding an organizational consulting practice. Similar to our career counseling and college advising practice, this organizational consulting practice was initially structured as a sole proprietorship. This meant income was readily folded into our individual tax reporting as a Schedule C. The career counseling and college advising practice was in my wife's name and the organizational consulting practice was in my name. The organizational practice flourished from the beginning due to the support of many people, colleagues and customers alike, and for that I am most thankful. Our initial financial goal was modest—earn enough so that we could stay in our house—and we knew that we could always move to a less expensive area if necessary. That goal was soon met and surpassed. Indeed, we were grateful that our income from nontraditional career components soon exceeded that from conventional employment.

While there was anxiety at first about the absence of a steady paycheck, this was quickly dispelled as the practice prospered, projects continued to surface, and the work was interesting and fulfilling. I was no longer beholden to a corporate bureaucracy and could make decisions based on my values and client and customer needs. After about three years, one of our customers was seeking a new on-site career services program and invited our bid along with bids from other organizations. We were most grateful to be awarded this contract. At that point we created an incorporated entity, changing the organizational practice from a sole proprietorship, and began hiring a tremendous team of people. This incorporated entity also prospered, and it led to other projects. The opportunity arose to create three more books. Editing the first two brought many additional voices forward. This is the third book. Woven throughout all of these steps there have been various volunteer experiences and experiences working in our private practice with individuals, which has been a continuing source of fulfillment.

Here is some of what I have learned:

- Be clear about your purpose.
  - An example of clarity of purpose is a focus on the relationship among individuals, organizations, and the community.
- Stay true to your personal beliefs.
  - Honoring personal beliefs has meant turning projects away on occasion.
- Expect unexpected supporters and barriers.
  - Supporters included customers and colleagues who were very helpful.
  - Creating the necessary infrastructure quickly highlighted administrative issues to address.
- Be patient and start the different components when the timing is right.
  - Patience meant waiting to start the organizational consulting practice until needed capabilities and relationships were in place.
- Develop needed skills before launching a nontraditional path and anticipate needing a broad range of skills.
  - Needed capabilities included interpersonal, sales, and organizational skills as well as content capability.
- Nurture relationships, for they are the lifeblood of the enterprise.
  - Relationships included those with customers and colleagues.
- Build portable knowledge and capabilities with commercial value.
  - Portable skills included diagnostic, analytical, strategic, and interpersonal skills.
- Enjoy the journey.
  - There's only one.

Comparing Nontraditional and Conventional Careers

We will now compare nontraditional and conventional careers from various perspectives to better understand both the differences and opportunities presented by a nontraditional career path. There are several aspects to this comparison, and we will look at each one:

- Locus of control and relationship
- Nature of work engagement
- Entry requirements and skill needs
- Practical aspects
- Outcomes
- Foundation for fulfillment and success

Locus of control and relationship. Central to the difference between a nontraditional and a conventional career is the locus of control and relationship. A nontraditional career is self-directed, though it also is supported by others as it takes place in an interdependent work world.[17] In conventional employment, on the other hand, others in an organization define values, objectives, priorities, progression, and work location. Furthermore, a nontraditional career is a focal point for relationships, for example, with customers, with partners, with other organizations, and with the community. In a conventional position, while some such relationships occur, they are filtered through the employing organization's lens, defining the nature, content, and extent of the relationships. It is like putting a membrane in the way that lets some things pass through while others are filtered out.

Nature of work engagement. One effective way of describing the relationship with work was created by Betsy Brewer.[18] It identifies the following four forms of our relationship with work in the context of the interior processes of discovering meaning (what), being (who), and doing (how):

- A job: based on material rewards, a transaction
- An occupation: involving greater meaning, but doing dominates
- A career: entailing continuity, requiring personal initiative but needing collective approval
- A vocation: calling in the service of a greater good

There is no implication that any one form of relationship is better than any other or that one form of relationship must precede another. I have found when exploring this with clients that many of us would like to move closer toward vocation, seeking deeper meaning in our work, by at least reaching the career stage. While this movement may be accessible through conventional employment, a nontraditional approach, constructed as it is around our individual needs and preferences, naturally leads us to a place of career or vocation on this spectrum, and it is more likely to be sustainable over time. Reflecting on the trajectory of your career so far and where a nontraditional career might fit in the future in the context of this framework can provide insights into what a nontraditional career might mean for you.

Entry requirements and skill needs. Entry requirements and skill needs are intimately interwoven. In a conventional position, it is primarily about bringing particular expertise to a position, fit with organizational culture, and enthusiasm about working for an organization. Fit speaks to the political component of aligning with an organization's culture and expectations. Indeed, interviewers in an organization will generally probe in these three

areas. However, while a nontraditional path also requires particular expertise, the other dimensions are different. The ability to relate well to customers, clients, and potential partners replaces the political skill of navigating a
corporate bureaucracy. In addition, skills and personal attributes replace
services that might have been provided by corporate staff groups in conventional employment. Enthusiasm is now about creating and sustaining a new
enterprise built around personal values and interests rather than fitting into
someone else's world view. We will explore the area of skills needed for a
nontraditional career in more depth in chapter 7.

Practical aspects. Practical differences include financial and logistical
factors. A conventional position provides defined income, at least while the
position lasts, sometimes provides access to benefits, and it usually requires
presence at a particular location at particular times. A nontraditional career,
on the other hand, may involve less predictable earnings with likely a significantly higher upside potential, and it requires taking personal ownership
for securing benefits. Health-care insurance is now accessible and manda-
tory in the United States due to passage of the Affordable Care Act, which
eliminates disqualification for preexisting conditions. This removes access
barriers for people working outside of a traditional, single-employer relationship. Provision for retirement in a nontraditional career can include savings using a one-participant 401(k) plan and an Individual Retirement
Account (IRA). It is no longer necessary to endure endless commute time
as nontraditional work can often take place at home or locally, and travel
can be scheduled for time efficiency. We will explore the practical aspects
of a nontraditional career in more depth in chapter 9.

Outcomes. How do outcomes differ? In a conventional position, progression and development depend on organizational culture and support.
Some organizations accept responsibility to commit resources to individual development; others place the entire responsibility with the employee.[19]
Employment is transitory, subject to the whims of others. In a nontraditional career, continuity is based on expertise and credibility that build
over time with successful completion of customer and client engagements.
Development can, and should, be naturally part of an evolving nontraditional career path. Fulfillment is a natural outgrowth of a nontraditional
career, which is closely tailored to personal attributes and aspirations. This
alignment with personal capabilities means that financial reward can also
be substantial, though this is typically a by-product rather a primary driving force.

Foundation for fulfillment and success. A nontraditional career path is
validated internally rather than externally as in a conventional position.

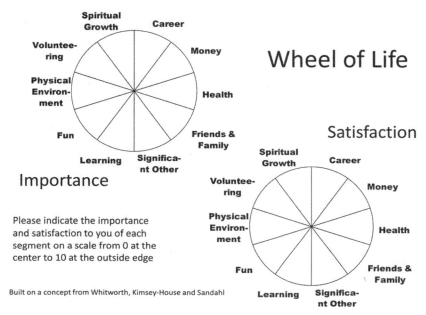

Figure 1.3

Performance reviews by others based on corporate needs are gone in a nontraditional approach. They are replaced by an ongoing personal assessment about how a nontraditional path is meeting intrinsic personal needs, contributing to meaningful relationships with others, and addressing practical financial and other considerations. We can capture one perspective about this in the wheel of life exercise, whereby one can assess importance and satisfaction in a range of areas, as shown in figure 1.3. This exercise is built on a concept from Whitworth, Kimsey-House, and Sandahl.[20]

The specific wheel segments can be changed by adding items that may be more significant to you or by deleting those less so. The wheel on the left in figure 1.3 is used to discern the relative importance of different areas, while the wheel on the right is used to show satisfaction in those same areas. Drawing an arc in each segment in each of the two circles, where the arc position corresponds to a self-assessed score from 0 (low) at the center to 10 (high) at the outside, will show you a pattern of importance and satisfaction. It is helpful to complete this exercise for the present time and then complete the importance wheel again showing how this would ideally look three to five years in the future. This provides one basis for reflecting on what changes might be needed to reach this ideal and what they might mean for a nontraditional career path. This brings us to creating a foundation of self-understanding for building a nontraditional career path.

Foundation for Building a Nontraditional Career Path

Much of our formal learning focuses on content knowledge in particular disciplines or crafts. This is an essential foundation for our contributions; indeed, investing our time wisely in such content learning is important in building expertise and credibility. This is our marketable capability, our core area, and it can take many and varied forms, from the creative, to the practical, to the analytical. For some of us, there may be more than one core area, while for others the different components of a nontraditional career will all relate to a single core area. There are also other aspects in addition to content knowledge that are important in allowing us to take control of our career destiny and build a nontraditional path.

A key aspect is building a strong sense of self-understanding to discern a nontraditional path that fits well. Components of self-understanding include values, personality preferences, interests, skills, and learning from past experiences. This sense of self-understanding is like a compass that helps guide our decisions and direction based on personal purpose. Personal purpose forms a cornerstone for a nontraditional career path.[21] It helps us address an important question: Why build a nontraditional career? For example, is that path for personal fulfillment, financial reward, mitigating risk of unemployment, giving back, or creating a legacy? Or perhaps it is a combination of one or more of these aspects or something else entirely. We may find that the primary purpose shifts over the course of our work life, with, for example, creating a legacy more prominent at later stages. Clarifying this sense of purpose affects the emphasis and structure of our nontraditional career.

This sense of purpose is intimately interwoven with personal values, one of the key components of self-understanding. Various tools have been developed to describe values. One particularly pertinent framework is that of career anchors, developed by Edgar Schein.[22] Career anchors result from the interaction between our values and needs and work context, as we find ourselves pulled into work that is more congruent with our values.[23] There are eight career anchors that can help us understand values we need to honor in finding a fulfilling nontraditional career path. We can tailor a nontraditional career path to address one or more of the following eight anchors:

- Becoming a recognized expert in a field
  - For example, expertise could mean leveraging and developing a particular area of content knowledge or service delivery capability.
- Becoming a senior manager
  - For example, this could mean building an organization that requires leadership and management.
- Being independent and able to set a work path
  - Being able to set direction is a natural aspect of a nontraditional career.
- Having a secure situation
  - Security results from the robustness of multiple income streams and from growing personal expertise and marketability.
- Having entrepreneurial focus/making a lot of money
  - Entrepreneurial success develops from the choice of income-producing career components and because there is no corporate bureaucracy to support.
- Being of service to a cause
  - Being of service is achieved by selecting nontraditional career components that address service causes having personal meaning.
- Being constantly challenged
  - Challenge is a natural outgrowth of a nontraditional career in a changing environment with the ensuing opportunities that unfold.
- Being able to integrate work and personal needs
  - Integration of work and personal needs is based on personal choice about how to best combine nontraditional career components and individual needs.

Schein's booklet, *Career Anchors Self-Assessment,* includes an assessment that identifies career anchor preferences.[24] Building an understanding of career anchor preferences provides insights into the nature of career components that will align well with our values.

Personality preferences, another component of self-understanding, provide insights into how we think and feel about, and prefer to interact with, the world. They suggest why we prefer to behave in certain ways and they offer insights into aspects of nontraditional career paths that will likely appeal. Various personality assessments have been developed, and one that is particularly helpful when considering career paths is the Myers-Briggs Type Indicator (MBTI), as described by Isabel Briggs Myers.[25] While it is beneficial to have MBTI feedback interpreted by a career counselor knowledgeable about the instrument, the assessment can be taken online and accessed directly in a version that provides explanation of the feedback.[26] The MBTI examines the following:

- Where we prefer to focus our attention and energy (Extraversion or Introversion)
  - Those who are more extraverted derive energy from being with others; those more introverted derive energy from being alone.
- The way we prefer to take in information (Sensing or Intuition)
  - Those with a stronger sensing preference take in information in concrete, tangible form; those with a preference for intuition take in information holistically, looking for patterns.
- The way we prefer to make decisions (Thinking or Feeling)
  - Those with a stronger thinking preference make decisions based on objective logic; those with a stronger feeling preference make decisions based on subjective values.
- How we orient ourselves to the external world (Judging or Perceiving)
  - Those with a stronger judging preference like to live with structure; those with a stronger perceiving preference like to live spontaneously.

Knowing personality preferences helps guide decisions about the nature and structure of nontraditional career components—for example, whether they involve extensive interaction with others or more work alone, whether they involve tangible, practical activities or are more conceptual, whether they focus on interpersonal feelings or are more analytical, and whether they are built around structure and repetition or spontaneity.

Interests, another component of self-understanding, address the kind of work activities that we find appealing and therefore would likely fit well in a nontraditional career path. Again, various assessments can help bring clarity; one that is particularly effective is the Strong Interest Inventory.[27] This provides insights into six general occupational theme areas and thirty basic interest scales within these theme areas, and it compares our work-related likes and dislikes with those of people in a wide range of occupations. The assessment, which can be taken online, does require administration and interpretation by a career counselor. However, simply reflecting on the six general occupational theme areas and understanding more of the associated career paths can provide insights into nontraditional components that may be appealing, and there is an online assessment available at no cost to help identify individual priorities in the six theme areas.[28] These six theme areas are as follows:

- Realistic: practical, hands-on interests such as mechanical construction, farming, operating laboratory equipment
- Investigative: interests aligned with research, analyzing, and synthesizing information, such as in scientific and mathematical pursuits or in social science research

- Artistic: creative interests that might include aesthetics and self-expression, such as visual arts or writing, though creativity is not limited to fine art endeavors
- Social: interests aligned with the helping professions that involve interaction with people and use of interpersonal skills
- Enterprising: interests in business and leadership positions that may be in entrepreneurial settings or in private or public sector organizations
- Conventional: interests aligned with detail and accuracy, comfortable with well-defined, structured activities such as office management or accounting

Skills, another component of self-understanding, are the things we like to do and are good at doing. An assessment that provides insights into this area, which can be helpful in considering what components to include in a nontraditional path, is SkillScan.[29] This instrument identifies sixty individual skills organized into six broad skill categories: management/leadership, relationship, analytical, communication, creative, and physical/technical. An online sort process identifies a self-assessed level of proficiency and preference for each skill while also grouping the skills into eighteen sets or subgroups of skills that are frequently used together. There are three sets or subgroups in each of the six broad skill categories. The skills assessment process is also helpful in highlighting skills we might choose to develop.

We have found that several more conceptual, narrative types of assessments are also helpful in building self-understanding. They can bring learning from past experiences to decisions about future nontraditional career paths. These assessments include the following, the first two of which look back in time, the second two of which look forward to the future:

- Life experiences: reflecting on ten past life experiences that were meaningful in terms of accomplishments and were enjoyable, then identifying from the top five experiences patterns of skills, interests, and values that emerge and what they may mean for a nontraditional career path
- Working environments: reflecting on enjoyable and negative aspects of past jobs and what they might mean for an appealing nontraditional career path in the future
- Ideal day: looking three to five years into the future, when a nontraditional career path is a good fit, and writing a description of an ideal workday from the beginning to the end of the day
- Questions for the future (from James Flaherty):[30] addressing the following questions for time frames of one year, three years, five years, and ten years into the future:
  - What do you want to be doing?
  - Whom do you want in your life? In what capacity?

Sweet Spot for a Nontraditional Career

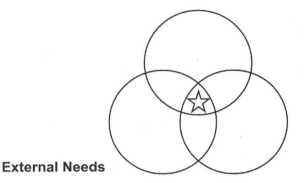

Passion/Interests

External Needs

Skills

Figure 1.4

- What resources do you want to have?
- What experiences do you want to be having?
- In what ways do you want to be growing/learning?
- In what other ways could you describe your life?

These four narrative assessments, when combined with information from the values, personality, interests, and skills assessments, provide insights that can help clarify the preferred path for your nontraditional career, the work environment you may choose to create, and how you wish to interface with clients, customers, and partners. Building a deeper sense of self-understanding from these assessments helps point to the intersection of the three circles shown in figure 1.4—passion/interests, skills, and external needs—which is the sweet spot for a nontraditional career. Cliff Oxford underscores the importance of understanding external needs and other business attributes.[31]

In my own situation, passion/interests included working with people on their development, working with organizations on strengthening workforce capability, and working in the community to help address the needs of those who are disadvantaged. The skills included career counseling and coaching, business understanding and experience, and strategic, analytical, and communication capabilities; external needs included individuals seeking support in their development and for-profit, nonprofit, and public sector organizations seeking to build workforce capability to enhance effectiveness. Let me suggest thinking about your own examples in each of these three circles and considering where the circles overlap for you. Evoking concepts of mindfulness and physical sensations and emotions that surface when reflecting on nontraditional options can provide additional insights.[32]

Table 1.1 is a template you can use to integrate the various assessments as you consider identifying and structuring nontraditional career components. You can base this on completing the assessments referenced earlier or on simply reflecting about your perspectives for the areas shown in the table.

There are some questions to consider when reflecting on the information in table 1.1:

- How can I best select and structure nontraditional career components to honor the values that are important to me?
- What are key characteristics of my personality preferences and what does this mean for nontraditional career components I might enjoy?
- Based on my interests, what nontraditional career components might appeal to me?
- Based on the skills I am good at and enjoy, what activities should I seek to include in my nontraditional career components?
- What has worked well for me in past positions and what has been a challenge? What might this imply for future nontraditional career components?
- What do my aspirations for the future suggest for nontraditional career components?

We have focused thus far on the inner aspects of who we are and what they mean for a nontraditional career path. There are other aspects also to consider. One key area is support from others. This can include family, colleagues, customers, and possibly a current employer. These relationships can encourage a fledging nontraditional career path to unfold. They can provide insights about strengths or blind spots that can be invaluable when setting out, and they can provide support as the path unfolds. Since decisions about your career path greatly affect significant others, for example, a spouse or partner, include those close to you in early thinking and decision making. I remember well how valuable this was with our first practice, which augmented traditional employment, and later, when we added our second practice, which meant moving entirely to a nontraditional career path.

Colleagues and customers with whom you may have built relationships can be vital supporters. They were instrumental in our ability to establish

Table 1.1. Integrating Assessments.

Assessment Area	Assessment Components	My Preferences (Where Applicable)	Implications for Nontraditional Career Components
Values	Functional expertise		
Values	General management		
Values	Autonomy		
Values	Security		
Values	Entrepreneurial/making a lot of money		
Values	Service		
Values	Challenge		
Values	Work and personal life integration/lifestyle		
Personality	Extraversion/ Introversion		
Personality	Sensing/Intuition		
Personality	Thinking/Feeling		
Personality	Judging/Perceiving		
Interests	Realistic		
Interests	Investigative		
Interests	Artistic		
Interests	Social		
Interests	Enterprising		
Interests	Conventional		
Skills	Management/leadership		
Skills	Relationship		
Skills	Analytical		
Skills	Communication		
Skills	Creative		
Skills	Physical/technical		
Narrative	Life experiences		
Narrative	Working environments		
Narrative	Ideal day		
Narrative	Questions for the future		

our practices, bringing work opportunities, as they did, along with encouragement. It might seem odd to include a current employer on this list of supporters, but in practice a current employer can play a pivotal role in the launch of a nontraditional career. I was employed by an organization that was supportive of my establishing a private practice on my own time. It turned out that there were benefits of knowledge transfer in both directions, insights that I could take from private practice to the organizational setting and vice versa. At a later stage, another employer was supportive of my launching a second practice, including maintaining existing customer contacts, as there was no competitive issue at this point, given a change in the employer's business focus. It helped provide an extended launch time and income during that launch phase.

The final aspect, also external, and instrumental in creating a strong foundation for a nontraditional career, is that of building financial resources both for the initial launch and to cover economic uncertainties. Income in a nontraditional career will vary from month to month, so providing a cushion for the slower times is important. For example, by starting our first practice while I was still in conventional employment meant that we were able to build a financial cushion that helped support the launch of the second practice.

In summary, a nontraditional career path offers an opportunity to integrate the emotional, intellectual, spiritual, and practical parts of who we are. It can be a journey with delightful surprises, with kindred fellow travelers, that brings personal fulfillment and community benefit. It can be an expression of who we are. Next we will look at some of the benefits and challenges of this journey.

CHAPTER 2

Benefits and Challenges

MANY OF US FEEL more comfortable and safe when driving a car rather than being a passenger in one. We know when we're driving that we have seen a car coming up from behind, or children or animals playing at the side of a residential street, and we adjust our driving accordingly. We know that we'll brake when needed to slow for traffic lights or a stop sign ahead. Yet strangely, when it comes to work, we sometimes ascribe greater employment safety to being in a large organization where employment decisions are made by others, and not necessarily made in our best interests. This perspective is changing with the reality that one Standard & Poor's (S&P) 500 company in the United States is now being replaced about once every two weeks in the S&P 500 index and the average tenure of an S&P 500 company in the index fell from sixty-one years in 1958 to eighteen years in 2012 based on seven-year rolling averages.[1] This is coupled with the recognition that large corporations are oligarchies with power concentrated in the hands of a few. They are not democracies, as "the governance structure gives ample opportunity to an almost medieval exercise of absolute power by management. It is no wonder that, under these conditions, companies can become like fiefdoms for the few, to be exploited like a machine."[2] So in one sense we have exchanged the tyranny of totalitarian political systems for the oligarchy of large organizations. Not surprisingly, we see people choosing to leave organizations when economic conditions improve. For example, the ratio of voluntary quits initiated by employees to layoffs and discharges that are involuntary reached 1.6 in March 2014 from a level of 0.7 in April 2009 toward the end of the recession.[3] When the ratio is 1.0, one person quits for each person laid off or discharged, and when the ratio is greater than 1.0, quits outnumber layoffs and discharges. This is supported by a survey showing that 83 percent of North American employees plan to pursue new job opportunities in 2014.[4]

In the world of small business, which is defined in the United States as an independent business having fewer than five hundred employees, about half of new establishments survive five years or more, about a third survive ten years or more, and about a quarter stay in business fifteen years or more (recognizing that specific sectors, time frames, geographic locations, and forms of small business will likely have different survival rates).[5] Small businesses have similar five-year survival rates in Canada, where they are defined as businesses with fewer than 250 employees in this study.[6] So we see that small business survival is not the catastrophic picture that is sometimes indicated—a pointer also for nontraditional careers.

In this chapter we will examine the benefits and challenges of a nontraditional career. We will first summarize perspectives that people have about different forms of employment, including nontraditional careers, and then we will focus on specific benefits and challenges associated with nontraditional careers. We will further illustrate this with descriptions from seven people who have successfully established nontraditional careers, and we will conclude by linking a nontraditional path to who we are and our vocation. Let us start by looking at perspectives of people about different forms of employment.

Perspectives about Forms of Employment

Perhaps not surprisingly, more than half (51 percent) in a sample of about twenty-five thousand people in twenty-three countries (mainly in Europe and North America) indicated that, given a choice, they would prefer to be self-employed rather than an employee.[7] There were significant variations among countries, with the United States near the top of the list with 71 percent of people indicating a preference for self-employment. How does this nontraditional, self-employment path work out for those who choose it? Studies from several countries show an enthusiastic embrace of self-employment. In the United States, self-employed adults were found to be significantly more satisfied with their jobs than other workers in a Pew Research Center study of 1,040 adults aged sixteen and older, 254 who were self-employed in a variety of areas (which could include small business owners, consultants, freelance writers, contractors, artists, construction workers, day laborers, farmers and agricultural workers, doctors, lawyers, and accountants) and 885 who were wage and salary workers.[8] Only 5 percent of those who were self-employed were dissatisfied with their employment situation, half the proportion of other workers who were dissatisfied, while 39 percent of self-employed workers were completely satisfied with their jobs compared with only

28 percent of wage or salaried employees.[9] Furthermore, 32 percent of those who were self-employed were self-employed mostly because they wanted to be, compared with only 19 percent of employees; whereas 50 percent of employees worked mostly because they needed the money, compared with 38 percent who were self-employed.[10] Those who were self-employed also placed a higher value on the intangible benefits of their work, such as feeling useful and productive, with 55 percent saying they were working to help improve society compared with 46 percent of employees.[11] These trends were true regardless of age. Income for those who were self-employed was slightly higher than for employees, but those who were self-employed reported feeling more financial stress, with 29 percent saying their family just met basic expenses (vs. 23 percent of employees) and 11 percent saying they didn't meet basic expenses (vs. 9 percent of employees).[12] These survey results were from 2009, before the Affordable Care Act was implemented in the United States. Implementation of the act likely reduces financial stress in self-employment. Similar positive responses in another study in 2013 showed high levels of satisfaction for those in nontraditional careers in the United States, with the vast majority planning to continue on the nontraditional path, based on a sample of more than two thousand people. Only those in contingent or agency settings, with little control over their work, showed lower levels of satisfaction compared with conventional employment.[13]

A study of 213 executives by the Independent Direction Directors Advisory Service (IDDAS) in the United Kingdom also shows a similar positive response to a nontraditional career path.[14] The majority (65 percent) said they were very satisfied or satisfied with their success in establishing an independent career, with key reasons being having control over work and time, the variety and unpredictability of the work, and freedom from corporate agendas and politics.[15] Here are two comments from the study: "There is no way I would go back to conventional corporate life" and "I am now 52 years old. I dearly wish that I had made the decision to work independently ten years ago."[16] Similarly, a study in Australia (IPro Index 2013) of 379 independent professionals across a wide variety of specializations and backgrounds (including mining/engineering, general project management, information technology, and banking and finance) showed that 87 percent were satisfied with their independent status and only 24 percent frequently think of ceasing this form of work.[17] Major reasons for engaging in this work were earnings capability, a sense of freedom, and the variety of work.[18] The survey showed high measures of well-being: 89 percent were proud of the work they do and 83 percent were enthusiastic about their job, with 98 percent feeling they could handle whatever came their way in their work.[19] There were also indicators of strong client relationships, with 66 percent experiencing a sense of commitment to their clients, 68 percent agreeing that their current client organization had come through in fulfilling promises made when they were engaged, and 99 percent agreeing that they are trustworthy and have their clients' best interests at heart.[20]

So we see a pattern of satisfaction with a nontraditional or self-employment path that crosses geographic boundaries and occupational areas. It is interesting to compare this with feedback from those in conventional employment. Only 60 percent of over seven thousand employees globally in a 2012 survey of engagement indicated a definite intent to stay with their current organization for the next twelve months.[21] The 40 percent not signaling a definite intent to stay (38 percent in Australia/New Zealand) is much higher than the 24 percent of those who were self-employed and think of ceasing this form of work as mentioned earlier in the IPro Australian study. Another study of more than 150,000 full- and part-time workers using different measures of engagement found that only 30 percent of American workers were engaged, though significantly, employee engagement levels were higher in companies with fewer than ten people (42 percent).[22] This low level of engagement in conventional employment is in contrast to the 87 percent of independent professionals satisfied with their independent status as mentioned earlier in the Australian study.[23]

One challenge in conventional employment is lack of alignment of individual aspirations with organizational purpose. Lack of feeling inspired about organizational purpose can lead to a reduced sense of affiliation and increased attrition as we found in our earlier work.[24] Indeed exit interviews showed that more than 70 percent of people left organizations due to dissatisfaction, not because of new opportunities or other reasons such as a spouse or partner moving.[25] We found that the primary reasons people considered leaving conventional employment were a negative work environment, lack of development opportunities in an existing role, and lack of promotional opportunities.[26] A nontraditional career addresses each of these aspects.

An example of the intentional negative work environment that some Japanese organizations create is the ostracism of employees the organization wishes to lay off but will not because of the social taboo surrounding large-scale layoffs. In this case, employees, often with many years of tenure, may be sent to a "chasing-out room" with no assigned task so they end up reading and browsing the Internet while filing a report of activities at the end of each day. Critics contend this is "to make employees feel forgotten and worthless—and eventually so bored and shamed that they just quit."[27] Unfortunately, I have seen disrespectful treatment from senior managers during my career. Approaches companies adopt to address problems with conventional employment include changing organizational practices, though efforts here appear to have stalled as employee engagement barely moved from 2000 to 2012, which is understandable as some organizations continue to adopt demeaning practices.[28] It is also not surprising given the problem of concentrated institutional power we discussed earlier. Furthermore, the decline in union representation, which we mentioned in the preface, has limited union presence as a mitigating factor. Given the climate in many organizations, it is understandable that conventional employment loses its appeal and a nontraditional career becomes ever more attractive. Indeed, it is a natural progression from the evolution of career relationships over time, as shown in figure 2.1.

The dependent relationship that existed from the 1950s through the 1980s, in which large organizations were prominent in the career landscape, fractured in the early 1990s as the concept of career tenure disintegrated in the private sector. Survival skills became those of career self-reliance—survival of the fittest. By the early 2000s, it was apparent that self-reliance alone can limit opportunities to give to, and draw from, others for mutual benefit. Instead, the concept of career interdependence surfaced, recognizing the strengths and benefits that come from integrating personal aspirations with communal resources.[29] On an individual level this means recognizing the

Career Relationships Over Time

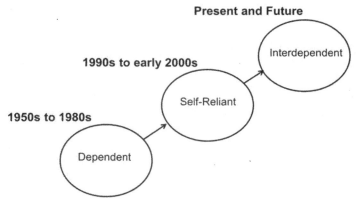

Figure 2.1

importance of building and sustaining a community of people with whom you are connected, a particularly important aspect of a nontraditional career. Now let's look at the benefits and challenges of a nontraditional career.

Benefits and Challenges of a Nontraditional Career

We'll begin with the benefits of a nontraditional career, which include the following:

- Autonomy
- Alignment with interests
- Contribution
- Financial attractiveness
- Improved risk/reward profile
- Personal meaning

Autonomy. The item that participants in the IDDAS survey found most rewarding about a nontraditional career was the "ability to control my own activities."[30] This sense of autonomy is inherent in a nontraditional career, bringing with it the opportunity to make decisions about the right path forward, values to honor, opportunities to pursue or drop, who to invite on the journey, and what to prioritize. For example, principles related to social justice and social responsibility that we adopted for our small organization are as follows:

- Being generous
  - With our clients, customers, and each other
- Honoring practices that create a safe, secure, supportive work environment
  - Aligning personal aspirations and organizational needs
  - Providing fair and equitable compensation
  - Offering vacation and sick time from the first day of employment
  - Supporting individual development
  - Providing time for community service
  - Offering schedule flexibility
  - Valuing inclusion
- Advocating for social and community causes that benefit all
  - Writing, speaking about, and addressing social justice issues
  - Supporting issues and candidates for office speaking to the needs of all in our society
  - Promoting universal, single-payer health care (Medicare for all)
- Participating in community service
  - Education, direct service, and financial contribution

These are principles that are personally and collectively meaningful. They address individual, organizational, and community needs, and they are nontraditional in their grounding in compassion, support, and giving rather than in competition and taking. I still remember the motto from the English equivalent of high school that I attended: "Work hard, play hard." While this has some merit, important values like compassion are absent. I learned something about what to be in this school, but I also learned something about what not to be. In a nontraditional career you can create what you choose to be rather than having to adopt someone else's what-not-to-be.

Alignment with interests. Of the independent professionals responding to the IPro Australian study, 89 percent indicated that they were happy when they were working intensely, and 83 percent indicated that they were immersed in their work.[31] Such high levels of satisfaction and connection with work speak to how well aligned their work is with their interests. A nontraditional career, consisting as it does of more than one component, can be fine-tuned to match personal interests in a way that is not possible in conventional employment, constrained as it is by organizational requirements and the need for specialization. Burnout typically results not from excessive workloads but from excessive time spent doing things that are not well aligned with our interests, skills, and who we are. Maximizing the time spent on work activities we enjoy doing and are good at while minimizing time in those areas that we find to be drudgery is important to our psychological health and accomplishment.

In chapter 1 we outlined six general occupational theme areas for interests, acknowledging that there are many interest subcategories within these theme areas. While the degree to which interests are focused on particular theme areas is different for each of us, most of us have interest profiles that are focused in one to three of the theme areas and have high expressions of interest in several of the subcategories within these theme areas. What is drudgery to one person is excitement to another. For example, writing can be an engaging, creative endeavor for one person and a chore to another. Public speaking can be energizing to one and intimidating to another. A nontraditional career offers an opportunity to select from a wide menu of options those that are appealing while leaving the others alone.

Contribution. There are many with whom we connect in our work lives, some directly through work while others, such as our family, are greatly affected by our work lives. We can think of these connections as constituencies or stakeholders in our nontraditional path.[32] Work-related constituencies may include employees, contractors we engage, investors, community members, suppliers, customers, and partners. The beauty of a nontraditional career is how it allows us to contribute to those close to us and broadly to these other constituencies. One of the delights is the opportunity it affords for giving as part of work. For example, it was such a pleasure early on in my self-employment bringing in a colleague to work with us as a contractor and being able to write that first check for the work. A nontraditional career also allows the time flexibility to engage in volunteer activities, and it may allow financial flexibility to contribute to causes that have personal meaning. I remember hearing a talk by Susan Packard Orr (the daughter of David Packard, one of the founders of Hewlett-Packard) in which she observed that Hewlett-Packard was giving to philanthropic causes when it was first founded in a garage in Palo Alto.

Financial attractiveness. Of the participants in the IPro study, 73 percent of those surveyed identified the perceived ability to earn more as one reason to engage in a nontraditional path.[33] However, as mentioned earlier, the Pew study found that financial stress can accompany self-employment in the United States, though we acknowledged that implementation of the Affordable Care Act likely reduces financial stress, providing, as it does, ready access to health care for those who are self-employed.[34] A nontraditional career is built on alignment of personal passion/interests, skills, and market needs. It is also built on the direct practice of these skills, unencumbered by a costly corporate bureaucracy. So the combination of a career path that matches capabilities well with an efficient cost structure provides a strong foundation for financial success that complements the intrinsic reward inherent in this path. Indeed, a nontraditional path is likely a primary vehicle for individual economic progress as real median wages in conventional employment in the United States were stagnant from 2000 to 2012 even though productivity increased by 25 percent.[35]

Improved risk/reward profile. It is a bit like the world of financial investing, where appropriate diversification into a range of investments enhances the risk/return profile. If the value of one stock or one bond fund drops, others may compensate, thus reducing dependency on a single item. Let us translate this into a work setting. A high-risk scenario is that of working for a single organization and having savings invested mainly in stocks of that same organization. If the organization fails, you lose your job and your savings. Unfortunately, the collapse of Enron in 2001 was an example of such a situation for some people.[36] While it is possible to readily diversify savings and reduce investment risk, the question of how to diversify the potential loss of earned income due to job loss or business failure remains.

A nontraditional career inherently addresses this form of risk since it is based on multiple income streams coming from separate work activities.

Let us compare nontraditional work with conventional employment in terms of risk and reward, recognizing that this is a subjective comparison and may look different to each of us. Reward can be intrinsic, as in, for example, a sense of fulfillment, or it can be financial or some combination of rewards. Risk is the likelihood that a job or position will go away. A conventional position, full-time employment in an organization, is low in reward and high in risk, since rewards go disproportionately to those at senior levels and employment continuity is defined by someone else.[37] Continued rounds of layoffs speak to this. Conversely, a nontraditional career with more than one income stream built around personal interests offers the potential for high intrinsic and financial reward as it is tailored to personal strengths and risk is mitigated by the breadth of activities. When I was giving a talk on this topic, one participant observed astutely that since the early 1980s the nontraditional career and a conventional position have reversed their profiles with respect to risk and reward. In the early 1980s compensation structures in organizations were more egalitarian since CEO pay had yet to accelerate to today's excess while there existed an implicit understanding that extended engagement with an organization was a reward for commitment and performance.[38] Neither egalitarian pay structures nor employer commitment to employment continuity hold today in most conventional employment situations.

What about other forms of work engagement? How do their risk/reward profiles compare to a nontraditional career with more than one income stream? When looking at a part-time position, contract/consulting work, or starting or buying a sizable business as standalone approaches rather than component parts of a nontraditional career, we find the nontraditional career with more than one income stream offering the most attractive risk/reward profile.[39] For example, starting or buying a sizable business offers the potential for substantial financial reward. However, financial and health risk are also high due to the likely size of the initial investment and the intense single-minded commitment required.[40]

Personal meaning. Personal meaning flows from each of the benefits we have reviewed, and, as we outlined in chapter 1, the form this takes depends on our values. This may mean for some people it's about building expertise in a particular area, for others it may be about leadership roles, autonomy, security, financial return, service, challenge, work/life integration, creativity, flexibility, or, more likely, some combination. A nontraditional career inherently aligns work with personal values, personality preferences, interests, and skills. It leads to making a contribution in a way that is personally fulfilling, and it offers the potential for significant financial reward at lower risk than in conventional employment. The rewards of a nontraditional career, both intrinsic in terms of fulfillment and extrinsic in terms of lifestyle choice, all contribute to work that has deep personal meaning.

Now let us take a look at the challenges of a nontraditional career and how to mitigate them. These challenges include the following:

- Breadth of skills needed
- Launch time
- Uncertainty
- Access to benefits
- Isolation

Breadth of skills needed. A nontraditional career is built on a broad range of skills that include not only content expertise in a particular discipline but also interpersonal skills to relate well with others, the ability to make our expertise known through sales and marketing, and organizational skills to build an infrastructure that includes, for example, forming a business entity, accounting, payroll, and information technology. We'll look at these skill areas in more detail in subsequent chapters, but for the time being let's just acknowledge the need for significant breadth of skills that we either develop ourselves or access through partnering. This contrasts with a conventional career, where specialization leads to the need for a more limited skill set focused on a given area of expertise likely also including interpersonal skills but probably not requiring the same skills in outreach as a nontraditional path or the same breadth of organizational skills and infrastructure building.

Acquiring this breadth of skills might seem like a daunting undertaking—though for those who prize variety, it may seem exhilarating—but two factors make this less daunting. First, today's information technology and communication tools make aspects of infrastructure building or marketing and sales readily accessible. These tools range from accounting software to track financial performance to outreach using social media or electronic distribution of promotional information. The second factor is the opportunity for partnering to supplement our capabilities in areas that are hard to develop or that we choose not to develop because we don't enjoy them. For example, payroll processing is readily outsourced, or perhaps we choose to partner in the marketing area with someone who specializes there.

Launch time. It takes time to start a nontraditional career, time to build needed expertise, time to build infrastructure, and time to secure customer engagements. Some aspects evolve on a medium- to long-term time horizon over a number of years, for example, building expertise throughout a career or engaging in extended educational activities. Others are short-term, such as creating a website or setting up an accounting system, which can take place in days or weeks. And some aspects may be a combination of both, such as long-term customer relationships that may have continued from conventional employment or customer relationships that rapidly develop after an outreach event such as a presentation. The financial consequences of an extended launch can be mitigated by gradually easing into a nontraditional path, as we'll explore later in the book, and by building financial reserves before launching.

Uncertainty. Some economists view uncertainty as applying to situations where we cannot know all the information needed to make accurate estimates.[41] For example, the profitability of a given business sector thirty years from now cannot be predicted due to the many unknown factors affecting it. Risk, on the other hand, in this view, is when the odds can be accurately defined, as, for example, in the probability of outcomes in a game of dice.[42] Some question the need for this distinction, suggesting that in practice, in the business world, we are usually dealing with significant complexity and therefore uncertainty.[43] Consistent with common usage, I use "uncertainty" and "risk" synonymously in this book. Let us consider two primary forms of uncertainty or risk associated with our work. The first is whether it will continue, and the second is what form it will take. We have acknowledged that with several career components making up a nontraditional career, while the uncertainty of continuing a particular component may be higher, like the uncertainty of being laid off from a conventional position, the risk of losing the overall portfolio is much lower than for any individual component. Regarding the form our work takes, our ability to select components to match our values, personality, interests, and skills means that changes can be driven by our preferences and linked to market needs rather than being subject to the whims of managers in an organization and the associated uncertainty.

Access to benefits. In the United States benefits typically linked to traditional employment in the past included vacation and sick time, provision for retirement, and health-care insurance. In a nontraditional career we can price our products or services to cover vacation and sick-time needs, aided by there not being a corporate bureaucracy to support. As we saw in the introduction, defined benefit pension plans are rapidly disappearing in the corporate world and being replaced by defined contribution plans. One-participant 401(k) plans offer similar tax advantages for sole practitioners in nontraditional careers to defined contribution plans. Historically, the United States was in the unusual position of tying health-care insurance to an employer. This made it potentially difficult to obtain insurance when in a nontraditional career due to the possibility of exclusion for preexisting conditions. Passage of the Affordable Care Act in 2010 and its implementation in 2014 removed this egregious preexisting condition practice of insurance companies and made health-care insurance readily accessible to those in nontraditional careers. Joining a community of people also in nontraditional employment, such as the Freelancers Union, is another potential avenue for accessing benefits and addressing employment rights.[44]

Isolation. One challenge of a nontraditional career is the isolation that can result from development and practice of individual expertise. This is mitigated by engagement with customers and clients, by partnering, and by connection with employees or partners who may be part of the team. It can also be mitigated by joining relevant professional organizations and participating in their local chapter meetings, conferences, and events. For example, in Northern California, the Northern California Human Resources Association (NCHRA) is an active proponent of the human resources profession, convening many local meetings. Similarly, the California Career Development Association (CCDA) provides forums to bring the career development community together.

Examples of Benefits and Challenges in Practice

To deepen our exploration of the benefits and challenges of a nontraditional career, we will now look at vignettes from seven people with successful nontraditional careers. Each person was kind enough to respond by e-mail to a series of questions about their nontraditional career experience, describing in their own words how their career has unfolded and the benefits and challenges that have arisen. Some people chose to be identified by their first name and others by their full name. Their career stages range from comparatively early, to mid-career, to the encore stage.

Alia. Alia is at a midlife stage. Her broad work background includes real estate leasing sales, entrepreneurship (starting small businesses during undergraduate work and in later life), administrative support, sales support, management and systems administration for an information technology start-up, instructional design and copywriting, marketing, resume writing, and mental health and career counseling. The aspects that connect Alia's career choices are psychology, writing, transformation, creativity, entrepreneurship, autonomy, novelty, and development of people and things. Alia also has successfully worked in multiple part-time jobs concurrently. She completed a master's degree in counseling psychology in 2004 then became licensed as a counselor. Building on this and reflecting the eclectic range of Alia's interests, recent components in her nontraditional career include career counseling delivered on a part-time basis in a health-care setting, mental health counseling in private practice and college settings, supervising other counselors, vintage/antique buying and selling, and estate sales.

Alia's concept of work has always been nontraditional, which caused stress when she was younger. The thought of doing one thing felt confining. Her nontraditional career evolved as she became excited about learning new skills and doing something new, and sometimes it evolved simply out of necessity. As she matured and felt less guilty about not following a traditional career trajectory, she realized there were common themes to the jobs she was drawn to, namely, transformation and development of ideas, people, and projects in venues that allowed personal autonomy and the ability to be her own boss.

Alia feels that the way she works has fulfilled her need for autonomy, the opportunity to use different skill sets and mindsets in a given week and experience continued novelty, as well as the ability to have more control over her life and time. Over time, she realized that she does her best work when she can exercise control over when she does her work, resulting in a better quality of work and life. However, working this way has required that Alia pay for her health-care insurance, and self-employment taxes on occasion, and retirement savings have been a challenge.

Anna Domek. Anna is married, in her early thirties, and the mother of two young children. Her educational background includes a bachelor's degree in psychology and history and a master's in counseling, with an emphasis in career counseling. She worked for five years as a career counselor at a state university and at that time also started teaching an online class in the evenings for a community college. After the birth of her first child, she left her full-time job and started a nontraditional career. This now includes teaching career development at a community college, teaching in a graduate career counseling program at a state university, and working from home in a part-time position on Internet-based and social media projects. Recently, Anna began work as an adjunct counselor at a community college, providing academic advising and career and personal counseling on an as-needed basis. When Anna first left her full-time position, she did not know if there would be sufficient work to provide needed income. It turns out that she was able to add career components and be selective about opportunities that arose. She was able to accept those opportunities in which she had a strong interest and that were a good fit in terms of time availability.

The benefits to this nontraditional path for Anna are many. It has allowed her to be home with her children the majority of the time while still providing professional fulfillment. She also feels a greater sense of control, adding and subtracting career components depending on her life situation. With multiple income streams, she is no longer tied to the whims of any one organization. Anna finds the regular changes in her work to be stimulating and is able to focus on the parts she enjoys, such as teaching and counseling, and bypass the less desirable parts, such as meetings, organizational politics, and administrative activities. Challenges of this path are benefits and security. Anna receives her health-care benefits through her husband's job, and each of her four positions could end at any time. This works well for Anna currently, with her husband employed in a traditional full-time job.

Beth Levin. Beth works in a job-share role as a part-time career counselor, providing services for employees within a large health-care system. Additionally, she provides values-based training and coaching based on neuroscience. She has a bachelor's degree in occupational therapy and a master's in counseling psychology, is certified in neurolinguistic programming, and is a board-certified coach. Her work includes developing and delivering career, leadership, and stress management coaching and training to corporations and nonprofit associations. Beth is in the maintenance phase of her career. She has established a strong competency in several areas and has built a local reputation for expertise in career development and training. As her husband begins to prepare for retirement, Beth is beginning to plan for reducing her work activities, which will include focusing on her favorite marketable skills while spending more time with her husband.

Her professional career began as an occupational therapist working in a nonprofit psychiatric system, and Beth quickly learned that occupational therapy patient care was not what she expected. The repetitive work and intense contact with seriously mentally ill patients was often stressful and dissatisfying. After considering various options such as sales, and going through a career assessment and exploration process, she discovered that her true areas of passion were training and career development. She returned to school to obtain a master's in counseling psychology. Beth explored a variety of roles while building her skills. Some roles were simultaneous as she began a nontraditional career path. These roles included volunteer leadership in the local chapter of the American Society for Training and Development, co-chair of an international career conference, a private practice career counselor, a training consultant, an associate outplacement specialist, and a part-time occupational therapist. She creatively utilized her mix of skills to build a bridge to this new career path. Her nontraditional career path had launched with her training consultant role at thirty hours a week. She was willing to take a risk and trade the security and benefits of a high-demand occupation, occupational therapy, for more personally satisfying work.

The benefits of choosing a nontraditional path for Beth include variety, autonomy, and stimulating and novel work. As an intensely curious person, she can translate her areas of new learning into practical assistance for her clients. Building on recent interests, she has developed webinars and training based on brain research that assist nurses in improving their stress management. Her nontraditional career also has drawbacks. She pays for the total cost of health-care insurance. She has a pension from previous employment which will be a component of her financial retirement plan, but she must plan well and be financially disciplined in order to meet future retirement goals. Working on a contract basis, she does not have the assurance that her skills will be in demand, but has contingency plans, such as teaching courses at local colleges and working as an occupational therapy consultant or educator. Although her nontraditional career carries risks, she is very satisfied with it, believes the advantages outweigh the risks, and is optimistic she can continue to develop fulfilling work.

Beverly Garcia. Beverly is beginning the second half of her career. She has been a professional in the workforce development sector for twenty-five years and expects to work for another twenty years. Beverly has a master's degree in human resource management and development. She recently left traditional full-time employment and, having established a nontraditional career for the past four years, can now claim success. She is the sole proprietor of a consulting business, serves as an independent contractor working for educational institutions, and is employed on a part-time basis providing career coaching in the health-care sector. She facilitates corporate training for private and public sector businesses and provides individual career coaching.

Beverly's nontraditional career unfolded without a business plan or marketing activity. Rather, it flourished by networking with colleagues and, perhaps, through some luck and good timing. In 2009 she connected with a colleague who offered to subcontract leadership development training as this colleague couldn't meet the business demands. While this was only two or three days of work a month, it was the catalyst for Beverly to open her consulting business, and within a few months additional projects were offered to her. It was at that point she resigned from her full-time corporate position. For the past three years, the majority of her assignments have been managing federally funded grant programs. However, welcoming change, for the past year Beverly has been working part time as an employee for a consultant to a major health-care provider. She also continues to work as an independent contractor on a variety of projects through her consulting business.

Although Beverly enjoyed working in corporate America, she appreciates the many advantages of her nontraditional career. She finds more autonomy, flexibility, and opportunities to be creative with a nontraditional career, and these are values high on her list for job satisfaction. There is greater variety of work and little chance of boredom or being limited by routine. Beverly has always thrived in a work environment where there are expectations of learning, growth, and development. Wearing various hats that feed her entrepreneurial spirit has been rewarding. Nevertheless, Beverly recognizes the challenges of a nontraditional career. Job security is a concern. Since income is variable, and there can be months between assignments, she has learned to be a careful financial planner. She also knows as a business owner that there are some less favorite administrative tasks she must attend to, such as bookkeeping. It takes self-discipline to accomplish these less preferred activities so she can focus on the enjoyable aspects of her work. Beverly expects to continue thriving in her nontraditional career. She enjoys the balance of seeing the results and accomplishments of her current work while simultaneously expanding her business. She finds her life in this career stage intriguing, as she continually wonders, "What else is possible? What new adventures are going to appear?"

Mary. Mary is forty-five years old and considers herself in a midlife stage. She started her career working as a human resources specialist for the Department of Treasury and discovered that she did not like the structure and details of her federal job. In 2000, she began working as a career coach, providing career support to federal employees in transition. She developed her own coaching business while working on various part-time federal contracts. She recently completed an independent consulting project with the Department of Veterans Affairs providing career coaching for a human resources program management team. Her recent activities also include part-time employment providing career coaching in the health-care sector and starting a career coaching practice again. The components to Mary's nontraditional career have included multiple income streams from working for a contract organization in the private sector, working as an independent consultant in the private sector, and owning her own private practice working with individual clients. Flexibility, creativity, and entrepreneurial skills such as risk taking, strategic thinking, and problem solving have all been necessary in order for her to work in this way.

Mary's part-time career started after she conducted a self-assessment study and determined that her work needed to involve helping people improve their lives and needed to include creativity, learning and growing, and autonomy, without the imposition of excessive details and deadlines. Mary completed her master's degree in counseling and development, which allowed her to begin working as a part-time career coach for the Department of Transportation's career center through a contract organization and then apply for other career coach positions as she gained experience. Mary's boss and mentor at the contract organization was extremely helpful in her career growth by providing her multiple opportunities to work at various federal agencies. Mary also gained contracts at other federal agencies through networking and referrals.

Developing a nontraditional career path has allowed Mary to create a customized career that fits well with her personality, strengths, and interests, such as being independent and creative rather than being more constrained in a traditional job by organizational structure and bureaucracy. She has valued the variety and personal growth and learning from different experiences associated with various part-time assignments and positions. Another benefit has been having more time and an improved work/life balance for other activities, such as volunteering, which has been very fulfilling. Mary had found that working in one place full time led to burnout because of lack of variety. The challenges of working in a nontraditional career path for Mary are associated with finances and benefits. Working full time can provide consistency, some financial security, and reduced health-care insurance costs. Part-time work has created an additional financial stress for Mary with sporadic employment and increased health-care insurance costs.

Michele McCarthy. Michele was born in Pittsburgh, Pennsylvania, and graduated from Duquesne University with a master's degree in counseling. She has delivered results coaching and career counseling for over thirty years and has pioneered five career centers in the Washington, D.C., area. Michele transitioned from a vocational adjustment therapist in a psychiatric hospital to a career counselor, initially delivering training for government agencies and most recently organizational career counseling in a health-care setting. Michele assists clients in reaching their potential and marketability through clarifying their aspirations and options, including pursuing advanced degrees and certifications. She is in a pre-retirement life stage and has several more working years before she formally retires.

Michele's nontraditional career has centered on working contract by contract since 1993, combined with her private career counseling practice. In addition to her counseling and training, in 2003 she took a certificate in expressive arts therapy. This was a multimodal program using visual arts, therapeutic writing, and sound healing. Upon ending the certificate program, she pioneered the first expressive arts program at the Montgomery County detention center working with youthful offenders to build self-esteem and make better life choices. As a hobby, she teaches intuitive art and collage and in general likes to explore new and novel directions.

Michele says the benefits of this career path are the variety and freedom it provides. This includes volunteering; for example, Michele teaches art for healing classes at a cancer survivor organization. Michele's motto is "If you feel free, you are free." She has always paid her own health-care insurance so that if she needs to change contracts, her coverage remains intact. She contributes to an IRA. The challenges of this path are the financial responsibility of contributing to a retirement fund and health-care insurance as well as practicing first-class networking when contracts end.

Wynn Montgomery. Wynn is seventy-three years old, retired, and a self-described workaholic. His conventional work life consisted of four phases: management in the automotive industry, management consulting, public sector management of an employment and training program, and management of a training institute associated with a state university in Georgia, the latter building on a master's degree in public and urban affairs. When Wynn was sixty-four, he retired from conventional employment and created a limited liability company to provide consulting services to a variety of clients. At this time he also began volunteering as a tour guide at the Oakland Cemetery in Atlanta, began writing memoirs for his grandchildren, and became active in the Southern Order of Storytellers. When he was seventy-one, Wynn and his wife moved to Colorado and Wynn closed down his consulting practice while continuing his ongoing efforts to polish his storytelling skills.

From a traditional career beginning as a college graduate who took a job in hopes of figuring out what he wanted to be when he grew up, Wynn moved to management consulting, which he hoped would give him further insights into career options. He was introduced to the employment and training field in which he soon became immersed, first as a consultant, then as a practitioner, then as leader of a training institute designed to enhance the knowledge and improve the performance of other employment and training practitioners, and eventually as a self-employed consultant working primarily with those same practitioners. During the latter half of his work career, Wynn began to explore volunteer activities that fed the teaching and telling genes that people had seen in him since childhood. First, he coached youth baseball and wrote instruction manuals, year-end letters, and season histories. When he discovered professional storytelling, he became a regular listener at local, regional, and national storytelling events. He also developed classes on using storytelling as a leadership tool. Wynn also developed courses on resume writing and career portfolios. As a volunteer guide at historic Oakland Cemetery, he began to hone his telling skills, which gave him the impetus to start sharing stories at gatherings of local storytellers. Upon arriving in Colorado, Wynn discovered Spellbinders, an organization that trains volunteers to tell stories in local schools and retirement centers. He has blossomed in that role and now tells regularly to second and third graders and senior citizens and is a member of two groups of professional tellers. He also facilitates classes in memoir writing.

In many ways, Wynn's employment history is conventional. He continuously worked full time from college graduation to retirement to self-employment, but he augmented that conventional approach with other activities that he was able to blend into his professional life, hence enriching his work and pointing him toward personally (and perhaps eventually financially) rewarding retirement activities. The challenge of Wynn's approach is that the activities which he was paid to do always had to take precedence over those he did for fun. Wynn observes that those who learn while young how to earn an adequate income doing what they love are truly fortunate. He perceives that, unfortunately, too many like him do not determine what they want to be when they grow up until late in their careers—or afterward. Wynn will tell you, however, that making that discovery late in life is better than never doing so.

We see common themes of personal fulfillment, making a contribution, variety, time flexibility, novelty, and love of learning woven through these vignettes. We see career components and time commitments tailored to match personal preferences, for example, allowing for volunteer engagement, and we see serendipity playing a part. Health-care reform addresses one of the challenges that surfaces, namely, securing health-care insurance. Addressing retirement funding is also a common theme. Perhaps more than anything else, we see the beauty and exuberance of the human spirit over and over again.

Finding a Fit: Connecting a Nontraditional Path to Who We Are

I remember taking an assessment a number of years ago that was supposed to show entrepreneurial attributes. Some parts, like managerial traits, were well aligned, but that was not the case when it came to personality. My more introverted, intuitive, feeling, and unstructured approach to the world in the Myers-Briggs framework (discussed in chapter 1) did not fit the mold of this particular entrepreneurial assessment. In fact, the entrepreneurial assessment would have consigned me to a non-entrepreneurial group, similar to the way I was consigned to a nonsinging group when young, in spite of having a love of music. While the nonsinging group may have saved many an ear and so been a public service, the entrepreneurial assessment would not have served me or our community well because my nontraditional, entrepreneurial path proved to be fulfilling, successful, and a blessing. Let me suggest viewing input you receive through your own lens of discernment. Here's Lauren Hargrave, whom we first met in chapter 1: "You will always have people who tell you, 'You can,' and people who tell you, 'You can't.' My advice: Keep the 'cans' on speed dial. . . . I picked up the advice I needed, left the rest on the table, and moved on."[45]

Nontraditional careers beckon to us from many directions and welcome people with wide-ranging backgrounds. For some of us, early educational experiences were challenging and discouraging. A nontraditional career path might offer either the opportunity to reengage with education as a foundation or the opportunity to pursue practical endeavors that build on different knowledge and skills, such as those needed for the practice of mechanical ingenuity or craftsmanship. For others of us, educational experiences may have been rewarding and so laid the groundwork for building expertise that can be a cornerstone of a nontraditional path in, for example, human resource, medical, legal, or financial areas. The benefits of a nontraditional career, which include autonomy, alignment with interests, contribution, financial attractiveness and improved risk/reward profile, and personal meaning, apply regardless of the context of the nontraditional components. They are there and accessible to all of us. Indeed, the beauty of a nontraditional career is melding it with our values, personality preferences, interests, skills, and life experiences to create something unique and special for ourselves, for those around us, and for our communities.

We have seen how to mitigate the challenges of a nontraditional career, which include breadth of skills needed, launch time, uncertainty, access to benefits, and isolation. The nature of the challenges and the mitigating approaches we use will depend on specifics of the nontraditional path we choose and who we are. For example, someone more introverted pursing a consulting path using specialized expertise may choose to develop and sustain in-depth relationships with a select number of clients and participate with colleagues in professional organizations to address isolation. Conversely, someone more extraverted creating and selling products may participate in public gatherings where there is an opportunity to connect with many people and promote product offerings. An intimate interweaving of personal characteristics with external needs and opportunities creates a nontraditional path that is a reflection of who we are. In the words of Frederick Buechner, "Vocation is the place where our deep gladness meets the world's deep need."[46]

So far we have explored what is meant by a nontraditional career, we have looked at examples, and we have examined how our individual characteristics can help guide what form this might take and components we might choose to include. We have examined the benefits and challenges of a nontraditional path and how that path connects to who we are. Now we will continue to lay the groundwork for moving forward. This means first addressing a number of strategic factors related to a nontraditional career path. These strategic factors are as follows:

- Whether or not to connect the components
- Finding differentiation
- Balancing the components
- Pace of entry
- Building needed skills
- Partnering

The first three factors look at how to structure a nontraditional career—connecting the components, finding the characteristics needed to become well established, and balancing components with different launch trajectories. The next three factors speak to implementation—pacing the entry, building the skills needed on a personal level, and partnering and the opportunities it presents. In the next section of our book, we will explore these six strategic factors, which influence the nature, evolution, and implementation of a nontraditional career, devoting a chapter to each of them.

PART II

Strategic Factors to Consider

Believe those who are seeking the truth. Doubt those who find it.

André Gide

CHAPTER 3

Whether or Not to Connect
the Components

RACHEL NELKEN, MENTIONED in chapter 1, created a nontraditional career with connected components: working as a relationship manager for an arts council, leading a youth music program, working freelance for music-funding organizations, and running action learning sessions. She comments, "I believe that each area of my work informs the other. Working with funders gives me a sense of the bigger picture, the national and regional context of my work and examples of best practice to learn from. Working on the ground gives me real-life, credible experience to draw on when talking to clients or commenting on policy issues. Running action learning sessions develops my leadership skills."[1] We also learned about Peter Bates, who switched from being an investment banker to a nontraditional career with disconnected components: a cabinetmaker, rare-breed pig farmer, and business adviser. He observes, "None of the jobs I do relate to each other. I have no background in farming. It would be difficult to find another person with such an odd range of jobs. . . . I enjoy greater security since I haven't placed all my eggs in the proverbial basket."[2]

These completely different approaches to a nontraditional career, both of which work for the people concerned, introduce a key strategic factor to consider, namely, that of deciding whether to build a series of strongly related components, for example, consulting and teaching, or whether to pursue components that have little in common, for example, administrative work and photography. Building related components makes possible the transfer of content learning from one component to another; for example, from organizational development to individual career counseling. It also provides infrastructure economies and marketing clarity and allows for a focus on areas of personal interest. On the other hand it may limit the range of options that can be included and it may increase risk. In this chapter we will examine the pros and cons of connecting nontraditional career components so you can decide which approach will work well for you. We will begin by examining what connection means, continue by exploring the benefits and drawbacks of connecting career components, and conclude with some questions for you to consider in reflecting on your situation.

Examining and Defining Connection

First we need to look in more depth at what is meant by "connection." Broad definitions of the word refer to aspects such as association, relationship, bond, and affiliation, and they also refer to the link between cause and effect. So an exploration of connection necessarily includes both what this might say about the nature of connection among different components of a nontraditional career and what it might say about outcomes from that connection. There are also degrees of connection to consider, ranging from tightly intertwined and permanent to tenuous and transitory. Are there parallels in other areas? Some that come to mind are an organization's portfolio of business units viewed from multiple dimensions, such as business performance (long a staple of business consultancy) or workforce characteristics, and mapping a research and development project portfolio.[3] In chapter 5, when we look at balancing the components of a nontraditional career, we will explore an approach related to research and development portfolio mapping. One important thing we can learn from these other areas, business and research and development, is that there is value in looking not just at the individual components but also at how the components in a portfolio connect with each other and what this means for decisions and resources. Our focus in this chapter is on the nature and form of this connection for a nontraditional career.

We'll start our exploration of component connection in a nontraditional career by looking at different forms of connection, recognizing that these are complementary rather than mutually exclusive:

- Content expertise
- Core functional skills
- Served customer base
- Geographic proximity
- Mutual influence

Content expertise. Education and work experience give us the content knowledge needed to practice and progress in a particular field. For me this includes career counseling and career development. Other examples we have seen or that come to mind include finance, engineering, music, carpentry, biotechnology, and medicine. If we delved deeper into each area we would find that each discipline has specialized subsets all with their own domain knowledge. Content expertise is usually described with nouns.[4] It can form a powerful connecting thread for a nontraditional career. For me this thread connects career development with individuals to workforce development with organizations to teaching and writing. In chapters 1 and 2, we saw examples of others where content expertise is an organizing principle for a nontraditional career.

Depending on the pace of change in a given field, adopting this connecting principle may favor either those with extended work experience or those new in the field. For example, in the area of finance or accounting, where core principles last a long time, work experience is a benefit in providing exposure to the many nuances that will likely unfold in practice. However, in high technology, with a knowledge half-life in months rather than years, exposure to the latest thinking is a significant benefit and knowledge of obsolete former technology of lesser value. Here past work experience may count for less.

Core functional skills. Some skills readily transfer from one place to another, one sector to another, one nontraditional work component to another. They can result from ability and aptitude, from training and development, or a combination of each of these. These skills are sometimes called "functional" or "transferable" skills and they are usually expressed in verbs, for example, "organize," "promote," "analyze," and "write."[5] Such skills can form a connecting principle for a nontraditional career. Writing, for example, can take various forms: technical writing, creative writing, copywriting, journalism. In this case the connecting principle, the focus of a nontraditional career, would be the skill associated with the process of writing rather than the accumulated knowledge contained in literature. Organizing similarly could be a focal point, for example, in offering a service to people that addresses their need to bring structure to possessions or to a living space or in offering a service to businesses to structure an office facility so it is inviting, comfortable, and efficient.

Served customer base. One challenge with a nontraditional career can be reaching customers so they know about the range of products or services you offer that would meet their needs. One means of addressing this challenge is to build deep customer relationships, spending time to understand evolving customer needs and communicating contributions you could make to address these needs. It is likely that your potential customers are not aware of the range of your capabilities; instead, they may view you through the lens of the limited products or services they initially receive from you. On several occasions I have begun a customer engagement with a given service, for example, exit interviews, leadership development, or a particular workshop, and then through conversations over time learned of additional needs that I could effectively address. This led to the extension of services to include these other activities, which frequently were greater in extent and duration than the initial engagement. So a connecting principle for a nontraditional career that complements the earlier two aspects is that of served customer base, where the emphasis is on extending the customer relationship for mutual benefit.

Geographic proximity. It might seem in this interconnected world that geographic proximity is irrelevant. Indeed, for some disciplines, such as the deployment of global technology solutions or Internet-based sales, it may not be significant. However, for a nontraditional career, which is built on relationships, geographic proximity can be a significant factor. Since in many situations opportunities come from referrals, or at least physical presence involving contact with potential customers, geographic proximity is important. For me, it led to referrals from one customer to another nearby, and to the ability to reach potential customers through presentations. Geographic proximity also affords the opportunity to build knowledge and understanding of a given local culture and adjust offerings accordingly. While my roots are in northern England, and I was brought up assimilating the culture of that region as part of who I am, I did not work there other than during school vacations, as I left the region to go to university. I would find it difficult to provide services there without experience of that business world. However, in northern California, where I have lived and worked for many years, I understand less of the social culture but have assimilated and now understand more of the business culture. This is also true of other regions in the United States where I have lived and worked. Geographic proximity can be an effective connecting principle for a nontraditional career, complementing the other connecting dimensions we have described.

Mutual influence. Mutual influence means that outcomes from one component of a nontraditional career directly influence another component. Perhaps an apt analogy here is that of vertical integration in a business enterprise. An example is an oil company possessing the exploration and production capability to find and distribute crude oil, the refining capability to produce and sell fuels for retail consumers, and the chemical production capability to take raw materials from crude oil processing and convert them to chemicals and items such as fibers, fabrics, and formed products. Integration brings the ability to generate demand for materials earlier in the chain based on those that are sold downstream. It can create challenges where customers of intermediate products become competitors and business challenges if there are depressed margins at certain points in the integration chain.

How might this look in a nontraditional career? An example would be the development of a career assessment instrument that could be delivered to individuals and provided to organizations as well as promoted through writing. Sale of the instrument could be one component of a nontraditional career, complementing other components that involve career work with individuals and organizations. A challenge in this case would be building sufficient expertise in development of the assessment so it offers advantages when compared with assessments created by specialist providers. The benefit to the nontraditional career portfolio would be the ability to incorporate learning from work with individuals and organizations into development of the assessment. For mutual influence to be effective, there needs to be content connection among the components and significant longevity of activity. A potential downside is that a change in one component, for example, ending work with individuals or organizations, would affect demand for the linked component, in this case the assessment.

Benefits and Drawbacks of Connecting Career Components

Having examined what connection means and the possible forms it can take, we will now look in more depth at the pros and cons of connecting nontraditional career components. To illustrate this we will consider two approaches, that of independence, where there is little or no connection, and interdependence, where components are closely linked.

Independence. The earlier example of a nontraditional career that includes cabinet making, pig farming, and business advising illustrates independence, where the components of a nontraditional career are unrelated. The only connection is personal interest in each component. Such an approach requires building content knowledge, expertise, credibility, and customers in completely different areas. As a result, the start-up and sustaining costs will increase in terms of both time and resource commitment. There may be potential infrastructure economies, for example, accounting systems, payroll processing, and insurance. One benefit of this approach is that exiting one component has little or no effect on the other components, so there is a reduction in risk. Likewise, component longevity is independent, as one or more components could be long term while others could be short term.

Interdependence. Interdependence implies significant linkage of the different components. As we have seen, this can occur through content expertise, core functional skills, served customer base, geographic proximity, or mutual influence among components. For example, in my case, with respect to content, knowledge developed initially and subsequent learning from delivering workforce consulting for organizations has been relevant and helpful when working with individual clients on their career development. Likewise, the foundation of knowledge needed to work with individual clients and subsequent perspectives from clients about their careers has informed organizational work, particularly career development initiatives at an organizational level, and leadership development in organizations. Foundational knowledge and subsequent learning from work with individuals and with organizations inform writing and teaching, where study in turn contributes back to the individual and organizational work. Activities in all of these areas benefit volunteer work when it builds on such content knowledge. Furthermore, skills developed in one component, for example, listening with individuals or public speaking to groups, are transferable to other components. Organizational skills associated with starting a practice are generally applicable to various components.

What about the rationale, benefits, and costs of the interdependent approach? Perhaps an apt analogy is that of a team in an organization. An effective team requires investment of time and resources, and there is a learning stage during its early development. The nature of activities the team is tackling needs to justify investing time and resources in team processes. For example, for a short-term task requiring minimal interaction, a limited tenure group rather than a fully integrated team may serve the purpose well. On the other hand, it is important to invest in building a strong team foundation when there is an extended engagement requiring close cooperation of multiple people. Similarly, with a nontraditional career where learning transfer among components will occur over an extended time, where value contribution from this transfer is significant, and where this matches personal preferences, interests, and passion, it is worth taking time to be sure knowledge and skill transfer happen among components. It is worth nurturing their interdependence. This is a small investment of time, mainly in organizing and storing information, as we carry this knowledge and experience with us. It is consistent with the nurturing of interdependent career relationships that we addressed in chapter 2.

Skills we develop in interdependent, nontraditional career components readily transfer. There may be a small investment in infrastructure in this case; for example, our delivery of assessments both for individuals and in an organizational setting required separate access portals. Where interdependence is a foundational aspect of a nontraditional career, the components are strongly connected, learning and knowledge are directly transferable among components, and this transfer occurs over a relatively long time period, in our case for more than ten years. We will see that interdependence brings significant potential benefits to the nontraditional career portfolio. The different foundations for interdependence are illustrated in figure 3.1. The center circle references interdependence as a core principle, and the five surrounding circles refer to the forms of connection we reviewed earlier.

Nontraditional Career Component Connections

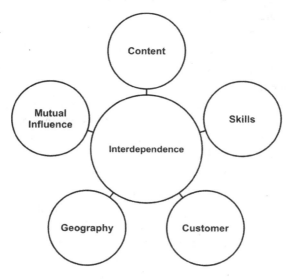

Figure 3.1

Now we will look in more depth at the benefits and drawbacks of the two different approaches to selecting components for a nontraditional career, examining both independence and interdependence. We will discuss the implications for development and sustainability from the following perspectives:

- Launch
- Scope
- Risk
- Marketing clarity
- Infrastructure economies
- Value creation for customers
- Intrinsic and material rewards

Launch. Let us consider how our two alternative approaches to career component connection affect launching a nontraditional career, both the rate at which it unfolds and the flexibility to adjust to changing early circumstances as options and challenges surface. Where components are independent, launch can occur on a different schedule for each component. Building expertise and credibility in disconnected areas will likely slow the rate of implementation, due to the need to engage with distinctly different customer and sector profiles, so that a phased approach may be preferable, constrained only by the need to generate adequate income. Tension may develop between the need to build experience in more than one area and the need to secure income. Flexing to adjust to changing circumstances can occur separately with each component, constrained only by the availability of time and financial resources if multiple needs surface simultaneously in different component areas.

Where components are interdependent, the launch process is affected by the form of connection. With content or skills as primary drivers of connection, launch timing of different components will be interconnected and determined by the time needed to build a foundational level of expertise. As core expertise builds, this enables the launch of multiple components, which with increasing experience and credibility in turn leads to surfacing additional opportunities. Where a served customer base or geographic proximity is the primary driver of connection, a launch of different components is more likely to occur sequentially rather than simultaneously. Success with one component at a given customer builds credibility for delivery of other components to that customer, or success with one customer in a geographic region builds credibility for working with others in the same region. Combining two connecting principles together, for example, content expertise and served customer base, can accelerate implementation by broadening the range of accessible opportunities. Flexing to address changing circumstances can focus on the component that is changing. For example, the extension of a teaching engagement does not influence workforce consulting, even though they may both be based on the same core content knowledge. When components are connected by mutual influence, with one supplying a product or service to another, the launch trajectory of the receiving component will influence that of the supplying component, as mentioned in the earlier career assessment example.

Scope. The scope of independent components that can be included in a nontraditional career is limited only by time available to build sufficient knowledge and capability in each area. In principle this means a broad array of components can be included. In practice there are significant limits to the breadth of diversification that is possible. It is similar to, though greater than, the challenge faced when an organization attempts to diversify in many directions, as time and financial resources are more limited to us as individuals. Indeed corporate diversification into unrelated areas is fraught with difficulties.[6] Likewise forays into unrelated components in a nontraditional career need considerable forethought and preparation.

Scope for interdependent components is necessarily limited to elements that are related to the core connecting aspect. So if connection is based on content expertise or skills, scope is configured around these aspects. The definition of scope can be broadened or narrowed based on personal choice. This can apply at the individual component level or at the portfolio level. For example, in working with individuals in career development, I intentionally welcomed clients from a broad range of sectors, at many life stages, and with varied educational backgrounds. It is helpful to bring breadth of work life experience to this range of client situations. I could have limited my served client population to only certain sectors or life stages; I chose not to do so. At the portfolio level, scope could be narrowly defined to include only career development activities with organizations or individuals or more broadly defined to include workforce development consulting with organizations, and career development and coaching with individuals and in organizations. I chose to adopt the broader definition. Customer relationships or geography can also play a role in establishing scope based on, for example, the range of needs of specific customers or the range of customers accessible in a geographic region. While the interdependent approach may at first appear to be scope limiting, in practice it can offer much breadth and flexibility for scope engagement according to personal preferences in defining breadth of scope. Conversely, the independent approach may at first appear to offer great scope flexibility, but practical considerations can place significant limits on the breadth of scope that is possible.

Risk. Let's consider risk from two perspectives: the risk of not being able to launch a nontraditional career successfully and the risk of not being able to sustain it once launched. When the components are independent, the risk of not being able to launch successfully is separate for each component and difficulty in one area will have little or no bearing on another area, unless there are financial considerations that come into play. On the other hand, when the components are interdependent, launch difficulties with one component may affect others. For example, in my case success in personally delivering workforce consulting was a necessary precursor to starting an organizational consulting practice that involved others as contributors. Consequently, from this perspective the launch risk is lower when components are independent. Conversely, success with one interdependent component may spawn success in other components, whereas when they are independent, success in one area will have little effect elsewhere. So the potential launch upside is greater when the components are connected and the potential downside of launch difficulties is also greater. Greater potential variability means greater risk. Offsetting this is the depth of knowledge coming from focus on a common connecting principle that is intrinsic to the interdependent approach.

Similar factors surface when considering sustainability of a nontraditional career. When components are independent, the performance and continuity of each is separate. Having distinct components provides diversification, which reduces variability. Furthermore, it may be easier to find components that behave countercyclically with changes in the economic situation and, as a result, lower ongoing variability and risk. "Countercyclicality" means that a rise in the performance of one component is accompanied by a fall in another, and vice versa. On the other hand, with the interdependent approach, external economic factors are likely to affect each of the components similarly, magnifying variability with changing external economic circumstances. So we anticipate greater variability, and therefore greater risk, with the interdependent approach.

Marketing clarity. Key aspects of marketing include understanding the needs of potential customers, how these customers might form natural groups with similar attributes, how product or service offerings can address customer needs in terms of what they consist of and how they are delivered, and how to reach and communicate with potential customers. The marketing process is iterative, moving from building understanding, to delivering products or services, to assessing results, and then continuing to refine the process. It benefits from focus in defining customer needs and communicating product or service attributes with respect to content and communication channels.

With this in mind, independent components in a nontraditional career each require a separate marketing strategy, addressing particular customer groups with associated product or service offerings, and access to the necessary communication channels for those customer groups. The culture and manner of interacting can vary greatly, for example, from the rapid, electronic communication and extensive use of social media typical of high technology to the paced, face-to-face contact typical of traditional sectors, such as much of the financial world and the public sector. Indeed, cultural differences among sectors are not unlike cultural differences among countries, where certain patterns of behavior and interaction have become ingrained. This presents a particular challenge for the independent approach to a nontraditional career, requiring as it does simultaneous engagement with disparate cultural communities. This is not easy. I recall, when first developing knowledge of the career counseling field, the challenge of embracing the interpersonal skills necessary for effective career counseling while simultaneously working in a competitive business environment operating with a different set of cultural norms. The challenge of dealing with sector differences suggests that partnering, which we address in chapter 8, to bring knowledge of different customer groups may be particularly important in developing a nontraditional career path based on independent components. It may, in fact, be a practical requirement for marketing of independent components to partner with others to effectively reach customers.

The interdependent career path, on the other hand, is naturally tailored to building a common understanding of customer needs and development of associated products or services, as it is centered on a core practice area and capability. Connection of products or services by content similarity or common skills simplifies communication of attributes and benefits. A common core message can be created that has broad applicability to customers; in fact, it is strengthened based on understandings gathered from various related career component areas. The cultural attributes of customers in different but related component areas are likely to be similar, which simplifies the communication process and the ability to achieve cultural alignment with customers. Communication channels will also likely be similar and may involve common professional organizations. For example, in the world of human resources, national and affiliated regional societies provide a forum for many in the profession regardless of the business sector in which they practice. When interdependence is built on a served customer base, then marketing can be tailored to specific customer attributes. Likewise, geographic proximity offers the opportunity to build on shared local cultural values. So we see that it is easier to achieve marketing clarity with an interdependent approach to a nontraditional career than with an independent approach.

Infrastructure economies. As a nontraditional career develops, the supporting infrastructure needs to grow with it. Infrastructure needs are basic in the early stages, when it is mainly about the personal contribution needed to deliver services or create products. Outreach is supported by a website, presence on social media sites like LinkedIn, and electronic and possibly hard copies of promotional material. Infrastructure support for generic business functions includes basic accounting software and securing the needed insurance, such as professional and general liability. As the scope of practice broadens over time to include, for example, engaging employees and possibly a changing business structure, infrastructure needs increase to include more sophisticated accounting software, payroll functions, more complex tax reporting processes, and possibly component-specific items such as particular software or physical facilities. We will learn more about this in chapter 9. When we look at infrastructure economies for generic business functions such as accounting or information technology platforms, there will likely be little difference regardless of how, or if, nontraditional career components are linked. In each case, when we add components we create infrastructure economies; for example, core accounting software doesn't change as additional career components access it. However, when infrastructure is tied to content-specific aspects of the career components, such as with communication materials, then infrastructure economies will be greater for the interdependent approach. Core content and materials can be used, or readily modified for use, by different though related career components. As a result, the interdependent approach to a nontraditional career offers greater potential infrastructure economies.

Value creation for customers. Value creation for customers comes from providing a service or product that is outside the domain expertise of the customer or that would be more costly for the customer to create internally than purchase externally. It can be helpful to distinguish between individual purchasers who also directly use the services or products and organizations where the purchasing unit may be separate from the service or product recipient. In the former case, the purchase decision is likely more rapid as the individual is the decision maker, though the scale is likely smaller and longevity of the relationship may be shorter. In the latter case, the purchase decision will likely be longer and involve multiple people, and the business relationship may have significant longevity. In both cases the purchase decision may include an assessment of whether to do this in-house as an individual or an organization or whether to purchase the product or service externally. Examples for both individuals and organizations might be accounting, catering, cleaning services, or career development support.

Let us consider the relative value contribution of our two nontraditional career approaches, independent and interdependent career components. When components are independent, it is necessary to create sufficient capability or domain expertise in each separate area so that the value contributed exceeds the cost to a customer of generating this capability internally. If this is a sporadic purchase, outside of a customer's core capability, then an independent component can add value based on initial know-how and augmenting it with learning from repeated product or service delivery to others. There may also be favorable infrastructure economies, for example, in access to facilities or meeting regulatory requirements for a catering service or in acquiring and staying current with know-how in an accounting area. However, value contribution must be demonstrated and maintained separately for each independent career component. This can be a challenge given that time is a key resource in a nontraditional career.

When components are interdependent, on the other hand, expertise created in one area informs other areas. There is a natural synergy, similar to the synergy sought in a merger or in an acquisition. Expertise gained from multiple career components can offer more value to a customer than would be possible from each component separately. For example, this might be learning and know-how transferred from in-depth study in one career component to another, or it might be infrastructure economies resulting from the acquisition of equipment or information technology capability in one component that can be used by other components. The potential for value contribution is substantial as expertise from the entire set of nontraditional career components can now be applied to a given customer in any single component. Having connected components may also help surface additional project opportunities with a given customer. Overall, the interdependent approach to a nontraditional career offers significant advantages in customer value contribution.

Intrinsic and material rewards. How do intrinsic and material rewards compare for the two approaches? Here we can refer back to the values that we discussed in chapter 1 and our primary motivations and purpose for engaging in a nontraditional career. We identified several possible purposes for taking a nontraditional path: personal fulfillment, financial reward, mitigating risk of unemployment, giving back, and creating a legacy. Both the independent and interdependent approaches can address each of these foundations of purpose. We can tailor either approach to a nontraditional career to address our particular needs according to our personal preferences.

As a means of integrating our thinking, let us now summarize perspectives about the benefits and drawbacks we have examined relative to the independent and interdependent approaches. Table 3.1 summarizes these perspectives.

The interdependent approach offers benefits in four of the seven aspects summarized in the table: launch, marketing clarity, infrastructure economies, and value creation for customers. The independent approach offers a benefit in one area, risk (with lower variability and therefore lower risk), with a potential benefit in one other area, scope, though this may not be accessible for practical reasons. Either approach can address the final category, that of intrinsic and material rewards, as this is tailored by each of us to meet personal preferences. In summary, when considering all of the various aspects together, we see that the interdependent approach is favored.

Table 3.1. Summary of the Benefits and Drawbacks of Nontraditional Career Component Approaches.

Aspect	Independent Components	Interdependent Components
Launch	Slowed by need to separately establish expertise for each component	Can be rapid once core expertise is built
Scope	Theoretically unlimited, practically limited by time and financial resources	Limited to components with connection to core theme, though this can be broad
Risk	Diversification lowers variability	Higher upside potential, though greater variability
Marketing clarity	Particularly challenging with unrelated components	Flows naturally from a core connecting theme
Infrastructure economies	Economies in generic business infrastructure	Economies in generic business infrastructure and in content-specific areas
Value creation for customers	Must create separately for each component	Significant benefit from component synergies
Intrinsic and material rewards	Personally defined	Personally defined

Questions to Consider When Determining Whether or Not to Connect the Components

Having examined the benefits and drawbacks of connecting components in a nontraditional career, and various forms of connection, here are some questions to reflect on when considering an approach that will work well for you:

1. What are the pros and cons for you in connecting components in your nontraditional career path?

2. If you decide to build links among the components and seek interdependence, what connecting principles might apply: content expertise, core functional skills, served customer base, geographic proximity, or mutual influence?

3. How can these connecting links be built?

4. If you decide to use an independent approach, what partnering may be appropriate?

CHAPTER 4

Finding Differentiation

IT SEEMED LIKE a daunting prospect. Respond to a request for a proposal to provide career services at multiple locations around the country for a large organization with many thousands of employees and the resources to purchase such services from any provider, including sophisticated, multinational corporations. How could we, small and emerging, possibly be considered for such a contract, given the likely competition? As it turned out, after a competitive evaluation process, we were awarded the contract. I am so thankful for this, and for being able to deliver services for many years, buoyed as a team by the knowledge that we were helping people make better lives each day. How did this fledgling operation of ours take on giants successfully? Why were we selected? We learned afterward that our proposal better met our customer's service needs and was more cost effective. We were able to differentiate our offering. As part of our exploration of the importance of differentiation, in this chapter we will explore why and how this was possible and what we have learned from this and other situations that can be helpful to you.

So far in our book we have considered how a nontraditional career relates to our aspirations, how it can be meaningful, and how it might be structured. We have looked at the interaction with customers and how different approaches to a nontraditional career might influence this. In this chapter we will introduce additional aspects into our exploration of a nontraditional career, namely, competition and where differentiation fits in. While we will acknowledge the classical view of competitive strategy as the interplay between competing organizations, potential new entrants, customers, suppliers, and substitute products or services, we will look more broadly at the issue.[1] General understanding has evolved from considering low cost and differentiation as two generic, mutually exclusive strategies to the value of simultaneously executing more than one approach.[2] Furthermore, instead of looking through the classical lens of scarcity, we will look through a lens of abundance, just as interdependence in career relationships comes from such a place.[3] This means that we will start by considering how a nontraditional career intersects with the broader community, the relevance of differentiation, what happens when changes occur in that broader community, and how our individual success is interwoven with the common good. From this foundation we will then examine how we might experience competition, the different forms it might take, and our options and choices to anticipate and respond. We will explore various sources of differentiation as a key response and how to protect and sustain them. We will conclude with some questions for you to consider in reflecting on your situation.

Differentiation and Community

At first sight it might seem that differentiation and community are antithetical. After all, isn't differentiation, by which we mean specialization and distinction, the polar opposite of community with its implicit sharing of common characteristics? Indeed, shared common characteristics are one important aspect of community. We can choose to define such characteristics narrowly, as in, for example, a shared geographic location or a shared viewpoint. However, sometimes such a narrow definition can lead to conflict and confusion with those who are excluded. Alternatively, we can think expansively about shared attributes of community, for example, our common humanity. This leads to aspects such as fellowship and living and interacting in ways that benefit each other. Then differentiation, the oneness we bring into community, becomes our contribution. And that oneness can be described in different ways. It can be described in spiritual terms as in our faith traditions, it can be described in how others experience us, and it can speak to our contribution through our work. It is here, in our contribution through our work in a nontraditional career, that we will explore the concept of differentiation.

How can we frame the broader community contribution we make through a nontraditional career? One parallel is with that of the organizational world and the breadth of organizational responsibilities. The social responsibility of organizations can be expressed in relation to constituencies or stakeholders, which include employees, senior leaders, investors, community members, suppliers, customers, and partners.[4] Similarly, we can think of constituencies or stakeholders in our nontraditional career path, as discussed in chapter 2. Let me expand on what this might mean. Figure 4.1 illustrates our personal constituencies, those who are important to us and are affected by our progress.

Four groups are shown in figure 4.1, family and people who are close to us; a broad work and social community that might include, for example, suppliers as part of the work aspect; team members, including employees and people we contract with, partners, and investors; and customers and clients. Our specialized, distinctive, and differentiated work capabilities contribute to each of these groups in different ways. Those close to us benefit from the fruits of our work, the broader work and social community benefits from our learning and experience, team members, partners, and investors find their own paths enriched by our work, and we meet customer and client needs as a result of our work.

Each constituency interacts with us in a two-way relationship: change in our career circumstances affects each constituency and vice versa. For example, our adding expertise in service delivery, such as coaching capability or in construction know-how, offers the prospect to a customer of a broader array of services. Changes in the external environment, such as licensing requirements for professional practice or construction codes for home improvement, directly affect our services or products. Similarly, change prompted by a customer, such as their moving a primary location, affects the

Personal Constituencies

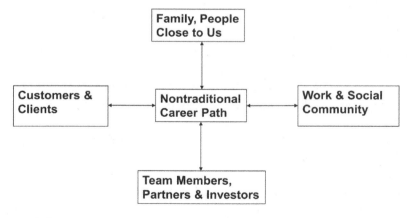

Figure 4.1

nature of our service or product delivery. Differentiation provides us with the wherewithal and a strong foundation to adjust to such changing circumstances—for example, having the requisite experience and background to meet evolving regulatory requirements or having the capability to serve new customers should an existing customer move or change direction. Differentiation also provides the basis for addressing competition.

Competition

Success in a nontraditional career is based on creating a differentiated approach with specialized, distinctive products or services. This is central to an effective initial career launch and for ongoing sustainability. It is a major reason why a customer will make a purchase and it is the basis for a viable financial position. Moreover, sustainable differentiation means that a customer would incur significant switching costs moving away from your product or service. Differentiation is grounded in the concept of abundance, since a differentiated offering creates its own market space rather than operating in an arena of scarcity, where an increase in demand for one person's offering leads to a decrease in demand for others. This also reframes the concept of competition from a tug-of-war over finite opportunities to the creation of new opportunities. It doesn't mean that competing products or services are absent; it just means that your offerings are distinct in customers' eyes. An understanding of competition is necessary to create and sustain differentiation. Competition can come externally from other organizations or individuals with similar offerings, it can come from customers considering backward integration, it can come from new developments that might render products or services obsolete, and it can come from new geographic regions with, for example, a lower cost structure.

Sometimes differentiation is subtle. An analogy is our perception and communication of our own skills. Often clients I work with are unaware of their special and distinct capabilities, the uniqueness that they bring and that we all possess. This realization of distinctiveness can surface with a process of exploration and discussion. An analogy on an organizational level is organizational culture, which may seem easy to replicate but is actually difficult to copy. Southwest Airlines, for example, has a culture focused on integrity and on valuing employees and customers, all built on many small reinforcing actions each day. Specialized and distinctive organizational presence is hard to duplicate, as are the subtle and special attributes we each bring to a nontraditional career. What does this mean for addressing competition through differentiation on a nontraditional career path? To introduce this we will first look at the example of the contract awarded in the face of stiff competition mentioned at the beginning of the chapter. This will give us some pointers for subsequently examining sources of differentiation. Let us first frame the customer relationship for the contract example, then examine the competitive situation, and finally look at the opportunities for differentiation.

The customer was a large, national organization, with over one hundred thousand employees and operations in a number of states. The customer was seeking delivery of on-site career services in multiple locations for a particular segment of its employee population. Career services were intended to support people in their development, thus benefiting individuals and helping secure workforce capabilities needed by the organization as it evolved over time. Our small practice had previously delivered other workforce diagnostic services successfully for this customer, and we had consulted on workforce development activities. As a result, there were extended relationships with people in the customer's organization based on trust and successful delivery of past services. We were invited to bid on a request for proposal (RFP) created by the customer. The need arose quickly and rapid deployment of services was sought. The customer's selection process included a panel with representatives from operations across the country. Our strengths in responding to this proposal included the following:

- Credibility in the field of expertise needed
  - In-depth knowledge of initiating, structuring, delivering, and pricing career services based on prior experience, supported by published content and a strong background in workforce and career metrics
- Alignment with the customer
  - Knowledge and understanding of the customer's workforce practices, including demonstrated ability to effectively partner across the organization and alignment with core values, particularly with respect to employee and community relationships
- Operational effectiveness and efficiency
  - Capability to rapidly assemble a team of experienced, committed staff both in career services delivery and in project management, complemented by economical operating practices
  - Strong customer service orientation, reflected in the ability to make timely decisions and customize services as needed

We brought much to this situation in terms of content capability and the alignment of our approach and operating practices with those sought by the customer. But so might others. So where was our differentiation? Let me suggest that it was in the following areas:

- The strength and depth of customer relationships and alignment with customer values meant that we could demonstrate an ability to work effectively in this complex organization in a manner that others in our field would find difficult to emulate quickly.
- The combination of experience in structuring and delivering in this content area, coupled with measurement expertise in this field, was distinctive. Small entities would be unlikely to have the range of skills needed, and larger organizations, while having broad general capabilities, would likely not have the specific domain expertise we possessed.
- The ability to bring a cost-effective solution that could be tailored to rapidly meet customer needs, unfettered by the high-cost overhead of a large corporate bureaucracy and the associated ponderous decision making.

In summary, we were able to bring the advantages of specialized capability, speed, responsiveness, cost effectiveness, and customer orientation of a small, fast-moving organization to differentiate our offering. Subsequent success in delivery validated the customer's decision. This project was a major undertaking for our practice so it received concentrated focus and dedicated resources. In a large organization it may have been incidental, receiving marginal commitment and resources. Many people were instrumental in this success: those on our team who contributed to the proposal; those ready to join the project, including potential partners; customer representatives who believed in us; and those willing to be references. To all of these people I am so grateful, for many have benefitted from that decision. We see abundance reflected in the example, flowing to the members of our team who delivered the services and to our customers and clients who received the services. Differentiation is apparent in the depth of customer relationships, in the nature of content provided, in customization, in speed and flexibility, and in cost effectiveness. It is in creating differentiation through these and other dimensions that differentiation's contribution and power emerge. Now we will step back from this specific example and explore potential sources of differentiation.

Sources of Differentiation

With interdependent components in a nontraditional career, linked by a common thread, we can seek differentiation through this core connecting thread. In addition, differentiation is accessible through the specific attributes of individual career components. When there is little or no linkage among the components, differentiation needs to be considered separately for each component, since each competes in a separate arena. Knowing that sources of differentiation may be addressed differently according to whether components in our nontraditional career are linked or not, let us now look at these potential sources. These are the sources of differentiation we will examine:

- Distinct product or service offering
- Operational excellence
- Focused, passionate engagement and commitment
- Team capability
- Nature, depth, and longevity of customer relationships
- Ability to tailor to customer needs
- Flexibility
- Speed
- Cost
- Component linkages

Distinct product or service offering. A distinct product or service offering is the most straightforward source of differentiation. In our earlier example, expertise in specialized career services and measurement formed the basis for a distinct, differentiated offering. This was further strengthened with continued delivery and learning. Other examples from my experience include workforce consulting services built on working in and with organizations, supported by publishing a book in a related area. Publication also helped establish credibility in the realm of workforce diagnostics, where accumulated experience coupled with development of analysis, synthesis, and communication approaches led to a distinct offering of value to customers.

Maintaining differentiation through a core strategic approach of continued innovation in product and service offerings can be highly effective. One organization that practices this on a large scale is DuPont. Innovation has been a central focus over the organization's history of more than two hundred years, as it has reinvented itself about every thirty years since the early 1800s.[5] Indeed, the latest metamorphosis has the organization focusing on supplying agricultural chemicals, seeds, nutrition, bio-based chemicals, and advanced materials where the company has "a unique ability to work beyond the boundaries of a single discipline and find novel, innovative solutions at the intersection," according to CEO Ellen J. Kullman.[6] This also means exiting traditional areas. So here we see the concept of differentiation practiced on a large scale and as integral to the path forward for an organization. We can adopt a related approach of ongoing innovation to maintain distinct product or service attributes as part of our nontraditional career path should we so wish.

Operational excellence. Operational excellence is primarily an opportunity for differentiation in product areas by providing attribute or cost advantages and in service areas by creating distinct and memorable customer experiences. Most products associated with a nontraditional career will be smaller-volume, specialized items whose product attributes are the main distinguishing characteristics. Examples include fabricated cabinetry, jewelry, or catered products. In these circumstances, differentiation through production is most likely to result from the specific features of the items and the know-how and creativity involved rather than economies of scale and manufacturing process. Works of art are an example; each item is unique. However, even with art there may be opportunities for production economies, as in the creation of limited edition replicas, for example. The opportunity for differentiation through production processes rather than product attributes will generally be associated with larger scale, and when that occurs, techniques such as lean manufacturing may be appropriate.[7]

When delivering services, on the other hand, core approaches to differentiating service delivery will be relevant at all scales. This starts at an individual level and extends to a culture of service excellence in a sizable practice that includes others. Service excellence results from establishing service as a leadership priority linked to core values, creating feedback systems to track progress and continually refine processes, understanding customer needs and where service can most effectively add value, implementing processes to refine service delivery, and engaging all involved in a team commitment to this area.[8] While specifics of implementation will vary according to circumstances, general principles of service excellence hold in wide-ranging situations, for example, from organizational consulting to teaching and from in-home health care to gardening. Differentiation through operational excellence can complement differentiation through a distinct product or service offering.

Focused, passionate engagement and commitment. Much of our exploration of sources of differentiation addresses knowledge or expertise, primarily thinking activities, or behavior, largely a doing activity. Another important aspect is the presence we bring into our nontraditional career, our sense of being (who) interwoven with meaning (what), which was mentioned in chapter 1. Thinking, doing, meaning, and being are incorporated to some extent into each of the sources of differentiation. Each source has a predominant orientation, and in the case of focused, passionate engagement and commitment, that predominant orientation comes from meaning and being. Meaning and being are most accessible when we engage in something well aligned with our values and purpose. Such alignment is a catalyst for engagement and commitment. This can be a significant source of differentiation when competing with large organizations, many of which struggle to connect on a deeper level with their employees, particularly as transitory employment relationships become the norm. Engagement and commitment are expressed in behaviors and are particularly evident in service endeavors where there is a high discretionary component. The maximum amount of energy and care that we choose to bring is quite different from the minimum amount required to avoid penalty.[9] So alignment of our nontraditional career with our values and purpose provides a springboard for strong engagement and commitment, which lead to excellent service. This is a significant potential source of differentiation, given the levels of service that will likely be forthcoming from many large organizations and the typically low levels of employee engagement that we saw in chapter 2.

Team capability. In a similar vein to engagement and commitment, team capability is associated with a state of being and can be a foundation for differentiation. It can take various forms, such as institutional knowledge coming from the expertise of team members, rapid integration and dissemination of new learning, or team behaviors that reinforce service excellence. Strengths of a small enterprise include the ability to build a strong sense of community and shared values and to rapidly share learning. Indeed, in one of our projects we could transfer learning much more readily among geographic regions than our customer with operations in those regions. Team capability coupled with small size can create an agile organization with a strong service orientation effective both in securing initial engagements and in delivering ongoing service.

This may be a permanently constituted team, as was central to our delivery of projects over extended time periods, where expertise, capability, and commitment of team members were instrumental in securing project continuity. Teams also may form quickly around specific time-bounded projects. An example from our organization was the delivery of exit interview projects. Here we rapidly assembled project teams from colleagues with strong interpersonal skills and relevant experience. We provided foundational training and created an information-gathering infrastructure. Skilled interviewing by team members, coupled with the information infrastructure and analytical capabilities, led to a differentiated offering of value to customers. The contribution of an effective team surpasses the sum of contributions from each person separately. It also may facilitate some specialization, for example, by content area, geographic region, individual capability, or interest. Rapid transfer of learning leads to the evolution and development of team capabilities, further strengthening differentiation.

Nature, depth, and longevity of customer relationships. We saw earlier an example of the value of extended and broad customer relationships in an organization in connection with career services delivery. Another example is that of a relationship that began with consulting on the delivery of career services in the public sector, evolved over several years to include the creation and delivery of leadership development for principals in an urban school system, and continued with consulting on a change initiative for a public sector agency. This occurred due to the support and foresight of Pearl Sims, who was initially with the public sector and then with Vanderbilt University. I am most thankful for Pearl's support. The evolution of these projects is an example of how a nontraditional career can start with components that are built on core expertise and then further develop as unanticipated client needs surface. Depth of shared understanding with a customer about work content areas builds over time leading to synergy from this shared understanding and to new initiatives.

Ability to tailor to customer needs. The ability to differentiate based on tailoring offerings to customer needs is related to the depth of customer relationships. Large organizations often have a primary goal of replicating products or services on a major scale. In fact, much of the wealth generated by high-technology companies in recent years comes from this. It requires a degree of standardization necessary for replication to be feasible and financially attractive. This is quite different from a small enterprise and a nontraditional career, where the goal typically is to create something of value for a small, select group of customers tailored to their specific needs. In our organization, while some core capabilities were shared across engagements, the final structure and customer-facing components were customized to each situation. The customer only needed to purchase what he or she wanted. If a customer needed additional items, we were pleased to provide them if we could (for example, the rapid provision of an 800 number as part of career services delivery). In general, differentiation through tailoring service or product delivery to meet customer needs may involve content, breadth of capability, timeliness, or geographic location, adjusted over time as necessary.

Flexibility. In today's work world, individuals and small organizations have ready access to advanced information and communication tools in a way that would have been inconceivable even in the early 1990s.

Information technology has democratized communication and analytical tools that at one time were the exclusive preserve of large organizations. Information and tools for gathering it are now readily accessible to individuals and small organizations. For example, we used an online platform for the collection of data that could be input from any location with Internet access. We could then readily aggregate, analyze, and synthesize information to inform service delivery and communicate progress to customers. In addition, our computer-based accounting system was current in real time, so we knew our financial status instantaneously. I recall working for one large and sophisticated organization that had a lag time of days before aggregated financial performance was known. An individual or small organization can more easily stay current with rapidly changing information technology than a large organization wrestling with outdated legacy systems that are costly to modify. This is a potential source of differentiation by providing timely analysis and synthesis of information for customers and by strengthening and flexing service delivery.

Speed. I have alluded to the agility advantage for an individual or small organization. This translates into rapid decision making, rapid implementation and marshaling of resources, and rapid provision of information needed by customers from state-of-the-art information technology tools. This is supported by communication from a lean organizational structure. Speed is important both in responding to opportunities and in adjusting service or product delivery to match changing customer needs or changes in the competitive environment. It has been said that senior leaders in large organizations only make a small number of key decisions each year, but the effect of these decisions is great. Similarly, with a nontraditional career or a small organization, there may only be a small number of opportunities that surface and need attention during a year. But each of these has special significance and needs a timely, thoughtful response; as a result, agility becomes an important differentiating attribute. Changing circumstances also require a speedy response. I recall one of our exit interviewers spotting quickly that our customer had mistakenly provided a contact list from the prior year. We quickly provided this feedback to the customer so the correct contact information could be generated, avoiding a costly and misdirected project launch.

Cost. Another singular advantage for a small organization, and one associated with a nontraditional career, is the ability to keep overhead low and, as a consequence, offer specialized products or services at lower cost than large organizations. When coupled with distinct product or service attributes, this is a powerful combination. When I first started a consulting practice, I looked into the possibility of securing separate office space. This turned out to be costly and complicated with respect to phone and computer access and, in one case, involved subleasing arrangements. This approach would have added both cost and complexity with little or no added value. On the other hand, working from home was cost effective, time efficient as commuting was eliminated, and, with modern communication and information technology, practical. It rendered the services more cost effective, it simplified the computing and phone infrastructure, and it saved a lot of time, which is a prized asset. Working with individual clients from a home office has also been effective, and meetings with organizational customers take place at their facilities. Such cost effectiveness can infuse a nontraditional career path and provide another valuable source of differentiation.

Component linkages. The final source of differentiation is that of linkages among the components of a nontraditional career, which works well when the components are interdependent. Combining knowledge and experience from different components results in component permutations that are special and distinct, strengthening offerings for customers. For example, the knowledge and capability resulting from a combination of workforce consulting, teaching, writing, working with individuals, supporting a team and practice, and volunteer work together are greater than would be the case from the sum of each separately. Learning in one area is immediately available to another area. Analysis and synthesis approaches developed in one component are applicable to other components. Infrastructure economies result from being able to use common support systems for multiple career components. Differentiation from component linkages applies both to customer-facing interactions built on accumulated learning and experience and to operational economies resulting from efficiency of infrastructure deployment.

Protecting and Sustaining Differentiation

Having examined potential sources of differentiation for our nontraditional career, we will conclude our exploration of this area by highlighting steps to protect and sustain such differentiation. Let me suggest the following:

- Establish and integrate more than one source of differentiation.
  - The more sources of differentiation, and the more complex their interaction, the more difficult it will be for others to imitate. For example, combining distinctive product or service offerings with cost advantages and deep customer relationships will be stronger and longer lasting than any

one of these separately. This is consistent with classical approaches to competitive strategy.[10] In our case, we sought to create differentiation and value through multiple sources that included content expertise, customized offerings, strong customer relationships, rapid response, cost-effective delivery, and flexibility.

- Emphasize continuous development and innovation as a means of revitalizing sources of differentiation.
  - For example, in providing workshops and webinars, we continuously sought to identify new areas of client interest and to create new content with offerings tailored accordingly.
- Use measurement to inform and refine service or product delivery.
  - For example, we gathered regular feedback from clients in organizations where we delivered career services about their experiences and the actions they took, providing ongoing learning that allowed us to strengthen and refine our service delivery.[11]
- Emphasize frequent communication of value contribution to customers to underline the nature, extent, and attributes of differentiated products or services.
  - For example, with career services delivery we created a framework and regular reports for customers that provided insights into aggregated client perspectives about their experience with the services, their knowledge gained, and their action steps, in addition to analyses that pointed to the contribution to organizational value.[12]

Taking such steps helps ensure that differentiation isn't a fleeting concept but that it becomes embedded in a nontraditional career, constantly revitalizing both the career path and the client and customer relationships that help sustain its vitality.

Questions to Consider When Addressing Differentiation

Having explored what differentiation means with regard to community and competition and examined various sources of differentiation and how to protect and sustain them, here are some questions to reflect on when considering how you might address this area:

1. What excites you in your work and interests and what might this mean for potential sources of differentiation?
2. How might those areas be valuable to others?
3. How might you combine multiple sources of differentiation?
4. What steps are needed to establish those sources of differentiation?
5. What barriers might stand in the way and how might you overcome them?

CHAPTER 5

Balancing the Components

IT SURPRISED ME THAT once our practice focused on working with individuals became established, the number of active career counseling and coaching clients I worked with at any one time was quite stable. This was over a period of about fifteen years and in spite of little formal promotional activity. As some clients completed their work, new clients came, often by referral, or previous clients in a new work/life phase returned. So while individual client situations changed, the overall level of activity was fairly predictable. This component of my nontraditional career, working with individual clients, was relatively stable partly because each client needed only a small portion of our total time and resource commitment. On the other hand, another component, writing books, was quite different in its predictability. This was episodic, each book requiring large and discrete time commitments so that variations over time were much greater than with individual client work. A further component, guiding and supporting an organizational consulting practice, evolved to significant scale, ultimately with relatively predictable financial performance, but this took an extended time to fully establish.

We can see several interwoven dimensions characterizing the career components: scale, predictability, and time to launch. They are all important in developing a nontraditional career. In this chapter we will explore the concept of balancing the components of a nontraditional career to meet time, intrinsic, and financial needs. Balance includes intentionally selecting a range of components or opportunities that vary in size, predictability, and length of time to secure. For example, a single, small-scale engagement with an individual client can occur rapidly and likely involves a small financial commitment and contribution. A large-scale organizational engagement over a longer time, on the other hand, generally involves greater financial commitment and contribution and longer preparation time, perhaps also including an extended delivery team. How should we balance these different components with each other? To address this and related questions, we will first explore what is meant by "balance" and why it is important. Then we will look at the example of my portfolio before examining different aspects of balance. Finally, we will look at implementation so you can apply these principles to your situation.

What Is Meant by Balance

Let us take two extreme sets of career components in a nontraditional career. For the moment it doesn't matter if they are interdependent or independent, and let's assume that all are equally fulfilling and have about the same level of financial predictability over time. In one case, all of the components provide only a small financial contribution but can be started quickly and comparatively easily, and projects or client engagements within each component have a short duration. In the other case, each component provides a major financial contribution but all take several years to launch, though once launched they have long duration. What are the implications of each case? In the first case, significant time is probably spent in marketing and sales to constantly replenish projects or engagements, and the financial contribution is meager because that is the nature of each component, but it is likely that there is continuous activity and a base of financial contribution that starts quickly. In the second case, time is spent in relationship building to create the basis for securing these major project initiatives. This period could last for many months or even years, during which time there is no financial contribution. At last, after a long time, these new components could launch and be sustained. Each of these two cases presents major problems. The first case never reaches the required financial level, and the second case has such a long gestation time that it never launches. So both cases fail. However, a combination of career components from each of these two cases can provide the firm foundation needed for sustained success. This is balance.

A parallel situation is that of a portfolio of research projects in an organization.[1] Here, similar issues surface: some projects offer rapid but low financial contribution with a relatively high likelihood of success while other projects offer a major financial contribution but with a lengthy research time and much uncertainty about ultimate success. Here, too, creating a balanced combination of both types of projects leads to a sustainable blend that is superior to either extreme. Another parallel is in the financial arena, where a combination of investments that balances profitability and the degree of variability among various items offers significant benefits in stabilizing long-term performance.[2]

This brings us to another significant aspect, namely, that of time horizon. Let us look at an analogy. Sometimes proffered advice in the world of financial advising and investing, when investment performance is poor, suggests considering an indeterminate long term. Unfortunately, when long term means almost ten years, this may exceed the time horizon of many investors, perhaps some at later life stages. This arose during the poorly performing U.S. stock market period from 2000 to 2009, over which time the S&P 500 composite index, the NASDAQ composite index, and the Dow Jones industrial index all declined.[3] At the other extreme, day trading, perturbations over short periods can be heavily influenced by the whims of current events and perceptions and, as a result, become uncoupled from inherent value. Both an indeterminate, long time horizon and an extremely short time horizon can present problems. Similarly, as we look at a time horizon for a nontraditional career we seek to avoid such difficulties. In considering a time horizon for our nontraditional career components we need to adopt an approach that is neither excessively long and therefore of little practical value or so short as to be meaningless. With that in mind, we recommend looking out in three- to five-year time increments, recognizing that this can be readily extended. Such time increments are both close enough to permit reasonable extrapolation and relevance and long enough in duration to be significant and meaningful.

How might the concept of balancing career components differ according to whether the components are linked and interdependent or are independent, as outlined in chapter 3? In both situations, generic aspects such as financial contribution, time to launch, and predictability are relevant. When components are linked, however, we can also consider additional factors such as balancing the commitment of scarce related resources and expertise among different components and balancing time spent transferring learning with that spent in sales or delivery.

Why seek balance at all? We seek balance because it strengthens the sustainability of a nontraditional career path. We can consider sustainability from two perspectives, personal and practical. Let us look first at the personal perspective. Sometimes I work with clients who have been in a series of conventional jobs, each lasting perhaps one or two years. On occasion the client initiated the departure, at other times circumstances led to the departure. Self-initiated departures often result from the client's interests and values being disconnected from the nature of their work. So the client moves from one place to another, trying to find a suitable work environment. These fleeting job encounters speak to a lack of career path sustainability. This is where an assessment process can assist in providing a foundation and a compass to navigate a future path. Similarly, with a nontraditional career, a well-balanced set of components begins from such a solid foundation of reflection, and as a result, it connects deeply with personal meaning being aligned with personal values, personality preferences, interests, skills, and aspirations. Balancing the components makes this degree of personal alignment possible and it leads to a sustainable path. We saw this in the examples at the end of chapter 2. The second aspect of balance is that of meeting practical financial needs that address both short-term launch necessities and long-term financial requirements. This is needed to create a firm financial foundation for nontraditional career path continuity. It is supported by continuously reappraising how the integrated career components are working together and adjusting as necessary. We will look at this, first as we map my nontraditional career components and then as we see how to apply a balance framework in general.

Balancing Nontraditional Career Components: An Example

We will now map the career components, which were introduced and described in chapter 1, in a new way, using the two dimensions shown in figure 5.1. The first dimension, financial contribution generated by a career

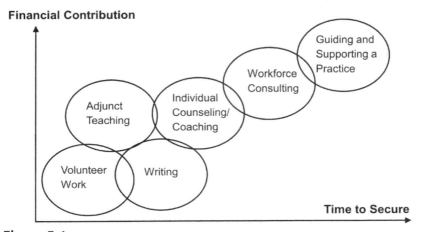

Balancing the Components: Example

Figure 5.1

component, is shown on the vertical axis of figure 5.1, the second dimension, time to secure, is shown on the horizontal axis.

For present purposes we will consider this representation as a qualitative, conceptual framework. It could be quantified and, in that case, financial contribution could be considered as either an annual peak number or an annual average based on actual or estimated performance over a multiyear period to take variability into account. Preferably this is based on income; revenue can provide a proxy for income if needed. Financial contribution could also be replaced with an index that combines intrinsic contribution with the financial aspect qualitatively, as we will describe later in the chapter. We will view figure 5.1 as a recent snapshot of past history. It can be constructed to represent various points in time, for example, addressing past history or future projections at different times. It also can show projects within each of the primary components, if needed.

There are six career components shown in figure 5.1. Let us discuss the placement of each. Volunteer work is shown in the lower left, as it made no direct financial contribution in my case and started quickly. In addition to the value contributed to an organization by volunteering, this component may make a significant long-term personal contribution by helping build skills or relationships. For example, my volunteer work with the Taproot Foundation hopefully contributed to three nonprofit organizations, and it led to Aaron Hurst, the founder of Taproot, and Barbara Langsdale, who was a senior leader with one of those nonprofits, We Care Services for Children, each writing a chapter for my last book, for which I am most grateful.[4]

Moving to the right along the time axis, we come next to adjunct teaching. It made some, though limited, financial contribution. I was fortunate to deliver classes in master's programs for two institutions in the Bay Area for about ten years, starting at one institution then continuing with another closer to home. This fit well as it was in a core area of interest and expertise and was an opportunity to contribute to students and the profession. It took a relatively short time to begin this work, just a few months in each case, and was quite stable once begun.

On the time scale we come next to writing. With an initial background in research, I have always felt drawn to publish knowledge and information that I have been fortunate to acquire through experience and practice. Writing has also offered the opportunity to provide a forum for the voices of others through edited volumes and to speak on topics that are personally meaningful. It also helped build professional credibility. Some aspects of writing, such as writing articles in journals or articles in newsletters for professional organizations, are relatively accessible and typically take less than a year to create and publish. Other aspects, such as writing books, take longer to complete and to secure. They usually require demonstrated capability on a smaller scale first. While those with celebrity or notoriety (or both) may generate substantial income from writing, it has been my experience that this contributes more intrinsically, and through building personal credibility, than financially.

Moving further along the time scale, we come next to working with individuals through career counseling and coaching. As mentioned earlier, once our practice became established, it was quite stable. It did make a financial contribution, though it would not have been sufficient to sustain us. As with the other components, this was meaningful and fulfilling work. The next item on the time line is workforce consulting. Here, credibility generated through the earlier writing component was helpful. For the first time this included engaging others on a contract basis. Projects in this component took longer to secure, now measured in months to years, but made a significantly greater financial contribution. It was also personally meaningful. At this point the nontraditional career was financially self-sustaining. The final component, guiding and supporting a practice, was also fulfilling. This entailed establishing a corporation and engaging a team of people employed by the organization. It was several years before the time was right for this. The financial contribution from this component was the largest and quite stable.

We can see how all of these components fit into a balanced whole that included smaller financial contributions secured more rapidly and larger financial contributions established over longer time periods. This portfolio proved to be enduring, sustainable, and fulfilling largely because of the balance that is inherent in it. Its strength in terms of the framework in figure 5.1 is the broad distribution of components. They are not all collected together in one part of the diagram, so the overall launch time is acceptable and the overall performance is strong. We will explore the pace of entry in our next chapter. Let us acknowledge now another aspect that is inherent in our component map, the ending of activities that are part of these components. Endings can range from completion of a career counseling or coaching engagement with an individual, which has a relatively small effect on the overall profile, to completion of a large workforce consulting project or completion of a book. For example, workforce consulting projects with defined time frames included exit interviewing or change leadership projects. Such project completions have significant time and resource implications. They speak to the importance of continuing to revitalize the components by introducing new projects. Having mapped this example of nontraditional career components, we will now look at various dimensions of balance.

Dimensions of Balance

We can consider dimensions of balance in two broad categories, first the practical aspects, and second, those aspects that are intrinsic or internal to ourselves. Let us first look at practical aspects, which we began to explore with the previous example. Practical aspects include the following:

- Financial contribution
- Time to secure
- Resource commitment

Financial contribution. One challenge facing many small businesses, which also potentially arises with a nontraditional career, is that of ensuring that cash inflows surpass cash obligations, even when both may be episodic. It is also true of conventional, full-time employment, given its tenuous nature today. There are steps that can help provide a cash flow cushion, for example, building sufficient operating reserve, such as several months of expenses, and building sufficient reserve to cover an extended launch period. These are important and necessary steps. However, they are like oil added to lubricate gears attached to an engine. Just like the oil, which is critical to help ensure uninterrupted operation, reserves are important, but they require that the core operation is sound, just as the engine and gears need to be in good condition for the oil to be useful. Let us address the cost and cash flow aspects of a core nontraditional career, since they are directly relevant to the issue of financial balance. One attribute of a nontraditional career, particularly when it is built on service-focused components, is that it has relatively modest capital needs. Start-up expenses are usually low. Infrastructure consists of basic communication and information tools that are comparatively low cost and readily available, while office expenses can be minimized by working from home. One study showed that 69 percent of nascent businesses in the United States conducted their business activities at home, and that 59 percent of established entrepreneurial businesses continued to be home based.[5] Many costs are tied directly to service delivery—for example, creating presentations, producing reports, and travel to customer sites—and these are variable costs. Some costs are fixed and independent of activity, for example, core telephone service and database access. Product-oriented career components will incur additional costs for production processes and inventory requirements. External funding may be needed to cover these initial and ongoing costs. Servicing interest payments for that funding becomes a fixed cost, along with other fixed costs such as utilities or rent for production facilities. A core, consistent cash inflow is required to cover fixed costs. Variable costs accompany service or product delivery and are covered by cash inflow from projects, recognizing that working capital is needed for timing differences between revenue received and disbursements.

How does this influence financial balance among nontraditional career components? An ideal cash inflow is stable, consistent, and delivers an appropriate level of profitability. As we saw in the earlier example, a combination of career components with varying levels of income, some smaller but rapidly implemented and others providing larger income streams but with a lengthy gestation time, can create such a favorable financial profile. Indeed, we could extend the level of detail on the financial-contribution/time-to-secure map to show projects within the components, each having its own financial contribution/time profile. With stable cash inflow, cash reserves are primarily needed for working capital. However, there may be situations in which annual cash inflow is favorable but seasonal because of large monthly variations. In such cases, for example, in agricultural activities or with product or service sales linked to particular times of the year, additional reserves are needed to cover seasonality. Again, a balanced set of career components or projects within those components, based on their income-generating profile during the year, can help smooth cash flow seasonality.

Time to secure. The second practical aspect of balance, which we saw in the earlier example, is the elapsed time needed to secure a given career component or projects within that component. This dimension is particularly important both because of its connection to the financial aspect we just reviewed and because time is such a valuable personal resource. We will examine the resource aspect in the next section. Let's focus now on the time-to-secure career components. On the surface it might appear that an ideal time profile for a given component would be an instantaneous start. However, such a start would bypass the relationship development needed for long-term sustainability. Conversely, an overly protracted time to launch a career component may speak to fundamental project difficulties. Again, a combination of components, some that start quickly though likely of short duration and others with a longer launch ramp and also longer duration, provides a more stable and viable nontraditional career foundation.

Resource commitment. The third practical aspect of balance is that of resource commitment: time, money, and personal energy. These are finite resources, so an allocation to one career component means that the amount already committed is not available to the other components. Various approaches to allocating such resources are possible. For example, the allocation could be based proportionately on the anticipated contribution from the component or on some estimated combination of importance and urgency. Indeed, there may be potential trade-offs, for example where a commitment of money could purchase support that would free personal time. Successfully balancing the nontraditional components with respect to financial contribution and elapsed time, as we reviewed earlier, is likely to create a favorable financial profile, which in turn allows greater flexibility in deploying resources. In the case of an interdependent career profile, a further aspect of time/resource balance surfaces. That is the amount of time committed to transferring learning among components relative to the time spent in delivery or in marketing or sales. It is likely that this balance of time will shift as the components mature, with marketing and sales dominating in early stages, delivery dominating in later stages, and learning transfer a continuous need. Tracking actual time spent in each of these three areas can provide a basis for determining if changes are needed and, if so, the nature and direction of those changes.

Now let's look at three intrinsic aspects of balance:

- Skills
- Interests
- Life and work

Skills. The opportunity to select and use skills that are fulfilling and rewarding is one of the benefits of a nontraditional career. It allows us to do things that we are good at and that we enjoy. Intrinsic to a nontraditional career is the need to engage in a broad range of skills. We will explore that range in chapter 7, but for now we can acknowledge that by selecting certain career components, projects within those components, and aspects of those projects we choose to deliver, we can balance the nature and range of skills we practice. This can include content-related skills, interpersonal skills, and strategic and business-oriented skills. It is in balancing the component and project selections that we tailor a nontraditional career path to match our strengths. (This was alluded to in the vignette about Beverly Garcia in chapter 2.)

Interests. With respect to interests, we can adjust career components to incorporate either a breadth of interests or focus on a few, in either case balancing activities so that they are well aligned with who we are. The independent nontraditional career is more likely to appeal to someone with widely disparate interests reflected in their being drawn to several of the six general theme areas for interests mentioned in chapter 1. An interdependent nontraditional career is more likely to appeal to someone with more focused interests, where breadth may come from practice of a variety of skills. Balancing interests according to personal preferences is therefore possible, practical, and desirable in a nontraditional career.

Life and work. Career counseling clients frequently express as one of their top values a wish to achieve a personally fulfilling balance of work with other aspects of life. In chapter 1 we looked at a wheel of life exercise that sheds some light on this issue, providing, as it does, an instrument to help examine the importance of different areas of life and current levels of satisfaction. Another complementary approach is to construct a pie chart that shows how time is spent currently in areas such as personal development, with a significant other, with family, in work, in educational activities, and in social activities. A second pie, for three to five years in the future, shows how time distribution would look ideally at that time in those same areas. The difference between the two pies can offer insights into changes that might be appropriate. A nontraditional career provides the opportunity to adjust the balance of life and work to match personal preferences, as we saw in some of the vignettes at the end of chapter 2.

Implementation

Having explored what balance means and why it is important, looked at an example, and reviewed dimensions of balance, we will now examine implementing concepts of balance in a nontraditional career. We will consider the following aspects:

- Developing criteria
- Selecting components
- Assessing performance
- Adjusting and refining

Developing criteria. In figure 5.1, we created a grid of financial contribution and the time to establish career components, acknowledging that our representation was qualitative and conceptual. Such a representation may suffice well for insights needed into the selection, resourcing, and integration of components. However, more structure is also possible. For example, as we mentioned earlier, we can quantify financial contribution by taking an actual or estimated annual peak value or an annual average value over several years, preferably of income, though alternatively revenue, if this is a snapshot before the component is well established. We can also extend the analysis and base it on the discounted value of estimated future cash flows to accommodate a longer time period and account for the time value of money. However, unless there is substantial initial capital investment, this is likely a more complex approach than is needed.

An alternative to a purely financial measure is to create an index of well-being, which includes both intrinsic and practical aspects, to replace the financial dimension. This index can include a qualitative assessment of financial contribution and other weighted items. Table 5.1 shows an example of this approach. You may choose to create your own elements for such an index. A template is provided as table A.1 in the appendix for your use.

The ten elements included in the example index are shown in the first column of table 5.1. The second column shows an example of how scoring might look for a particular career component. Scoring is based on a 7-point scale, with 7 being the most positive response. In prior information gathering, we have found that a 7-point scale provides the needed degree of differentiation.[6] The scores are multiplied by the assigned weight for each element, which is shown in the third column. The weights total 100 percent. This generates a total weighted score in the fourth column. Since

Table 5.1. Example of a Weighted Index of Well-Being for Assessing Career Component Contribution.

Index Element	Value for a Given Career Component on a 7-Point Scale (7 Is the Most Positive)	Weight (%)	Score (Value x Weight)
Anticipated degree of personal fulfillment	6	15	90
Alignment with values	6	15	90
Fit with interests	5	10	50
Opportunity to use skills	5	10	50
Fit with other career components	5	15	75
Geographic proximity	3	5	15
Low initial capital investment	5	5	25
Strong financial contribution	5	15	75
Low variability or seasonality	4	5	20
Likelihood of extended duration of the component	5	5	25
Total			515

this example uses a 7-point scale and a total weighting of 100 percent, the maximum possible total score is 700. The example shown generates a score of 515, or 74 percent of the maximum total possible. We can create a similar analysis for each career component to facilitate mapping on a grid like that in figure 5.1, where the vertical axis would now be the index score rather than financial contribution. In some cases the primary distinguishing elements among career components may still be those of a practical and financial nature, if, for example, intrinsic aspects are similar among components. The index of well-being can also offer insights, when looking forward, as a screening tool to suggest either including or excluding career components from further consideration based on the components meeting a predetermined minimum score, for example, 50 percent of the maximum possible, or 350.

We can represent the other dimension of the grid, time to secure an established position, as an actual or estimated number of months or years to reach a given level of financial performance or to achieve a particular index score. If this is based on financial performance, then, as before, we can select the time to reach a certain milestone, to reach a peak level, or to reach an average over an extended time. We can also identify separate subsets or projects that are part of the primary career components. When we identify individual projects within a career component, the component may then be represented by an oval shape rather than a circle on the grid in figure 5.1, reflecting the differing characteristics of projects in that particular career component. The mapping approach, based on either a financial measure or an index, provides a basis for characterizing a nontraditional portfolio, from which we can derive insights about resource and time commitments as well as components to exclude or include. We will now focus on this latter aspect, specifically selecting components to include.

Selecting components. Mapping the components of a nontraditional career on a financial contribution or index versus time grid, as they begin to take shape, shows where there are gaps. We seek to create balance so that some components, or projects within components, generate smaller returns quickly and others generate large returns that may require a lengthy gestation time, with a range of components in between. As potential new components emerge then an important evaluation step is to place them on the grid, to assist in deciding whether to proceed, delay, or shelve each new component based in part on its contribution to balance. We may also construct a search process for new components with particular balance attributes in mind to fill gaps. For example, if the existing balance is skewed to short-term/smaller components, our focus on new additions would be on those offering a larger contribution though taking more time to secure. The reverse would be true if the existing balance were skewed to projects with a lengthy gestation time. So far in this section we have focused on assembling the nontraditional career components. How do we assess ongoing performance? Let us look at that now.

Assessing performance. In the same way that an organization has multiple constituencies or stakeholders and performance assessment needs to consider all of these constituencies, so with our nontraditional career we need to assess performance as it affects our broad constituencies. We looked at four groups in chapter 4, family and those who are close to us; a broad work and social community; team members, partners, and investors; and customers and clients. The index we constructed earlier in this chapter and the financial measures and time measures are primarily built around our personal focus and needs. These are central to the sustainability of a nontraditional career. The first level of assessment for a nontraditional career is whether it meets our intrinsic, financial, and elapsed time expectations.

However, there are additional aspects affecting the four groups of constituencies that may not be reflected in our person-centric index. For example, those close to us may particularly value our time availability, our work and social community may value our purchasing decisions and volunteer contributions, our team members may value continuity and stability of employment, and customers and clients may value our evolving and deepening expertise. While such aspects are implicit in our person-centric index, they are not explicit. So one step we can take is to reach out to our constituencies to clarify their priorities and then create an additional index that reflects these priorities. This complements our person-centric index as a measure of ongoing performance in our nontraditional career and in helping guide decisions about resource commitments and component engagement.

Adjusting and refining. Maintaining the health and vitality of a nontraditional career means continually adjusting and refining the components in it. Much of this occurs at the project level within each component, as projects are completed and new projects emerge. For large projects, mapping using the approaches we have described in this chapter may be helpful. It is also important to reassess and rebalance time and resourcing for the components themselves on a regular basis. This may be prompted by changing internal priorities, changing needs of constituencies, changing project attributes within components, the emergence of new opportunities, or the need to either reinvigorate or drop components as they reach the end of their life cycle. For example, in my nontraditional career, teaching on an adjunct basis was a fulfilling means of contributing to the profession, though it occupied a relatively small amount of time. Closure of the department with which I was connected, as the institution refocused on only larger programs, offered the opportunity to either seek alternative teaching options or redirect time into other components. I chose the latter step, as needs in other components were growing and local teaching options were limited in the field where I specialized. Here an external event led to component rebalancing. Conversely, an interest in social justice and social responsibility over a number of years led to the emergence of a book editing opportunity and therefore an increase in my time committed to the writing component. In another area, guiding and supporting a practice, needs changed as we adjusted project management resources to reflect changing circumstances, which led to adjustments in my time for this component and some rebalancing of time. So we see that rebalancing components to reflect changing priorities and external circumstances is fundamental to the continued evolution of a nontraditional career. Part of the beauty of this nontraditional path is its ever unfolding and developing nature. To complete the exploration of balance, here are some questions that may be helpful when considering your nontraditional career.

Questions to Consider When Addressing Balancing the Components

1. What is the likely annual financial contribution and the time needed to establish a position for each of the components you are considering for your nontraditional career?

2. When you map your nontraditional career components on a financial contribution/time-to-secure grid, what gaps do you see in the profile and how might you fill those gaps?

3. What elements would you include in an index of well-being that characterizes the contribution of nontraditional career components?

4. When you map your nontraditional career components on a well-being index/time-to-secure grid, what gaps do you see and how might you fill those gaps?

5. What priorities of your family and those close to you; the broader work and social community; potential team members, partners, and investors; and potential customers and clients would you include when evaluating your nontraditional career path? What does this mean for the resources and time you would commit to each of your nontraditional career components?

6. How is your time spent today in areas of life that are significant to you and how would you like this to be in three to five years? What might this mean for how you structure your nontraditional career components?

CHAPTER 6

Pace of Entry

I REMEMBER MANY YEARS AGO, before starting my nontraditional career, being told that success depended on committing everything to such a path and immediately jettisoning all other conventional work responsibilities. But is it really that way? Is full, instantaneous immersion the only approach that works? In my experience the answer is unequivocally no. It is possible, and often preferable, to begin gradually. Where nontraditional career components are built on existing expertise, gradual development may be preferable. This can permit integration with conventional employment for a period of time and provide financial support during ramp-up. If, on the other hand, the new components represent a major departure, or require significant capital expenditure, full, immediate engagement may then be preferable. One of my son's colleagues, who studies animal behavior, once told of his experience with small animals when presented with an exercise wheel. Most happily run on the wheel. However, some press their legs against the sides of their cages so they don't have to run on the wheel. I am not proud to admit to relating to this latter group. The animals have different individual preferences. And so do we. Our different personal preferences, along with practical considerations and support from others, determine a preferred pace of entry into a nontraditional career.

In the three preceding chapters on strategic factors, we have focused primarily on how to structure a nontraditional career. In this chapter and the two that follow, we will focus on the implications for implementation. In this chapter we will look at pace of entry and whether to gradually build a nontraditional career path while maintaining more conventional employment or whether to take a single, immediate step. After describing these two alternative approaches, we will explore factors to consider when deciding on the preferred approach, discuss how to determine the best approach, and finally touch on the emotional roller coaster of starting up.

Describing the Alternatives

There are two broad pace-of-entry options we are considering. The first option is abrupt cessation of conventional employment, either current or an employment search process, with immediate, full-time engagement in a nontraditional path. In this case, upon launch, the entire focus is on the nontraditional path. Financial resources need to be secured in advance to cover the launch period and then rapidly supplemented by income streams from the nontraditional path. For those currently employed, this case would mean leaving conventional employment when starting on the nontraditional path. For those who are not employed, it would mean ending an employment search when beginning the nontraditional path. This case does not mean lack of preparation prior to launch, rather it means that extensive prelaunch preparation is required.

The second option is to gradually assemble the components of a nontraditional career over an extended period of time while continuing in full- or part-time employment or while continuing an employment search process. In this case focus gradually shifts from that of conventional employment to that of a nontraditional path, paced according to individual preferences. This gradual shift allows for both some learning unfolding over time and the option to test and adjust delivery of nontraditional components. Resources for launch and initial development of the nontraditional path include income from conventional employment that is gradually augmented by income from the nontraditional components over time. Eventually conventional employment ends and the nontraditional path becomes the entire focus of work. As in the first case, extensive preparation is beneficial prior to the initial launch, though in this case preparation continues over time as the nontraditional components phase in. This second case, of gradual pace of entry, describes well my nontraditional career path. I gradually added components over time while continuing in conventional, full-time employment. Over time the nontraditional path grew, and when events supported ending conventional employment, the nontraditional path became my entire work focus. It has been a blessing.

Factors Affecting the Decision about Pace of Entry

Let us examine some factors affecting the decision about pace of entry and the launch approach. Current circumstances materially affect this decision. For example, if you are currently employed and have built strong customer relationships and a credible body of expertise in a service area, then you likely have flexibility about launch timing. In this case the broad range of factors we will examine can inform your decision. On the other hand, if you are currently unemployed and face financial constraints, practical considerations may dominate decisions in the early stages of your nontraditional career. If you are just entering the workforce, factors related to building expertise and credibility are most significant; at mid-career, work/life integration questions and support from others may be particularly important; and at a later career stage, giving back and creating a legacy may assume greater significance. So the influence of the various factors we will examine and how they affect pace of entry will vary according to current circumstances. We will look at these factors in three broad categories: personal preferences, practical considerations, and support from others.

Personal Preferences.   Personal preferences refer to internal perspectives that inform our decisions. They are shaped by our natural disposition leavened by life experiences. These experiences in turn are shaped by our immediate environment, our family and local community, and events on the broader national and global stage. For example, growing up during the economic disruption created by the Great Depression of the 1930s, or the great recession more recently, leaves an indelible impression about financial security. Entering the workforce at a time of high unemployment is costly in terms of a longer term employment trajectory and will influence personal perspectives about employment risk. I remember well how economic conditions in the early 1970s led one not atypical, on-campus recruiter from a large company to comment to me that there were about one hundred applicants for his one position. How different this would be three years later, after graduate school, receiving multiple job offers and almost surprise from some potential employers that they were even being considered. The first experience probably led to a high degree of personal risk aversion for me during my initial years of employment, which was not eradicated by the second experience. So in addition to the importance of our inherent disposition, our personal preferences are molded both by current circumstances and by our life experiences and the environments that shaped those experiences. With that in mind we will look at various personal preferences and their influence on the decision about pace of entry. They are as follows: personal purpose, sense of urgency, development, and risk tolerance.

Personal purpose. How does personal purpose affect a decision about pace of entry? Let us look at different ways this can happen. If we are considering a nontraditional career because we are drawn to it by a sense of vocation, because it matches our values, personality preferences, skills, and interests, rather than considering it as an escape from an undesirable alternative, we are likely to modulate our pace of entry so that it integrates into rather than disrupts our life. If, on the other hand, current employment is intolerable, we may seek a more rapid approach to a nontraditional career. Moreover, external economic disruption and unemployment may lead to reappraisal of personal priorities and perspectives that accelerates the desired pace of entry. One specific value related to personal purpose that influences our pace of entry is that of integration of work and life. Here the perspectives of some of the constituencies we reviewed in chapter 4, for example, family and those who are close to us, will likely have a significant influence. Ascribing greater importance to work/life integration as an important personal value will likely lead to a more gradual pace of entry that is less time disruptive. At later work/life stages, when considerations of personal legacy may grow in importance in the absence of financial distress, pace of entry may also be more measured to fully integrate personal meaning into nontraditional career components. Financial hardship at this stage, on the other hand, may lead to an accelerated pace of entry. At early work/life stages, where much time remains to experiment, pace of entry may accelerate to provide insights, through experimentation, into a meaningful personal path. So we can see that personal purpose is an important foundation on which decisions about the pace of entry rest. Closely linked to some of the items we just explored is the second aspect of personal preference, sense of urgency.

Sense of urgency. Our sense of urgency is affected by life circumstances such as a need to experiment at an early work/life stage with different options, the immediacy of unpleasant current employment, concerns over financial security, or external disruptive events such as unemployment, all of which heighten our sense of urgency. However, our wish to maintain work/life integration, bolstered by the views of those close to us and the importance of finding strong alignment with personal preferences, may soften an otherwise heightened, self-imposed sense of urgency. Furthermore, the extent of our established work relationships, our expertise, our content base, and the credibility on which we can build are important aspects that affect our ability to act on a sense of urgency. Based on an appraisal of these aspects, in conjunction with our evolving sense of urgency, we may conclude either that rapid movement into a nontraditional career is possible, with an abrupt cessation of conventional employment (or ending a search for such employment) or that a more gradual approach to a nontraditional career is appropriate, allowing time to build capabilities and resources. This brings us to the third aspect of personal preferences, development.

Development. The need to develop skills, knowledge, or expertise can materially affect pace of entry. This involves personal judgment about when capabilities are sufficient to launch, when they may need augmenting, and, if so, what form this augmentation might take. With paced entry it is possible to build expertise in different career component areas at different rates. In my case a newly acquired educational background coupled with hands-on career counseling experience and earlier work experience provided a foundation to launch a career counseling practice. Acquiring that needed educational background and experience took several years of elapsed time. My organizational consulting component required additional skills and expertise to establish credibility before launching. While I had some prior experience with responsibility for a sales and marketing group in a product-oriented business, and had observed the tremendous skills of this team, I did not have the hands-on consultative sales experience needed to launch a consulting practice. I was fortunate to receive consultative sales training and experience while employed. Writing a book helped materially with credibility. I also needed time to assemble a critical mass of content knowledge about organizational interventions that would be valued by customers. These aspects together—building skills, writing a book, and building a critical mass of content knowledge—took several years. At each stage personal judgment guided decisions about when and how to proceed with the nontraditional career component launches.

Immediate full-time engagement in a nontraditional career path, on the other hand, requires concentrated learning prior to launch. This is a significant challenge and can be aided by a partnership or consultation with others having prior experience to create the needed foundation. This is one reason franchise approaches are sometimes considered, as learning from the experiences of others can help provide the initial content base. Development in such a situation can then focus on skill building since content knowledge is secured from others. With immediate full-time engagement, personal judgment, as before, determines the timing of entry. Needed skill development is an important part of that consideration. One aspect of personal judgment, and the final personal preference area we will examine, is risk tolerance.

Risk tolerance. While on the surface assessing risk is all about the risk of different options associated with new, nontraditional paths, in practice it is about assessing not just the risk of the potential new paths but also the risk of continuing on a path of conventional employment. Conventional employment, as we saw in chapter 2, carries high risk. So our pace of entry is conditioned in part by our risk tolerance with respect to three alternatives: immediate full-time engagement in a nontraditional career path, phased entry over time into a nontraditional career path, and continuation in conventional employment. Immediate full-time engagement in a nontraditional path presents risk challenges since other income streams are removed, and it may require substantial initial capital. This risk is mitigated to some extent by a full-time commitment of energy and resources to the nontraditional endeavor. On the other hand, phased entry mitigates risk by providing income from other sources during an extended launch and allows for experimentation and refinement over time. In the case of a conventional position, employment continuity is largely determined by others and risk mitigation comes from ensuring personal marketability. Each path poses different risk challenges; each is addressed by different risk-mitigating approaches. The preferred approach to pace of entry is based on individual judgment taking risk into account.

Practical Considerations. We will now look at the second broad group of factors that affect pace of entry, namely, practical considerations. These are factors external to ourselves that influence when and how to proceed with launching a nontraditional career. We will look at three elements: financial base, infrastructure needs, and geographic aspects.

Financial base. Regardless of the career component details, a sound financial base is needed to cover much of the launch period, as income from nontraditional components is still building during this time. Moreover, there will be an ongoing need to provide working capital that covers the timing difference between receipts and disbursements, as well as inventory requirements if there are product sales. In the case of product sales, there will also likely be an additional initial investment required for manufacturing capability. Key questions influencing pace of entry are (1) how much money is needed to finance the launch phase? and (2) how will it be obtained? Financial needs include both business-related expenses and living expenses. During the launch phase, business-related expenses for service components can be minimized by, for example, working from home. Estimating business costs for the first year of launch is a useful exercise, along with estimating living expenses for a similar time period based on recent past history and incorporating any anticipated changes. This provides an insight into core financial needs for the first year after launch. It then becomes a matter of personal judgment as to how much time the initial financial base should be designed to cover. This in turn is determined by personal risk profile, the estimated time needed to generate significant income from nontraditional career components, and whether entry is gradual, in parallel with continued conventional employment, or consists of immediate full-time engagement.

The second financial aspect is how to secure needed funding, as the nature of such funding also affects pace of entry. There are two broad approaches. The first is to build a personal financial base over time, just as in saving for any undertaking, and the second is to seek outside sources of funding. This second approach is most feasible where specific assets that can be used as collateral are involved. The first approach is well suited to service-oriented career components with lower capital requirements and is particularly effective when pace of entry is gradual and income from initial components can help fund the start-up of later components. This is how my nontraditional career developed. Income from our practice, where my wife and I worked with individuals, grew over time, in parallel with income from my conventional employment, which then formed part of the financial base for later launch of an organizational consulting practice. So both the financial requirements of the launch process and the nature of sources of funding significantly influence pace of entry. Related to this aspect are infrastructure needs.

Infrastructure needs. I remember receiving wise advice from a colleague when starting out with a nontraditional career: economize on initial costs where possible and minimize infrastructure investments. This was good advice. Capital investment and infrastructure requirements affect the pace of entry. As they grow in scale, so does the complexity of securing funding and the time required to launch. Indeed, if funding requirements grow too large, it may be necessary to cede some decision making to funders, at which point the dynamics of a nontraditional career change fundamentally. This would happen with venture capital funding, for example. Such funding decisions should not be taken lightly but, rather, involve deep reflection on the personal purpose of engaging in the nontraditional path.

When nontraditional career components include sale of products, initial investment is unavoidable, even if it is just for inventory. Additional investment for production capability will likely be required, though this may be handled through renting facilities, for example, if starting out with a winemaking business as a career component. In a service-oriented area, infrastructure will likely consist of computing and communication capabilities as well as office space, which may be available at home. As we observed earlier, computing and communication tools are available today at comparatively low cost. There may be some additional aspects of infrastructure that are needed, such as a website or database capabilities, but such elements can be added during the buildup to launch to accelerate the pace of entry. In our case, we created and maintained our own website. This was initially for our practice working with individuals and was later extended to include an organizational consulting practice.

Geographic aspects. The final practical factor we will mention in connection with pace of entry is the geographic factor. In today's connected world, the geographic reach of nontraditional career components can readily range from local to global. However, as we observed earlier, relationship building is a key part of creating a successful foundation for a nontraditional career. This occurs most effectively in person, through shared community, which may be a local community or a community connected by interest or expertise. A launch process that begins with such core communities to build credibility leads to a firm foundation on which later geographic expansion can occur if this is desirable. Such an approach, which can be viewed as paced growth, is also more likely to lead to a sustainable, long-term nontraditional career path rather than one that begins explosively then fades.

Support from Others. The third broad set of factors affecting pace of entry is support from others. Particularly in preparation for launch, and in early stages after that, supportive others can be instrumental in accelerating successful entry. Indeed, support from others is a key factor influencing the personal preferences and practical considerations we just reviewed. It affects both the income-producing and cost and infrastructure aspects in those early stages. We will look at three potentially supportive groups: family and friends, customers, and a current employer.

Family and friends. Bringing family members early into a decision to pursue a nontraditional career lets them understand the rationale and be present to offer practical and emotional support. This might mean being available for conversations in thinking through decisions, affirming decisions, and in helping make practical aspects like working from home feasible. Family and friends are those most likely to provide unconditional support. My wife's support was critical when we began our first practice, in terms of both her expertise in a complementary area of practice delivery and her patience during my earlier educational activities while I was also working full time. It was critical later, when deciding to embrace organizational consulting as an additional component. Such support is important whether the approach is instantaneous immersion, which can significantly affect others quickly, or the approach is gradual engagement, where patience and perseverance are important. Family and friends can help accelerate the pace of entry by providing affirmation, encouragement, and practical support.

Customers. It might seem strange to include customers as a factor. After all, isn't our relationship with a customer simply that of being a supplier? In fact, strong customer relationships speak to longevity of the relationships and shared values. Such customer relationships can profoundly influence the pace of entry leading to the purchase of products or services at an early career component stage, when such support is vital, and can provide encouragement. Customers can also provide valuable insights into product or service offerings as the components of a nontraditional career take shape and develop. These insights in turn accelerate the pace of entry. In starting the organizational consulting component of my nontraditional career, several customers with whom I had extended prior relationships as a supplier were very helpful with their purchase of services and, in one case, input and thoughts about launching, which were also extremely valuable. Customer support directly affects income generation, which accelerates growth and is indirectly beneficial in providing product or service insights that translate into accelerated entry.

Current employer. In the forming stage of a nontraditional career, it is important to remove as many impediments as possible to this path forward. Such impediments can include restrictions on the ability to operate in certain content areas or to work with certain customers. A current employer may seek to impose such restrictions if permitted by local law and if the new career direction is seen as a competitive threat. On the other hand, a supportive current employer will not limit areas of practice and will encourage use of personal intellectual capital that in turn can accelerate pace of entry. This will most likely happen when new career components do not compete with a current employer's business. I was fortunate, when moving forward with an organizational consulting practice, that this did not compete with the employer's evolving direction and therefore I was able to proceed without impediments and with support. This helped accelerate my pace of entry. This is an area to consider as you are determining an appropriate pace of entry.

Determining the Best Approach

Now that we have looked at the factors that influence the pace of entry, let us integrate them in such a way that helps guide decisions about the best approach. I will use the example of my nontraditional career to show how these factors can lead to a natural alignment about a preferred approach. Table 6.1 summarizes the three broad category areas that affect pace of entry, personal preferences, practical considerations, and support from others, and the ten factors within these three broad areas.

Table 6.1. Example of a Grid to Guide a Decision about the Pace of Entry into a Nontraditional Career.

Category	Factor	Immediate Full-Time Engagement	Gradual Launch
Personal preferences	Personal purpose		Gradual addition of components over several years matched personal aspirations
Personal preferences	Sense of urgency		Initial paced entry, later more rapid component addition fit well
Personal preferences	Development		Built needed skills mainly from conventional employment over time
Personal preferences	Risk tolerance		Risk mitigated by phased entry
Practical considerations	Financial base		Financial base developed before entry and augmented by phased approach
Practical considerations	Infrastructure needs		Minimized infrastructure investments and spread over time to support entry
Practical considerations	Geographic aspects		Gradual expansion from local to national
Support from others	Family and friends		Supportive family at each stage, gradual entry minimized disruption
Support from others	Customers		Customer relationships led to unfolding project opportunities over time
Support from others	Current employer		Support for own practices in parallel with conventional employment

The two columns to the right of the factor column in table 6.1 refer to the two approaches, immediate full-time engagement or gradual launch. Consider this as a template to use when evaluating preferred pace of entry. A blank template is provided as table A.2 in the appendix for your use. Table 6.1 is completed in retrospect for the example shown, the gradual launch of my nontraditional career. Let us review this example, first noting that all of the factor preferences point to a gradual launch approach, as this is what transpired. When looking forward, not all factors may point to the same pace of entry path as they do here. The four personal preference factors are personal purpose, sense of urgency, development, and risk tolerance. Regarding personal purpose, the gradual addition of career components over time aligned well with my purpose as this was a vocational direction to which I was drawn. It was not an immediate response to moving away from something undesirable. Similarly, regarding sense of urgency, this was an important move, but it was not driven by an external event, so paced entry fit well. Later external changes helped accelerate the addition of further nontraditional career components. The opportunity to build skills over time, such as sales skills, fit well with the development factor, and my risk was mitigated by phased entry that allowed for experimentation and income continuity.

As with personal preferences, for the example shown, gradual launch was the best fit with all three factors related to practical considerations. Building some financial base before entry and then augmenting it with income from conventional employment and income from initial components during phased entry eliminated financial upheaval. Infrastructure investments spread over time supported the needs of new components as they were added. These investments were relatively small. Starting locally and extending nationally over time also fit with the gradual launch approach. Similarly, the gradual launch approach aligned well with the final three factors related to support from others, namely, family and friends, customers, and current employer. A patient and supportive family enabled this transition to unfold over an extended time. This gradual approach minimized the disruptive effect of these changes on them. As customers learned more of how I might meet their needs, I was most grateful that they continued to offer new opportunities. This, too, fit well with the gradual approach. Finally, two employers at different times were supportive of my development of nontraditional career components in parallel with conventional employment, as this was not competitive with their work. This helped facilitate a gradual launch.

We can see from this example how each pace-of-entry factor points to gradual entry as a preferred approach. As you look forward to a possible future approach, some factors may point to immediate engagement and others to gradual launch. This could happen, for example, if there is an external triggering event such as job loss. Personal judgment will then determine the preferred path, weighing the relative importance of the different factors and the resulting benefits and challenges of different pace-of-entry approaches. This also speaks to the importance of preparing well in advance for a nontraditional career transition so that pace of entry is determined by personal preferences rather than being imposed by external circumstances. One approach that can aid with advance preparation is testing career components on a small scale. For example, this might include small production runs of a product in the concept stage, or delivery of selected workshops on a volunteer basis, or complementary coaching with one or two clients. Such small-scale testing can provide valuable insights into the best launch approach based on customer or client needs as well as insights into potential scale implications. Related concepts also surface in perspectives about start-up businesses.[1] This testing step fits with reappraising the path based on experience and learning. It blends naturally with the gradual launch approach, in which incorporation of ongoing learning is relatively straightforward, but it is also important in refining a path over time that may have begun with immediate full-time engagement.

Starting

The decision is made, the launch approach determined. Now it is time to take the plunge and dive in. What might this feel like and why is this important? Up to now we have examined what our nontraditional path might look like, the basis for choosing components to include, and how to proceed. Metaphorically, we have looked at identifying which tree to climb and how to climb it. We have emphasized values that shape our decisions, that help us decide on the right tree so we don't reach the top, look out, and realize we picked the wrong one. We have only mentioned emotional aspects in passing; let us now acknowledge their significance. How we feel about a particular path will materially affect our enthusiasm for embracing that path and the pace at which we move forward. If we like climbing a particular tree we'll be enthusiastic, and if it doesn't feel so good, we might not start the climb at all or we might abandon it. So our emotional response is important; it can accelerate or derail the launch process.

While the intensity of our emotions will vary according to whether this is an abrupt change or whether it is more gradual, and whether it is self-initiated or imposed by external events, the same basic framework will hold. In the introduction we mentioned a framework created by William Bridges that distinguishes between change, which is an external event prompted by something new, and the subsequent gradual internal transition we experience that begins with an ending.[2] In Bridges framework, this internal transition proceeds through three stages, an ending, a neutral zone or a time of uncertainty, and a new beginning. Endings are when we disengage from old ways of doing things and let go of who we were in that situation. This may mean letting go of past expectations and the expectations of others and embracing new expectations. The neutral zone is a confusing in-between state in which we are not who and where we were and not yet who and where we will be. In music there is a concept known as a modulating bridge, which is a short section of music that makes possible a transition from one key to another. The neutral zone is like that modulating bridge. We then come to new beginnings, where we embrace the new reality that change brings. There are different emotions associated with each stage. We may experience anxiety and fear in the endings stage, confusion, stress, and creativity in the neutral zone, and acceptance, hope, energy, and enthusiasm in the new beginnings stage.

Embracing a nontraditional career path is a change event that will probably trigger these internal transition stages, just as job loss or change in a conventional position will trigger them. For example, simply moving from a schedule of having to be at a particular conventional work location at certain times to having flexibility and freedom about time and location may feel slightly disorienting before it becomes comfortable. The emotions associated with the transition stages are natural. An abrupt move into a nontraditional career will put us straight into the transition stages with a greater intensity than would be the case with a gradual move that causes the stages to unfold over an extended time. It is important to acknowledge and honor the different stages, accepting the emotions that come with them. Sometimes there is a tendency to try to move straight to the new beginning stage and the positive emotions that come with it. However, this will likely be simply a temporary bypass. Be kind to yourself during the transition. This means taking care of yourself in little ways, for example, by talking with supportive others or by allowing time for activities and relationships that are nurturing. By acknowledging and embracing the emotional aspects of this transition to a nontraditional career path, they become a rite of passage to a new, more exciting and fulfilling place.

Questions to Consider When Addressing Pace of Entry

1. What does your purpose in beginning a nontraditional career path say about the pace at which you might choose to engage in this path?

2. What is driving your sense of urgency for engaging in a nontraditional career path and what might this mean for your pace of entry?

3. What skills do you need to acquire before starting and how will that affect your pace of entry?

4. How does your perspective about risk affect your preferred pace of entry?

5. What financial base do you need initially and in the early stages and what does this mean for your pace of entry?

6. How might you minimize infrastructure requirements to accelerate entry?

7. What geographic scope are you considering and what does this mean for pace of entry?

8. How might support from family and friends, customers, and, if appropriate, your current employer, affect your pace of entry?

CHAPTER 7

Building Needed Skills

IT WAS ONE of those cold, late winter days in Chicago in 1978. About twenty of us filed rather sheepishly into a hotel conference room. We were there for a three-day workshop on public speaking for engineers. A root canal paled in comparison to the prospect of this experience. My supervisor at the time encouraged me to go. He recognized the need, though probably not the extent of my fear about public speaking. That lack of confidence started many years earlier in elementary school in England and was reinforced by the English equivalent of high school. I had reached the erroneous conclusion by that winter day in Chicago that public speaking was a gift given to only a few at birth and the rest of us, particularly me, would never be able to do it. However, I was willing to give it one last, painful try. Three days later, after what was a transformative workshop experience, I was able, for the first time in my life, to approach the prospect of public speaking not with terror but with the knowledge that this was possible and might even be enjoyable. This transformation was no mean feat and a great credit to the workshop facilitator, Jeremiah Goldstein.

Since then I have given many presentations, and I may even have informed and moved some people in those audiences. Presentations have become energizing experiences for me, and while I am no Socrates reincarnated, it seems that these presentations have been enjoyable for most participants. This is good because public speaking has been an important part of my nontraditional career, whether in a workshop, classroom, or seminar, or during volunteer activities. It was central to launching and sustaining my nontraditional career. And it is a good example of developing a skill needed for a nontraditional career that might have seemed out of reach but was actually close at hand.

In this chapter we will examine the range of skills and personal characteristics needed for success in a nontraditional career. These skills and characteristics include a blend of content knowledge and consulting capability to deliver products or services meeting customer needs; sales, marketing, and business skills to reach prospective customers and ensure viability; personal attributes to relate well and communicate effectively with others; time management skills to balance service or operational needs with those of marketing and sales; and organizing skills to create infrastructure, such as information technology and human resource capabilities. We will identify the skills and characteristics needed and some potential approaches to building them. Our purpose is to raise awareness about those skills and characteristics that make a difference to a nontraditional career path. This is a foundation for identifying strengths and opportunities for development. Skills and personal characteristics are the basis for navigating around possible barriers that may surface and for building a vibrant and sustainable nontraditional career.

Defining and Understanding Skills and Characteristics

Skills and personal characteristics that form a strong foundation for a fulfilling nontraditional career integrate elements of working well autonomously with organizational and small business leadership because a nontraditional career includes all of these aspects. For the purposes of this exploration, we will distinguish broadly between skills, the things we do, and personal characteristics, who we are, recognizing that both are equally important and that they are intimately interwoven. Our values frame how we prioritize and approach skills and characteristics. There are many approaches that describe skills and leadership attributes, each bringing a different perspective.[1] For example, Lombardo and Eichinger, in *FYI: For Your Improvement,* describe sixty-seven competencies or skills grouped into six broad factors: strategic skills, operating skills, courage, energy and drive, organizational positioning skills, and personal and interpersonal skills.[2] As mentioned in chapter 1, SkillScan identifies sixty individual skills organized into six broad skill categories: management/leadership, relationship, analytical, communication, creative, and physical/technical.[3] Meanwhile, just a few of many examples of perspectives about effective leadership include Daniel Goleman and his colleagues identifying personal competence (how we manage ourselves) and social competence (how we manage relationships) as key domains of emotional intelligence for leadership, Robert Greenleaf's perspectives on servant leadership, Sam Intrator and Megan

Scribner's poetic views about leadership, and my perspectives about the roles of effective leaders.[4]

This multiplicity of views underlines the subjective nature of these perspectives and the fact that there is no single best framework. Rather, coupling insights from the body of prior learning with personal experience provides a basis for defining and exploring skills and characteristics that are important for nontraditional career fulfillment and success. Relevant skills are those things that we do well, that we like doing, and that make a difference in our work. Relevant characteristics are those attributes of who we are that infuse our work, bring meaning to it, and enable us to be most effective. Perhaps most important is acknowledging that we can develop both skills and characteristics as we see fit according to the needs and direction of our nontraditional career. Indeed, the brief discussion about learning public speaking skills, which introduced this chapter, is an example of doing just that. We will begin our exploration by looking first at individual skills, and then we will address personal characteristics. These skills and characteristics will be relevant to each of the components of a nontraditional career.

Individual Skills

We will examine individual skills in four broad categories: foundational, interpersonal, customer facing, and support service management.

Foundational. Foundational skills are fundamental to the creation and development of a nontraditional career. They include the following:

- Content knowledge
- Business and financial acumen
- Time management

Content knowledge. A central skill for a nontraditional career is content knowledge in a particular discipline, which is developed through education and work experience. Examples of content areas mentioned in chapter 3 include finance, engineering, music, carpentry, biotechnology, and medicine. Content expertise, as in these examples, is usually described with nouns.[5] This is concrete information that is generally objectively verifiable. Credibility is built by demonstrating both foundational knowledge and the ability to apply this knowledge effectively. Content knowledge in a given discipline or craft develops over time, underlining the importance of remaining current by staying connected to evolving practice. This happens through relevant professional or trade organizations, through published material, or through continuing education. The half-life of knowledge varies greatly, from months in the high-technology arena to years in some areas of public policy. Regardless, maintaining a vibrant nontraditional career means staying current with emerging practice and knowledge. Since the nature and source of content knowledge is specific to a particular focus area, I simply acknowledge here the importance of establishing and maintaining a strong and credible base of content knowledge.

Business and financial acumen. Creating a nontraditional career is similar to starting and guiding a business or a nonprofit organization in that it requires clarity of purpose, the ability to identify and meet customer or client needs, an efficient operating infrastructure, and a viable, sustainable financial framework. Not surprisingly, these are many of the same issues that concern the CEOs of large organizations.[6] Our focus here is on the skill of creating a viable business foundation and financial framework. This means understanding the basis for competing successfully, with differentiation a key component, as described in chapter 4. It means assessing risks and making decisions about strategic direction with incomplete information. It means staying current with evolving trends and practices in the relevant sectors. It means managing costs so that products or services in each career component are competitive, and it means pricing to balance the need to attract customers while generating sufficient margin over costs to provide an acceptable profit. It also means establishing approaches to track financial performance, which we will look at in chapter 9. Various means to build business and financial acumen include on-the-job business experience in an organization prior to launching a nontraditional career, educational activities such as an MBA or certificate programs, and use of the U.S. Small Business Administration resources, including mentoring options such as the SCORE program, which provides small business advisers and mentors at no fee as a community service.[7]

Time management. I remember a period in general management in an organization when my calendar was fully scheduled a month in advance. Is this an example of effective time management? Not at all. There was little flexibility, and any small schedule change created difficulties. I have since learned from this and other experiences how important it is to build in open time. This may seem counterintuitive, but it is one of the most effective time management skills we can embrace, leading to flexibility and effective time utilization. Time management is an important skill for a nontraditional career, for time is such a valuable resource. There are competing time demands and there is a continual need to address multiple tasks over time. For example, there is a need to balance delivery of services with the marketing of services and administration. Effective time management means being able to concentrate on important priorities, not just urgent tasks that surface, and being able to problem solve as needed. It also means being able to attend to a broad range of activities over time and being well organized and able to say no to requests when necessary. Educational resources can provide tools and techniques for refining time management skills, for example, tracking actual time spent as a helpful precursor to more effective time utilization. One example is Inscape Publishing's (a Wiley Company) Time Mastery Profile, which provides insights into effective time management behaviors.[8] Effective time management is a complement to planning, which I will discuss in chapter 10.

Interpersonal. The following three interpersonal skills are particularly relevant to a nontraditional career:

* Communication
* Negotiation
* Networking and connecting

Communication. Many elements fit under the broad umbrella of "communication," ranging from written communication to verbal communication, which in turn can occur individually, in a small group, or to a large audience. Moreover, the purpose of communication can range from informing to persuading, and from listening, a critical skill, to being assertive and resolving conflict. Different skills are needed for each form of communication and for each purpose. For example, studies estimate that body language accounts for 65 percent to more than 90 percent of face-to-face communication, an aspect missing from written communication.[9] This is a complex area, and a particularly important one, as a nontraditional career involves much interaction with others, whether customers, employees, or partners. Communication is intrinsic to service or product delivery, to customer satisfaction and development, and to relationships with employees, contractors, or partners (if the latter are involved). In short, it is central to success. General behaviors that distinguish effective communicators are that they are clear, concise, able to tailor their message to their intended audience, and able to achieve the outcomes they are seeking. Rather than attempting to outline development of communication skills, any one of which could occupy an entire book, I will just acknowledge the importance of this skill area and the availability of many resources, such as continuing education programs, numerous books, and the

Small Business Administration, all of which can contribute to skill building in this area.[10] For example, Robert Bolton provides practical insights into developing effective listening skills, assertion skills, and conflict management skills in his book *People Skills*.[11] Perhaps most important is assessing your current skill level in the various communication areas by seeking feedback from those with whom you interact. This can then provide guidance in determining communication areas you may choose to develop. It is easy to be blindsided here. I recall one client, at a senior leadership level, declaring his readiness to engage in interviews for a job based on conducting many such interviews himself. A short exercise consisting of recording a simulated interview with him quickly demonstrated the need for skill development in this area.

Negotiation. It may seem that negotiation is an esoteric skill reserved only for those engaged in high-stakes corporate contracts. In fact, negotiations occur frequently, in personal settings, such as how many green vegetables little ones in a family will eat at a given meal, to work settings, such as when a future assignment will be delivered. In a nontraditional career there are likely to be contract negotiations with customers around project deliverables and pricing, there may be negotiations with employees around compensation and benefits, and there will be negotiations with partners concerning strategic direction and decisions. So negotiation is a fact of daily life. It affects both the revenue-generating aspect of a nontraditional career as well as the cost basis of a service or product component. As a result, negotiating skills are important, recognizing that the outcome of successful negotiation is an agreement that is embraced by both parties and is successfully implemented. The attributes of skilled negotiators include planning in advance and establishing clarity of goals, testing understanding and summarizing during negotiations to resolve ambiguities that might cause problems in implementation, asking questions to clarify the other party's position, providing input about internal feelings rather than being poker-faced, and introducing proposals with prior notice while avoiding irritators such as "this is a reasonable offer," avoiding excessive counterproposals which signal lack of listening, and not diluting positions with too many reasons as the weaker reasons dilute the stronger.[12] Again, building skills in this area through the many resources that are available will pay significant dividends.[13]

Networking and connecting. When those in nontraditional careers were asked in the IDDAS study introduced in chapter 2 what they would do differently if they could turn back the clock and start again, the single greatest response was to network and self-market more actively.[14] The response underlines the importance of networking and connecting as a means of building relationships, identifying and developing opportunities, and connecting with potential customers and partners. Networking and connecting are defined well by a nonprofit organization I once worked for as building relationships to prosper in today's fluid world. This is not selling, using people strictly for personal gain, or manipulating, or badgering others.[15] It is about mutual benefit. Here is Robert Muller, former assistant secretary general of the United Nations: "Network through thought. Network through love. Network through the spirit. . . .You are a free, immensely powerful source of life and goodness. Affirm it, spread it, radiate it. . . . And you will see a miracle happen: The greatness of your own life. Not in a world of big powers . . . but of five and a half billion individuals. Networking is the new freedom, the new democracy, a new form of happiness."[16] While these stirring words speak to the global significance of networking, they are also relevant to us locally.

If networking is so good, why don't we do it more? Some perceived barriers include concern over rejection (in fact, most people welcome mutually beneficial exchanges), lack of time (networking and connecting need to be accorded a high priority), the sense that it is impersonal (not when it is viewed in light of building relationships), and the perception that it means having to be pushy (in reality, it is not about forcing anything). It might seem that those who are more extraverted have natural advantages in this area. Indeed, extraversion brings with it an outgoing presence and the ability to be energized by groups. However, the challenge is to open space for listening so that others can talk. Those who are more introverted will likely be more reticent in large groups, but they bring a natural ability to listen and connect most effectively with a smaller number of people, likely in greater depth. So it is fine to adjust your approach to networking and connecting according to your personality, either reaching out more broadly if you are more extraverted or reaching out to a smaller number of people where there may already be a connection if you are more introverted. In either case, it is important to be clear about the purpose of networking, to nurture your network with acknowledgments, to listen and be interested, to be visible (for example, by volunteering), and to be trustworthy (for example, with sensitive information). As with the other skills, information and resources are available about building networking and connecting skills.[17] While networking and connecting are most effective when conducted in person, the tools of social networking, such as LinkedIn, can be powerful supports for this process.[18]

Customer Facing. Now we will look at three customer-facing skills:

- Marketing
- Consulting
- Selling

Marketing. Marketing is the process of determining what products or services to offer that address customer needs, how to price those products or services most appropriately and how to differentiate them, what approaches are most effective to reach customers and promote the products or services, and how to distribute the products or services. It means building an understanding of how potential customers might naturally group together, for example, by geography or by industry sector, and what is important to each group. For a large organization, marketing is a complex business function built on a body of knowledge in this discipline. For a nontraditional career, marketing is more straightforward, consisting of clearly defining product or service offerings, determining why they appeal to customers, and identifying how to reach customers. While much has been written on this subject and there are numerous courses addressing it, the Small Business Administration provides straightforward perspectives about marketing that are helpful in this context.[19] Perhaps most important is acquiring a basic knowledge of marketing principles through reading or workshops, thinking through marketing issues when establishing a nontraditional career, and revisiting these issues on a regular basis.[20]

Consulting. An important aspect of a nontraditional career is the ability to build relationships with customers and potential customers, first to understand their needs and then to link these needs to products or services that you can provide. Service delivery may include a classical consulting assignment, with a diagnostic component, delivery of a solution, and tracking of outcomes. However, consulting capability in our context is broader as it includes any customer interaction, regardless of the nature of the product or service, since the emphasis is on building a bridge between customer needs and product or service capabilities. The significance of consulting capability varies according to the nature of products or services, ranging from a central role in an organizational consulting practice to a more limited role in a consumer-oriented product area. Consulting capability blends interpersonal skills with business acumen, which was mentioned earlier, so it is inherently more complex, involving, as it does, a combination of other skills. Skill building in this area can be tailored to the specific attributes of your nontraditional career. Various professional organizations representing the consulting field provide insights into emerging practices and resources.[21]

Selling. It was only after becoming responsible for a sales and marketing team that I came to appreciate the breadth and depth of skills involved in effective selling. I also came to appreciate that the sales process, just as with consulting, is primarily about understanding customer needs and meeting those needs. It is not about foisting an unwanted or unnecessary product or service on an unwary consumer. However, this latter image could explain our reticence about engaging in sales activities. This reticence is not surprising if we couple concerns about the nature of the process with misperceptions that only those who are aggressive and highly extraverted are effective in sales. In fact, effective sales behaviors are those of reading cues such as body language, tone of voice, and relying on intuition to sense what is really happening.[22] This is coupled with an ability to turn around difficult situations, a focus on outcomes while paying attention to people's emotional needs, and an aura of confidence.[23] Aggressiveness is not a primary driver of success.[24] Primary characteristics of effective sales managers are analyzing the future implications of their decisions, communicating clearly, maintaining expertise in their area, and engendering energy and enthusiasm in others.[25] The important interpersonal and strategic attributes associated with sales match well those of a nontraditional career, where sales are essential to success. A customer must purchase a product or service for this career path to be viable. As in other areas, there are many resources available to build skills.[26]

Support Service Management. The final two skills are in the area of support service management:

- Information technology
- Human resources

Information technology. Computing and communication tools are central to creating an effective and efficient operating infrastructure supporting a nontraditional career. This includes core activities such as document preparation, presentation preparation, spreadsheet processing of data, and outreach activities such as use of e-mail and social media, and it may include use of online database systems for gathering and synthesizing information as well as website creation and maintenance. This means being sufficiently knowledgeable to acquire, use, and maintain the relevant hardware and software or to purchase the appropriate support services. Outsourcing some aspects can be helpful, for example, purchasing software and hardware maintenance contracts provides ready access to assistance with initial installation and system learning as well as ongoing problem solving.

Human resources. Initially it is likely that a nontraditional career will be a solo undertaking if career components focus on service delivery, so human resource aspects will not surface. It is also likely that over time there will be a need to engage other people to support opportunities that develop. For a product-focused career component, other people may be engaged at the outset. Success is dependent on engaging people with the right skill sets, who fit well, and creating an environment that builds their affiliation and supports their development. This is the domain of human resources. It includes understanding and abiding by appropriate employment regulations, supporting employees in meeting performance expectations, and building a culture that creates a positive work environment. This blends strategic workforce practices, interpersonal sensibilities, and practical implementation considerations. Support resources in this area are accessible through payroll service providers, through professional organizations such as the Society of Human Resource Managers (SHRM), and on a contract service basis.[27]

Personal Characteristics

Now we turn to the area of personal characteristics:

- Integrity
- Tenacity
- Self-awareness (or self-understanding)
- Empathy
- Comfort with ambiguity

Integrity. Integrity means being honest and fair, ensuring that what is done matches what is said, being consistent, and following through with commitments. As Max DePree states poetically and appropriately regarding integrity, "At the core of becoming a leader is the need always to connect one's voice and one's touch."[28] Integrity is demonstrated by behaviors and it is the foundation on which personal characteristics are built. It is fundamental to effective decision making. Here's an example in contrasts. A helicopter carrying executives from the MBNA Corporation crashed into the East River while taking off from Manhattan in June 2005. Everyone on board escaped alive. Some were pulled from the choppy waters by everyday New Yorkers, including one grocery service delivery worker who leapt into the water to rescue people. When invited by the police to speak to the news media to take credit for his quick action, Miguel Lopez said he could not because he had to hurry back to work.[29] This quiet, selfless act of courage and integrity is in contrast to the behavior revealed in a nearby trial reaching its conclusion at the same time. In this trial, Dennis Kozlowski, the former CEO of Tyco International, was found guilty on fraud, conspiracy, and grand larceny charges for a combination of stealing and covertly selling "artificially inflated" shares with a combined value of $580 million.[30] This fraud was equal to the annual compensation of more than ten thousand U.S. households at a median level in 2005.[31] In stark contrast, Miguel Lopez is a shining example of integrity. Such integrity needs to permeate all aspects of a nontraditional career, and it will result in building trust and a strong foundation with others.

Tenacity. While projects sometimes materialize quickly and the path from initial proposal to implementation is measured in days or weeks, those projects with broad ramifications and significant scope can take more than a year to materialize. During that time there may be periods of quiet and other periods of active discussion. There may also be a time of heightened activity when rapid response is critical to meet a timing deadline that surfaces. Holding fast through this process, staying tenacious, is essential. Rarely will a customer's time frame follow a predictable path, and patience and perseverance are needed to stay the course. A benefit of extended project preparation time is the opportunity this affords for strengthening and further building relationships. It also provides time to refine and develop a project approach, and to gain support from others.

Self-awareness. With self-awareness comes the confidence to pursue a path of personal meaning and to stay true to personal values along that path. In the past world of corporate dependency, self-awareness was perhaps less important. Others defined organizational values; and the organizational path forward, in the parlance of those times, the top, by virtue of their administrative positions, told the middle what to do to the bottom. Why would self-awareness matter in such a setting? After all, it's about getting on an already moving bus, or getting off. With a nontraditional career, on the other hand, self-awareness is central to setting direction, selecting career components that matter, and behaving in a way that matches personal values. In chapter 1 we looked at elements of self-awareness: values, personality preferences, interests, skills, and learning from life and work experiences. We examined how to build understanding of who we are and what that means for our work. Building such understanding is an important precursor to a nontraditional career. Continuing to reflect on such aspects over time and adjusting the nontraditional path as needed, sustains the vitality and personal relevance of a nontraditional career.

Empathy. Empathy is a distinguishing attribute of successful leaders in many sectors and contexts.[32] It means listening with your heart as well as your head in order to understand and imaginatively enter into another person's feelings.[33] Empathy is about genuinely caring for others, being available to provide support when needed, and recognizing that work and life are intimately interconnected. It is a foundation for meaningful and effective work relationships whether they are with customers, those who may join you on your nontraditional career path, or with others in the community. It reinforces integrity and leads to openness, which in turn reinforces building trust. It is the glue that holds relationships together within an organization and it is an important part of team effectiveness. Demonstrating and modeling empathy on a personal level in work interactions encourages others to do likewise.

Comfort with ambiguity. It won't be a surprise that a nontraditional career can be unpredictable. New projects surface unexpectedly and others end, taking their natural course. New career components are added over time and others reach the end of their natural life cycle. This means that dealing with ambiguity and uncertainty are intrinsic to a nontraditional career. An example of this is the need to hold simultaneously the ability to work autonomously and the ability to work collaboratively. Both of these seemingly contradictory aspects will be called upon. Moreover, flexibility and creativity are woven into comfort with ambiguity. Comfort with ambiguity is an important attribute to develop and bring to a nontraditional career path. It means being able to cope with change, shift gears comfortably, decide and act without the total picture, and maintain equilibrium when there is uncertainty.[34] Lombardo and Eichinger identify approaches to developing skills in this area.[35]

Summary

The eleven individual skills and five personal characteristics that we reviewed are summarized in table 7.1. The table can be used as a template to note those skills and characteristics that are strengths, those that are development opportunities, and those where partnering may be appropriate. This can help in focusing development efforts and in considering partnering. In some cases the individual skills and personal characteristics may develop while in conventional employment as part of that employment or in parallel with it, through a combination of on-the-job experience and relevant courses.

For example, over a number of years, I was fortunate to attend courses in the (for me) foundational areas of career development, business strategy,

Table 7.1. Summary of Individual Skills and Personal Characteristics.

Category	Skill or Characteristic	Strength	To Develop	To Partner
Individual skill/ foundational	Content knowledge			
Individual skill/ foundational	Business and financial acumen			
Individual skill/ foundational	Time management			
Individual skill/ interpersonal	Communication			
Individual skill/ interpersonal	Negotiation			
Individual skill/ interpersonal	Networking and connecting			
Individual skill/ customer facing	Marketing			
Individual skill/ customer facing	Consulting			
Individual skill/ customer facing	Selling			
Individual skill/support service management	Information technology			
Individual skill/support service management	Human resources			
Personal characteristic	Integrity			
Personal characteristic	Tenacity			
Personal characteristic	Self-awareness			
Personal characteristic	Empathy			
Personal characteristic	Comfort with ambiguity			

managing strategic innovation and change, and economic evaluation of projects; in the interpersonal areas of public speaking, communicating with the media, business writing, negotiating, interpersonal skill development, and leadership and management; in the customer-facing areas of marketing, consultative sales, and customer service; and in the support service management area of effectively using core computer programs. At the time each course fit a particular need. In retrospect I can see that this accumulated learning was helpful in launching and establishing a nontraditional career, even though in most cases that was not the original intent. Some of your development may happen naturally in conventional employment, while you may plan and develop other aspects over time. Think of yourself as the CEO of your nontraditional career, guiding its development. There are some skills not mentioned here that nevertheless are important, for example, legal- and tax-related skills. These typically will be outsourced, or they may involve partnering. Exploring partnering, whether customer facing or operationally oriented, is the subject of our next chapter, which addresses the final of our six strategic factors. Before we move to this topic, here are some questions that may be helpful when reflecting on skills and personal characteristics.

Questions to Consider When Addressing Building Needed Skills

1. What strengths do you bring to individual skills in the foundational, interpersonal, customer-facing, and support service management areas? Which skills would benefit from additional development and in what order of priority? Which skills might benefit from partnering?

2. What strengths do you bring to the personal characteristics of integrity, tenacity, self-awareness, empathy, and comfort with ambiguity? Which personal characteristics would benefit from additional development and in what order of priority?

3. What will your development plan consist of to provide the foundation of skills and personal characteristics you need?

CHAPTER 8

Partnering

IT SEEMED LIKE a simple question. How do organizations benefit if they provide career development support for their employees? Surely there must be a great deal of evidence to support that benefit. This was a key question for me, new to the career field in the late 1990s and responsible for career services delivered into organizations by a nonprofit I worked for at the time. It turned out there was much useful anecdotal evidence from employees about the value to them, but little else. How to fill the gap? Here is where partnering comes in. One of the organizations receiving career services was Sun Microsystems. Seema Iyer, who was with Sun at the time in human resources, welcomed the opportunity to work together on this question. So we partnered. I was able to identify the data needed, Seema obtained it so I could analyze it, and together we synthesized the results, which showed significant organizational benefit.[1]

This is an example of the value of partnering. Pooling resources and knowledge led to an outcome that would not have been possible without it. Partnering is the focus of this chapter. We will examine when partnering can be helpful in a nontraditional career and the different types that are possible. This includes customer-facing partnering to enhance revenue and partnering designed to build internal capabilities. Partnering offers the opportunity to collaborate for mutual benefit, built on clarity about the rationale for it. In exploring this, we will look at what we mean by partnering and the different kinds that fit with a nontraditional career, the benefits and challenges of partnering, and, finally, the attributes of successful partnering. We will conclude with thoughts about implementation.

Partnering in a Nontraditional Career

Partnering is defined as sharing or being associated with another in some action or endeavor. When applied to business, this may include sharing in the risks and profits. There are specific attributes associated with the legal entity of business partnership, which we will explore in chapter 9 when looking at the different legal structures associated with a nontraditional career. However, in this chapter we are not looking at the legal entity of formation; rather, we are focusing on the activity of partnering with others as a means of supporting development of a nontraditional career. Let's look at three forms of partnering in this context: informal, customer facing, and building internal capabilities.

Informal. *Business Behaving Well,* the last book I edited, had fourteen additional contributing authors.[2] *Building Workforce Strength,* the previous book, had fifteen additional contributors.[3] Each contributor brought special expertise and knowledge. Each brought a passion for their area of interest and an enthusiasm to contribute their writing. The two books were deeper and broader as a result of these contributions. Moreover, the books provided a platform for the voices of these contributors to be heard. The contributors were included because they brought particular expertise and either I had known them for an extended time or they came by referral from someone I had known for some time. They became partners in each of these two writing endeavors. While writing a book is time bounded, with a definite beginning and end, the relationships that led to this partnering extended over a much longer time. This is an example of one type of partnering, namely, informal. While there were some specific agreements from the publishers about the book contributions, my relationships with the contributors were built on trust and common interests; they were informal.

Other examples of informal partnering come from volunteer work. Some presentations on the subject of social justice and social responsibility in *Business Behaving Well* benefitted from partnering with a colleague who covered areas she knew well while I covered aspects with which I was most familiar. The combination was stronger than either one of us alone. Another example is contributing to volunteer projects for the Taproot Foundation, which places teams from the for-profit world into nonprofit organizations. Here the team structure was designed to facilitate partnering and focus resources for the benefit of the nonprofit organizations.[4] Informal partnering has been a meaningful aspect of many components of my nontraditional career, such as the writing and volunteer components, and it will likely be important in your nontraditional career.

Customer Facing. Customer-facing partnering typically brings more structure to the partnering relationship. We will consider three forms. The first form is based on two or more individuals or organizations with complementary skills coming together so that they bring more value to the customer than they each could separately. The second form is when a supplier relationship evolves into partnering with a customer, built on shared expertise that addresses significant customer needs. This is distinct from a purely vendor relationship based on meeting contractually defined deliverables. The third form is a hybrid of the first two based on the partnering of more than one provider with a customer. Customer-facing partnering relationships are often based on contractual agreements that outline how the parties will work together. As a result, this introduces more structure than we saw in the earlier informal situation. Let us look at examples of each of the three forms of customer-facing partnering to illustrate how they can unfold. The first is an example of two entities coming together with complementary skills, the second addresses supplier/customer partnering, and the third is an example of more than one provider working with a customer in partnership.

We will begin with partnering based on complementary skills. Much of my initial organizational consulting was focused on the affiliation of individuals with organizations and the implications for leadership and workforce practices.[5] A related activity, though not a focus for me, was the deployment of performance management processes. More closely connected is the deployment of processes for guiding individual development activities viewed from an organizational level. As part of my consulting activities, I came across a small organization with a strong infrastructure for delivering computer-based performance management and development planning that could be aggregated at the organizational level. The offerings from this organization and my consulting focus were complementary. So we created a basis for collaboration and outlined it in a letter of understanding. It had much flexibility and was built around ad hoc collaboration as opportunities might arise. The primary benefit turned out to be a mutual exchange of learning rather than customer engagements. This exchange was particularly helpful at an early stage in my nontraditional career. Similar opportunities will likely arise for you as you develop your nontraditional career.

Now let us look at the slightly more complex example of supplier/customer partnering. Career services supporting the development of individuals in organizations provide a foundation for fulfilling individual careers. Moreover, organizations benefit by gaining a more capable and flexible workforce, and communities benefit from enhanced individual and organizational prosperity. Delivery of such services is based on content expertise. This is often interwoven with organizational workforce planning and development activities, and it can involve broad geographic coverage to reach a dispersed workforce. Given the strategic relevance and breadth of scope, such initiatives are typically multiyear undertakings requiring effective ongoing measurement to refine delivery and demonstrate continued value contribution.[6] With a somewhat complex and extended engagement such as this, there is much to be gained from supplier/customer partnering. It allows service delivery to evolve based on ongoing learning, changes in the external environment, or shifting organizational priorities. Such partnering also facilitates close coordination about communication of outcomes and outreach activities that encourage participation in the services. Examples of such partnering in delivering career services in health-care and high-technology settings are provided in *Building Workforce Strength.*[7] In these situations, supplier/customer partnering took place in the context of defined contractual agreements, bringing more structure to the arrangement than in the informal partnering we covered earlier. These contractual agreements addressed both required deliverables and financial aspects. Such partnering enables service delivery to become more effective over time, and it is an approach to consider as your nontraditional career develops.

Now we will look at examples of the third and most complex form of customer-facing partnering, that of more than one provider with a customer. These examples consist of two assignments that were spearheaded by an academic institution to provide services in two public sector settings, as mentioned in chapter 4. One assignment was to deliver leadership development modules for an academy created to support principals in an urban school system. The second was to provide organizational and change management consulting to a public sector agency operating at the state level. In both cases the educational institution was the primary contractor and I was engaged as a subcontractor, one of several for the second assignment. In each case, from the outset, the project leader from the educational institution based the approach on partnering among those providing services and the public sector entities. This was possible in large part because of the long-term relationships that existed between the project leader and many people at both public-sector entities, as well as with the providers. Indeed, it was as a result of previous smaller projects that these assignments came about for me. Most important from a partnering perspective is the long-term relationship of trust that led to these endeavors. This is another example of a form of customer-facing partnering that might emerge for you.

Building Internal Capabilities. Let us now look at how partnering can support building internal capabilities. We will look at this from two perspectives: first, engaging with others to support service delivery, and second, partnering with suppliers, acknowledging that there is some overlap in these two designations.

Between operating as a solo provider and creating an organization with employees, there is an approach that works well for time-bounded, sizable projects that are too brief to justify hiring people. This is where partnering with others can be beneficial. One form this can take is engaging people on a contractor basis for the duration of a project. In the United States there are legal requirements that stipulate when someone can be hired on a contractor basis rather than as an employee.[8] An example of where this form of partnering worked well for me was in the delivery of exit interview projects, mentioned in chapter 4. I began this work when part of an organization then continued it in my own practice. For the interviewing part of the projects, we needed people with a career or human resources background skilled in interviewing in sensitive areas. Each project was of a limited duration, typically one to three months, and the contractor arrangements allowed people to set their own part-time interviewing schedules. Based on contacts in the career field, I was able to identify people with the needed expertise who were pleased to engage on a project basis. We had a simple written agreement defining deliverables, confidentiality provisions, and compensation for a project. It worked well for those who participated and it made possible a broader project scope than I could have handled alone. Similarly, for some projects I engaged a colleague to conduct data analysis, building on her skill set and interests. This was particularly valuable during intense project engagement periods. You may find similar opportunities to engage others on a contractor partnering basis as your needs grow.

The second area of partnering to build internal capabilities is that of working with suppliers. We will look in more detail at practical aspects of building an infrastructure with suppliers in chapter 9. However, we can acknowledge here that suppliers of services such as payroll, tax, legal, accounting, and information services bring specialized expertise that in some cases can lead to partnering beyond transactional activities. For example, I engaged a colleague with expertise in information technology on a contract basis for selected exit interview projects to help establish online infrastructure needed for data collection and aggregation. This engagement included discussions about broader aspects of the data collection process. In the legal area, various questions arose from time to time, primarily either of a contractual nature or related to employment law. Having a relationship with an employment law firm that extended over time was beneficial. It was possible to rapidly address questions that surfaced in the context of shared knowledge about past practice. Here, too, you may find partnering opportunities.

There are parallels with partnering in other situations, such as organizational workforce partnering, as proposed in *Affiliation in the Workplace.*[9] In the area of social responsibility, as described in *Business Behaving Well,* parallels exist with partnering between for-profit and nonprofit organizations, between organizations in the public and private sectors, and between labor and management.[10] Such examples can provide additional insights for partnering in a nontraditional career.

The Benefits and Challenges of Partnering

Why engage in partnering? After all, it can be time consuming and may divert attention away from other activities, such as delivering services or products, building customer relationships, deepening expertise, or creating infrastructure. This question is similar to the one that arises when considering the formation of a team: Do the benefits outweigh the costs?[11] Can we justify the time and likely financial resources that will be needed for partnering to be successful? We will explore this by examining the potential benefits and challenges of partnering. Potential benefits include the following:

- Strengthened capabilities
- Enhanced revenue
- Lowered costs
- Broadened learning
- Accelerated innovation

Strengthened capabilities. As we saw in the earlier example of complementary skills, partnering can broaden the range of capabilities available to address customer needs. This works well when skill sets are complementary rather than overlapping and when the combination of skill sets is valued more by customers than the separate skill sets. Such a combination can provide access to customers who would otherwise be inaccessible. The test of this benefit is customer reaction. We also saw earlier that partnering to expand the scale of internal capabilities through contractor relationships can provide access to projects of greater scope or scale than would otherwise be possible. Strengthened capabilities enhance revenue, the next potential benefit of partnering.

Enhanced revenue. Enhanced revenue can come from a broadened customer base or increased scope or scale in service or product delivery. Implicit is a synergistic gain from partnering and the combination of capabilities that result. It is clear how this occurs when engaging others on a contractor basis, since the parties separately do not have the scale needed for project delivery but together they bring necessary scale. When partnering is built on complementary skills from each party, synergy comes from expanded offerings or expanded customer access, resulting in a broadened customer base.

Lowered costs. Cost reduction will come primarily from partnering with suppliers. For example, partnering with an employment law firm, mentioned earlier, can potentially reduce costs since accumulated knowledge and understanding can reduce the time and, as a result, the cost of addressing questions as they arise. Another example is partnering with a supplier of assessment instruments to lower costs based on the scale of purchases.

Broadened learning. While not necessarily having immediate financial effect, broadened learning strengthens the foundation on which a nontraditional career is built. Partnering is likely to offer significant opportunities for mutual learning based on complementary capabilities. Indeed, a key question to ask as a partnership unfolds is whether it continues to contribute to shared learning. If this is happening, it is a strong indication of continued viability. If, on the other hand, shared learning is absent, this is an indication that the partnership may no longer meet the needs of participants.

Accelerated innovation. If we consider innovation as developing an idea from conception to implementation, as opposed to creativity, which is generation of an idea, then creativity typically resides with individuals and innovation resides with teams. Individuals are creative, teams are innovative.[12] Partnering increases the size of the team, the opportunity for new ideas to surface, and potentially the likelihood that such new ideas are implemented. The test of this potential benefit is the number of new ideas that come from partnering and are successfully implemented.

Now we will look at some potential challenges of partnering:

- Unbalanced
- Too demanding
- Cultural disconnect
- Misaligned objectives

Unbalanced. I recall attending my first (and last) meeting of a particular professional society in Silicon Valley a number of years ago. It was an awards ceremony. All was on track when suddenly, at a particular point in the ceremony, one recipient chose to be carried from the back of the large hall, with much fanfare, to the stage on a thronelike dais supported by several people. He was the CEO of a company. This is the kind of organization or person with which it would be difficult for me to develop a balanced partnership. One of the challenges with partnering is achieving an equitable balance so that each party perceives equal commitment and reward. Continued open communication is important to maintaining balance, both perceived and actual.

Too demanding. Partnering works well when it complements the core activities of each party and does not interfere with these endeavors. It becomes challenging if the demands of partnering, in terms of time or financial resources, create conflicts with other core activities. Open communication and setting appropriate boundaries are important steps in addressing such concerns.

Cultural disconnect. Recognizing that there are many views about what constitutes organizational culture, there is consensus that it exists and that it is important.[13] It is particularly important when we consider partnering, which involves close interaction at an individual level, organizational level, or both. One challenge in partnering is finding other individuals or organizations that are culturally aligned. This can mean alignment of fundamental values such as integrity, social justice, and social responsibility, or alignment in how things are done, with planning and forethought or in a reactive manner. Assessing cultural alignment before partnering is important to avoid an early and unanticipated demise of the partnership.

Misaligned objectives. If one partner seeks growth and the other stability, if one seeks to reinvest and the other seeks to withdraw profits, objectives will be misaligned and sustaining partnering will be challenging. Clarifying objectives at the outset is important to avoid such misalignment. Objectives will likely diverge over time, which may at some point lead to dissolution of a partnering arrangement. However, initially at least, it is important to ensure that objectives are aligned.

In summary, when considering partnering opportunities, weighing the potential benefits and the potential challenges we have reviewed can provide insights into whether and how to proceed with a particular partnering relationship.

Attributes of Successful Partnering

We have examined what is meant by partnering and the benefits and challenges that it offers. Now we will look at eight attributes of successful partnering:

- Cultural alignment
- Aligned objectives

- Complementary contributions
- Mutual respect
- Mutual benefit
- Shared information
- Value created through synergy
- Innovation

Cultural alignment. We noted previously that cultural disconnect can derail partnering. Now let us look at how cultural alignment is indicative of successful partnering. How do we characterize culture in a work context? Many dimensions have been proposed.[14] Here are some examples relevant to partnering in a nontraditional career:

- Perspectives about responsibilities to multiple constituencies
  - For example, valuing generosity toward employees, community members, customers, suppliers, and partners in the broader context of social responsibility[15]
- People or task orientation
  - The balance between support for individuals and task accomplishment
- Approach to decision making
  - The extent to which decision making is participative rather than directive
- Support for innovation
  - The embrace of new ideas and degree of risk tolerance
- Emphasis on cooperation
  - The extent to which operating practices favor collaboration rather than competition, for example, the approach to performance management
- Time horizon
  - The relative influence of short- and long-term considerations on resource decisions
- External service orientation
  - The emphasis placed on understanding customer needs and delivering customer service excellence
- Communication
  - The degree of openness in communication
- Learning
  - The extent to which learning is emphasized and supports adaptability

For example, if a prospective partner adopts a broad view of relevant constituencies, operates in a participative decision-making style with open communication, and emphasizes collaborative practices with continual learning, and these are principles that you share, then there exists a strong foundation for successful partnering. Conversely significant gaps in one or more areas would suggest that partnering could be challenging.

Aligned objectives. Objectives can be broad, such as exploring opportunities in a particular domain of practice, or specific, such as establishing presence at a given customer with a certain level of sales or achieving particular revenue in a given time frame. They can include long-term aspirations and short-term steps toward meeting those aspirations. For example, one partner may have an objective to facilitate an introduction of the other partner to a particular customer to extend the range of products or services that are provided. The other partner may have an objective to understand the needs of this customer, learn where delivery capabilities could contribute, and consummate an agreement to supply services. Such objectives are complementary. A sign of successful partnering is consensus about objectives that are acknowledged as important, consensus about their relative priority, and consensus about the resources needed to accomplish these objectives. Reaching such consensus is based on discussion and cooperation; the resulting aligned objectives are an indication of effective communication. They are a sign that partnering is on a firm footing.

Complementary contributions. Successful partnering is built on each person or entity bringing specific and complementary contributions. Examples in client-facing partnering include complementary

- content knowledge, for example, knowledge of a given assessment instrument partnered with a related workshop framework;
- functional strengths, for example, delivery of a particular organizational diagnostic partnered with a sales relationship;
- infrastructure, for example, in-person delivery partnered with online delivery; and
- skill areas, for example, flooring installation partnered with painting and decorating.

The specifics of how these complementary contributions will be deployed will vary according to individual customer needs and the nature of the partners' contributions. For example, in one situation one partner may bring a customer relationship while the other brings particular expertise relevant to that customer. With another customer these roles may be reversed. Successful partnering incorporates such complementary contributions, and the vitality of partnering is maintained in part by their continued development.

Mutual respect. Respect can be described as a general attitude about how to view others, perceiving and responding to someone else as an equal so that each person feels they are properly seen and considered.[16] It is related to Carl Rogers's client-centered approach to counseling, which is based on unconditional positive regard and therefore encompasses respect.[17] When partnering is shorter term, focused on a specific event, for example, informal partnering contributing to a book, as mentioned earlier, or partnering to build internal capabilities, as with the exit interview projects mentioned earlier, then a relationship of mutual respect needs to be established quickly as the partnering launch time is short. Ideally this respect is based on an existing relationship. When partnering is longer term, over several years, for example, the supplier/customer relationships mentioned earlier, mutual respect needs to strengthen over time in order to sustain this extended relationship as it develops. In our exit interview studies, people frequently cited lack of respect as a major factor in their leaving an organization. Lack of respect can lead to fractured relationships. Maintenance of respect is an important foundation for extended partnering. So a signal of successful partnering is that each party exhibits respect for the other(s).

Mutual benefit. Another signal of successful partnering is mutual benefit. This can take both tangible and intangible forms, as we saw earlier in the chapter. Tangible benefits include equitable financial compensation, likely based on an agreement created at the outset that reflects the contributions of each party. Tangible benefits may also include the opportunity to engage in new projects as they emerge, exposure to different operating practices that might suggest efficiencies, and the opportunity to use resources more effectively based on increased utilization resulting from partnering. Intangible benefits include shared learning, the potential to innovate by identifying and developing new ideas, development of new customer relationships, and the opportunity for shared thinking and brainstorming about future directions. This range of benefits is substantial and the extent to which they are captured is a signal of successful partnering.

Shared information. Since effective partnering is based on complementary contributions, each party brings breadth and depth of knowledge that is mostly new to the other party. Application of this knowledge is built on cultural norms and operating practices that uniquely reflect each person or entity. This limits the potential business risk of sharing information. Indeed, with partnering based on mutual respect and an expectation of mutual benefit, sharing information is a practice that will strengthen rather than threaten partnering. It is central to the successful initiation and evolution of a partnering relationship. Concerns about maintaining proprietary information confidential are mitigated with an appropriate confidentiality agreement established at the outset. An example of shared information is in exit interview partnering, where key operational aspects of the data collection infrastructure were shared with the contractor partners who were conducting interviews so they could use the system infrastructure to record exit interview results for later aggregation, analysis, and synthesis. Thus another signal of successful partnering is sharing of information.

Value created through synergy. Partnering is successful when value is created through the synergy that emerges from the relationship. This means that the value contribution to a customer from the tangible and intangible attributes of each party, when combined, is greater than the sum of each separately. Synergy does not happen by accident; rather, it happens by the intentional actions of each party to complement the other. When partnering is based on shared information and complementary contributions, it becomes possible to create additional customer value from enhanced product or service offerings or from operational economies. For example, one of the projects described earlier involved an organizational consulting activity with a public sector agency. In the course of assessing organizational needs, it became clear that a diagnostic survey component would be useful. Having prior experience delivering such diagnostic approaches, it was possible to readily construct the needed information gathering and synthesis protocol and incorporate this component. Many of the interviews conducted as part of this information gathering were conducted by staff from one of the partnering organizations, which simplified deployment and lowered the costs. Synergy in this case, through shared capabilities, led to enhanced operational efficiency. Such synergy is an indication of successful partnering.

Innovation. The final attribute we will examine as an indicator of successful partnering is that of innovation, the identification of new ideas and their implementation. This is the lifeblood of emerging enterprises. Innovative ideas are particularly susceptible to suppression in fledgling stages as they are still evolving and are vulnerable. Successful partnering helps foster the emergence of new ideas and protects them during these early stages while they are being refined. This was needed in the successful career services analysis project that introduced the chapter, since one external adviser quipped at the outset that the project was unworkable. So our final attribute, indicative of successful partnering, is the ability to foster innovation.

Implementation

Prior to implementing partnering relationships, it is important to understand the intended strategic contribution and the outcomes sought. This is similar to the initial steps in workforce partnering.[18] Recognizing that perspectives will change over time as a nontraditional career evolves, a starting place is mapping potential career components, as described in chapter 5. This helps identify components that might fit well but where there are aspects missing that might hamper implementation. These are potential partnering opportunities. A second step is to review skill needs and skill gaps building on the approach described in chapter 7, as this can also highlight areas where partnering would be helpful. Once the potential contribution of partnering is established, we can look at the mechanics of implementation.

Implementation is affected by the anticipated duration of partnering and likely trajectory of the partnering relationship. Short-term partnering supports immediate event- or project-related needs such as in the earlier exit interview example. In such cases, given a likely rapid launch time, it is beneficial to have already established a relationship so that project-related partnering can be readily initiated when needed. A simple contract outlining the nature of the partnering relationship and any confidentiality considerations can provide the legal framework. Longer-term partnering typically has significant resource and time implications, so it benefits from a more extensive definition of intent and contract scope at the outset. For both short- and long-term partnering relationships, it is advisable at the beginning to define anticipated outcomes and to identify measures of success. Assessing progress in these areas over time then helps guide future decisions about partnering continuity and adjustments that might be appropriate. For both short- and long-term partnering, it is important to define at the outset what will happen on dissolution of the partnering relationship, for example continued confidentiality provisions. With short-term, project-based partnering, dissolution will occur naturally on project completion. With long-term partnering, objectives may diverge over time. Defining an approach to dissolution in the initial agreement makes the separation straightforward while maintaining amicable relationships.

When considering a partnering relationship, the attributes of successful partnering that we reviewed can provide a framework for evaluation. Table 8.1 summarizes these in template form for use when considering partnering opportunities. The attributes are listed in the left column. Inserting a check in one of the next two columns indicates whether a particular attribute is a strength or a concern. The final column is for observations related to each attribute.

When built on a strong foundation of alignment, partnering can complement a nontraditional career. It can potentially enhance both customer-facing

Table 8.1. Attributes to Consider in a Partnering Evaluation.

Attribute	Strength	Concern	Observations
Cultural alignment			
Aligned objectives			
Likelihood of complementary contributions			
Extent of mutual respect			
Likelihood of mutual benefit			
Openness to sharing information			
Likelihood of value creation through synergy			
Catalyst for innovation			

and operational aspects of a nontraditional career, valuable both during initial launch and once it is established. Now that we have completed our review of the six strategic factors, we will address in more depth practical steps and the path forward. In the next chapter we will look at the nuts and bolts of creating an infrastructure. Before addressing that topic here are some questions to consider when addressing partnering.

Questions to Consider When Addressing Partnering

1. Which of your career components would benefit from partnering and why?
2. What skills areas could you enhance through partnering?
3. What benefits might accrue by developing partnering relationships?
4. What challenges do you anticipate in approaching partnering and how might you mitigate those challenges?
5. What partnering opportunities would you consider?
6. How do these opportunities align with the attributes of successful partnering?

PART III

Practical Steps and the Path Forward

There is nothing like returning to a place that remains unchanged to find the ways in which you yourself have altered.

Nelson Mandela

CHAPTER 9

Creating an Infrastructure: Nuts and Bolts

THERE WERE ABOUT TWENTY of us in the room, each with a computer. The quiet, purposeful air helped as we learned how to use a payroll system for our respective organizations. Most people, other than me, specialized in payroll activities. I could see why from the large, complicated book of instructions that came with the session; seven years later, it is still on my bookshelf. Near the end of the session, we were each asked when we would first use the payroll system in our organizations. Answers were mostly a month, two months, or three or more months in the future, after people translated their knowledge to their organizations. Then it was my turn to observe, with some embarrassment, that I would have the first payroll run the next day. There was a slight gasp of disbelief from the others, along with words of support. Words you might offer someone on a mission with no hope. I realized, for the first time, that there was a big challenge ahead. And while it was in fact a challenge, all eventually turned out well, though with some foreknowledge there might have been less nail biting and the path forward would have been smoother. So in this chapter we will look at how to be more prepared. We will look at the practical, nuts-and-bolts stuff of launching and building a nontraditional career. We will begin with an exploration of different forms of business structure, look at regulatory and related considerations, address internal infrastructure, and conclude with customer-facing considerations.

Business Structure

Why is business structure important or necessary? Why not just start delivering services or products without any such structure when embarking on a nontraditional career? There are two primary reasons. First, a structure is needed to meet regulatory requirements, for example, for filing federal and state taxes or to obtain business licenses and permits. Second, some structures can provide protection for personal assets should problems arise in the business entity. This is known as limited liability, meaning that a partner or investor cannot lose more than the amount invested and is not responsible for the debts or obligations of the entity in the event that these are not fulfilled.[1] In this chapter we will focus on the regulations of the United States, recognizing that domestically some requirements vary by state and locality, and that while specifics will vary by country, general principles are likely relevant in other places. We will focus on structures of for-profit entities since a nontraditional career needs to provide financial support. If nonprofit structures, which carry a 501(c)(3) tax-exempt designation in the United States, are of interest, there is much information available for this area.[2] There are various forms of for-profit business structures that can fit at different stages of a nontraditional career. Let us look at each of these forms, beginning with the simplest then moving on to those that are more complex. I will summarize the following forms of business structure:

- Sole proprietorship
- Partnership
- Cooperative
- Limited liability company (LLC)
- S corporation
- Corporation (C corporation)

Sole proprietorship. The most straightforward structure is a sole proprietorship.[3] It is an unincorporated business owned and run by one person who benefits from the profits. As the sole owner, you are personally responsible for any business liabilities, which is one of its primary disadvantages. On the other hand, having minimal set-up requirements is a significant benefit. It is necessary to establish a fictitious business name (this terminology simply means business name), sometimes known as "doing business as" or DBA, if the business operates under a name other than that of the sole proprietor. This is readily obtained through application, usually to the local county clerk, and publication in a local newspaper. It is also necessary to obtain the necessary licenses and permits; for example, our local township requires a business license that is renewed annually. Another advantage of a sole proprietorship is that tax filing is straightforward, using Schedule C to report income or losses, which is integrated into the standard annual tax return form 1040. Information about filing U.S. federal taxes and associated tax forms in connection with a sole proprietorship is provided at the Internal Revenue Service website.[4] One example of information for state tax filing is described by the California Franchise Tax Board.[5] Given its simplicity, a sole proprietorship can be an excellent way to start a nontraditional career, as long as the disadvantages of the difficulty of raising external financing and personal liability for business debts are not critical. This is how I began my nontraditional career, first in our career counseling and college advising practice and then with an organizational consulting practice. This worked well for a number of years before the scale and complexity of the organizational work required that we adopt a different structure for this practice.

Partnership. A partnership is a single business where two or more people share ownership.[6] Each person contributes to all aspects of the business and shares in the profits. A legal partnership agreement is recommended at the outset. A slightly more complex form is a limited partnership, which is structured with general partners who manage the business and are personally liable for debts and limited partners who contribute capital, share in the profits, but have no operational role and have limited liability. A limited partnership needs to be registered with the state, usually the secretary of state's office. As before, a fictitious business name and business licenses and permits will likely be required. Information about filing U.S. federal taxes for a partnership and associated tax forms is provided at the Internal Revenue Service website.[7] One example of information for state tax filing is described by the California Franchise Tax Board.[8] Partnerships are relatively easy to set up and they draw on complementary skills of the founders. Disadvantages are the personal liability for actions and debts of the partnership and the challenge of maintaining harmony as circumstances change and contributions of partners differ.

Cooperative. A cooperative is an organization generally owned and operated for the benefit of those using its services.[9] In a worker cooperative the workers own the business and decide what to do with the profits, as opposed to consumer cooperatives, which are stores owned by members who shop at a discount.[10] It may be of interest to you under certain circumstances as part of a group endeavor. Profit and earnings are distributed among the members, who are known as user-owners. Usually an elected board runs the cooperative, with regular members through their votes controlling the direction. Members join by purchasing shares; the weight of their vote is not determined by the number of shares. Forming a cooperative is different from forming other business entities. A group of potential members agree on a need and approach and then, through a consensus process, develop a business plan. Many cooperatives choose to incorporate, though it isn't required. Those cooperatives that do proceed to incorporate go through a process of filing articles of incorporation, creating bylaws, creating a membership application process, and obtaining licenses and permits. In general, cooperatives register with the Internal Revenue Service and state and local revenue agencies. Cooperatives do not pay federal income taxes as a business entity; they follow the regulations of the Internal Revenue Service's subchapter T cooperatives tax code. Members pay federal and state taxes on the margins earned by the cooperative, with requirements varying by state. The advantages are a democratic structure, continuity when members change, scale that creates purchasing economies, and no taxation at the entity level. Disadvantages include the challenge of raising capital and difficulties if members do not actively participate.

Limited liability company (LLC). A limited liability company is a hybrid structure that provides the limited liability features of a corporation but with the tax and operational efficiencies of a partnership.[11] The owners, referred to as members, may be a single individual, two or more individuals, corporations, or other LLCs. Profits and losses flow through directly to members' tax returns. Information about tax implications for an LLC is provided at the Internal Revenue Service website.[12] Though there are some variations from state to state, there are certain general principles, such as the need to choose a business name, filing articles of organization, creating an operating agreement, and obtaining licenses and permits. Benefits of an LLC include the limited liability aspect, less recordkeeping than corporations, and straightforward profit sharing. Disadvantages include the need for dissolution in many states when a member leaves an LLC, and some tax considerations, for example, in California LLCs are taxed based on gross revenue, not profits.[13]

S corporation. An S corporation (S corp) is a special type of corporation with a subchapter S designation from the Internal Revenue Service. With an S corp, profits and losses pass through directly to personal tax returns and the business is not taxed itself at the federal level. Information about filing U.S. federal taxes for an S corp and associated tax forms is provided at the Internal Revenue Service website.[14] The approach to state taxation varies by state. To create an S corporation, it is necessary to first file as a traditional C corporation in the state in which the business is headquartered, then elect to become an S corporation. An S corp is considered a unique and separate entity apart from those who own it, which limits the liability of you as owner or "shareholder," though with certain limitations.[15] Any shareholder working for the entity must pay himself or herself reasonable compensation at fair market value. As before, business licenses and permits may be needed. An S corp provides limited liability, tax advantages, and continuity of operation with change in ownership. Disadvantages include administrative aspects such as annual reporting requirements. We found the S corp structure worked well when our organizational practice expanded in scale and scope.

Corporation (C corporation). The C corporation structure is usually for larger companies with many employees. It is an independent legal entity owned by shareholders, and the corporation, not the shareholders, is legally liable for the actions and debts the business incurs.[16] C corporations operate based on complex tax, legal, and reporting regulations. They are formed under the laws of the state within which they are registered and state laws vary. As before, business licenses and permits will likely be needed. Corporations are subject to income tax on their profits, and then dividends distributed to shareholders may also be subject to tax. Information about filing U.S. federal taxes for a corporation and associated tax forms is provided at the Internal Revenue Service website.[17] Corporations offer the benefit of limited liability to shareholders. Raising additional capital through the sale of shares is relatively straightforward, for example, through a public offering. Disadvantages include the costs and complexity of start-up and ongoing reporting requirements. This structure is likely most relevant if the objective is one of achieving large scale.

These are the primary business structures to consider as you launch and develop your nontraditional career. The structure you use may evolve over time, for example, progressing from a sole proprietorship to either an S corp or an LLC as new career components are added and scope grows.

Regulatory and Related Considerations

Now we will look at various regulatory and related considerations in the following areas:

- Business registration and taxation
- Engaging contractors and employees
- Insurance

Business registration and taxation. Let us briefly further address the topic of registering a business as this arises for several of the business structures. At the federal level an employer identification number (EIN), also known as a tax identification number, is used to identify a business entity. It can be obtained readily online from the Internal Revenue Service's website.[18] An EIN is required if you have employees or are operating as a partnership or corporation. In many cases it may not be needed if you are a sole proprietor.[19] There is additional registration at the state or local level. The Internal Revenue Service website provides links to each of the state websites where there is information about state and local requirements.[20] For example, in California, where my nontraditional career has been based, information is given at the California website about applying for a business license in the local township, applying for a fictitious business name at the county level, registering at the office of the secretary of state as a corporation, obtaining a seller's permit at the state level to authorize the sale of tangible property (for example, products), obtaining the appropriate state income tax forms at the state level, and workers' compensation insurance requirements.

While this may sound daunting, some of these requirements only surface when moving beyond the stage of sole proprietorship into a more complex business structure. In moving to this more complex stage, it can be helpful to engage an attorney specializing in small business and employment law. This attorney can file appropriate documentation at the federal and state levels as well as address questions that may arise. We needed to incorporate quickly when we formed the S corp to address rapidly growing business needs. Here our attorney proved excellent and the relationship worked well on an ongoing basis to address questions that came up, for example, those related to contract preparation. As scale increases it may also be helpful to engage an accountant specializing in small business to handle tax submissions, as we will discuss later. Again, we were fortunate to find an accountant with in-depth knowledge of this area who provided valuable support as needed.

Engaging contractors and employees. Now we will look at various forms of employment relationship for engaging people, including addressing some practical issues. As business opportunities grow, engaging others will probably be necessary to support this growth. As mentioned in chapter 8, one option is to engage people on a contractor basis for specific projects, being sure that this approach meets Internal Revenue Service guidelines.[21] Reporting contractor earnings is straightforward using form 1099-MISC, which is required for each contractor receiving $600 or more from you in a given tax year.[22] Hence engaging someone on such a contractor basis is sometimes known as a "1099 relationship." It is necessary first to provide each contractor with form W-9, a request for taxpayer identification and certification.[23] This ensures that you have the correct information for then completing form 1099-MISC. We used a simple written agreement with each person we engaged as a contractor, outlining project expectations, defining compensation, and including confidentiality provisions.

As project needs increase, it will likely be necessary to engage people as employees, possibly in addition to contractors. Engaging employees is sometimes called a "W-2 relationship." in reference to form W-2, which must be filed with the Internal Revenue Service each year for every employee receiving compensation of six hundred dollars or more in that tax year.[24] Engaging employees is a significant step, since the relationship is now continuous rather than being defined by specific project deliverables as in a contractor situation. Creating a supportive work environment now includes organizational responsibilities such as encouraging individual development.[25] It means adopting a philosophy related to performance and communicating this. For example, larger organizations frequently use performance management processes that try to link (small) pay raises to results. Such an approach theoretically creates alignment of objectives within the organization. However, such approaches can also be demotivating to many and of marginal value to the few who receive slightly higher pay raises. In our organization we took a different approach. A primary need was for collaboration, not internal competition. So our practice was to hire experienced, effective individuals and not differentiate based on subjective performance but to offer ongoing development support. This worked well and helped create a supportive work environment and excellent results due to the commitment of team members. Ensuring that resources are in place when employees start and establishing a warm and welcoming environment are critical to creating a strong foundation for future relationships.

On a practical level, engaging employees means determining whether it is appropriate to use an exempt basis, meaning they are not eligible for overtime compensation, or a non-exempt basis.[26] It means developing effective hiring and onboarding processes. This includes creating job descriptions; defining the nature of employment, whether full or part time; defining appropriate compensation and benefits; and establishing onboarding activities. Behavioral interviewing can contribute to effective hiring. This is a powerful and widely used tool to assist in the selection process, based on the premise that past behavior is the best predictor of future performance.[27] Behavioral interviewing is built on first identifying skills needed for a given position and then structuring questions to probe for examples of behaviors demonstrating those skills. Establishing appropriate compensation and benefits can be aided by accessing resources that provide examples of compensation levels, reviewing similar job postings, and basing them on personal knowledge.[28] Benefits to consider include the amount of vacation time, sick time, community service time, time for jury duty, and the nature of support for individual development. Enactment of the Affordable Care Act now makes it possible for individuals to readily obtain health-care insurance independently, as we will reference later in the chapter, while also offering options for small businesses. Given the greater complexity of the employment relationship relative to contractor engagement, it is important to use a comprehensive employment offer letter that outlines the conditions of employment, responsibilities, compensation, and benefits, confidentiality considerations, and various other factors associated with employment, such as procedures to resolve disputes, and at-will provisions that specify the circumstances under which employment can be terminated. An employment attorney can provide a template based on your situation.

A further practical consideration is verifying eligibility to work using form I-9 from the Department of Homeland Security.[29] It is also necessary to clarify employee preferences with respect to withholding allowances using Internal Revenue Service form W-4.[30] This provides a basis for determining appropriate federal withholding taxes, which will be handled through payroll administration. Payroll administration also addresses payment of state withholding taxes, state unemployment insurance (SUI), and Social Security and Medicare taxes, and it can include payment of workers' compensation insurance. Payroll administration can be handled effectively by engaging a payroll service provider, as I will describe later in the chapter.

Insurance. Of the several forms of business-related insurance connected with starting a nontraditional career, some are discretionary and others are required. Moreover, some insurance needs will vary according to the nature of career components. For example, we purchased professional liability insurance for our career counseling services, both for working with individuals in our private practice and to cover team members in our organization providing career services on site for other organizations. We used an insurance provider that specialized in related areas, Healthcare Providers Service Organization (HPSO), having found the provider through membership in a professional organization.[31] Exploring content-specific insurance needs when starting a nontraditional career is important. Professional liability insurance is an example of insurance that some customers may require and that for others is discretionary.

That is also true for other forms of insurance, such as general liability (and additional umbrella insurance providing higher coverage limits) and automobile coverage for business use, which may be discretionary or required by some customers. General liability insurance covers property damage and/or bodily injury arising from negligence. If you plan to work from home, explore the possibility of modifying and extending the provisions of your current insurance to accommodate business coverage. If establishing a separate location, new insurance coverage will be needed. Our existing insurance provider was able to offer general liability coverage when we added a private practice from home. However, this initial provider was not equipped to handle the expanded scope of our business after we formed an S corp. We needed to change providers to one specializing in the business area and selected Travelers Insurance Company, which was affiliated with our payroll provider, Automatic Data Processing, Inc. (ADP).[32] This worked well. Indeed, Travelers also provided our workers' compensation insurance, which is one of the other required areas of insurance for an organization with employees and may be needed by a sole proprietor in some occupations.[33] Workers' compensation insurance provides for wage replacement and medical benefits to someone injured in the course of employment. Regulations vary by state, and insurance premiums vary according to the insurance provider, the occupation and industry sector, and an organization's past history of work-related injuries. The benefit of integrating workers' compensation insurance with payroll processing is simplified administration and premium payments being distributed throughout the year rather than being concentrated in a single lump-sum payment.

Some additional insurance items can also be integrated with payroll processing. For example, another form of required insurance for an organization that has employees is state unemployment insurance, mentioned earlier.[34] This is part of a joint state-federal program that provides cash benefits to eligible workers who are unemployed through no fault of their own and meet other eligibility requirements defined by state law. Each state administers its own program within federal guidelines. The program is primarily paid for by a tax on employers. Rates vary by state and by past unemployment insurance claims for a particular organization. When registering a business, it is necessary at the outset to register with the appropriate agency for each state in which you are doing business to obtain SUI rates and registration numbers.[35] SUI payments can then be handled along with payroll processing. A related item, federal unemployment tax act (FUTA) tax, which in conjunction with SUI contributes to payments for workers who have lost their jobs, can also be handled with payroll processing.[36] This is also true for other state-specific insurance programs, such as California's state disability insurance (SDI). SDI provides short-term benefits to eligible workers for loss of wages when they are unable to work due to non-work-related illness or injury or the need to take time off to care for a seriously ill close family member or to bond with a new child.[37] It provides partial wage replacement and is funded through employee payroll deductions. In some cases insurance premiums for these various programs are capped based on state or federal requirements. The number and variety of items, particularly when multiple states are involved, speaks to the complexity of administration in this area. This administrative complexity can be considerably simplified by using a payroll service provider as we will describe later in the chapter.

The final insurance item is that of health care. The significant changes accompanying implementation of the Affordable Care Act in 2014 are summarized at the HealthCare.gov website.[38] In most states, provision of health care by employers for employees is optional for organizations with fewer than fifty full-time employees. Be sure to check the regulations for the states in which you will be operating. Hawaii, for example, requires that employers provide health-care insurance to employees. The Affordable Care Act enables individuals to purchase health-care insurance independently and prohibits exclusion for preexisting conditions. It also provides options for small business purchase of health-care insurance. This area is likely to continue to evolve, hopefully to an effective Medicare-for-all, single-payer system that would greatly increase health-care system effectiveness and efficiency while significantly reducing costs.[39]

Internal Infrastructure

Now we will focus on the internal infrastructure to address the various requirements. We will look at the following aspects:

- Payroll
- Accounting
- Tax return preparation
- Financial considerations
- Information technology

Payroll. We mentioned several times the contribution of a payroll system to easing administrative burdens when the organization has employees. This leverages your capabilities so that you can focus on your areas of core content expertise. One important question that surfaces immediately is which payroll provider to use. Let us examine some factors that influence this decision. Expertise of the provider is a primary factor given the breadth of connected considerations, such as federal and state tax payments, quarterly wage and tax reporting and filing, new hire reporting, and various insurance aspects, all linked to payroll administration. Selecting a provider with in-depth expertise in such payroll-related areas, and with the ability to service multiple states, should this be needed, are key factors. Ready access to the provider when questions arise, an easy-to-use system, and the ability to set up a payroll run at any time are also important, along with reliability of service access, flexibility to adjust payroll frequency to match needs, and flexibility to accommodate different state or locality requirements. For example, some places will levy local taxes, while others may have taxes for special situations such as transportation. Further considerations are the costs for using the service and the ability to provide access to related services, for example, workers' compensation insurance, human resources information such as hiring protocols, background check capability for new hires, provision for direct deposit to employees' bank accounts, and the ability of the payroll provider to directly mail employees' W-2 forms annually. Moreover, the ability to directly extract financial information from a payroll run to import into an accounting system is a valuable time saver. Think of the payroll administration system as a virtual administrative assistant that automatically takes care of many administrative tasks. Given the range of payroll activities, it is not surprising there was a gasp from others in the payroll training session, as I mentioned at the beginning of the chapter, when the first payroll processing was due the next day.

We used ADP as our payroll provider once we formed our S corp. We chose to run our payroll every other week using an online portal to input information; telephone input is also available, as is telephone support. All the features mentioned in the preceding paragraph were available. We used direct deposit for our employees and received ready-to-mail hard copies of pay stubs for each person each payroll period. The system worked well, even for a payroll neophyte like me, providing much needed administrative support.

Accounting. Accounting is another area where computing tools offer a major administrative benefit. Accounting software provides a straightforward means of tracking financial performance and generating information that simplifies annual tax return preparation. When we started our second sole proprietorship, we began using Intuit's QuickBooks Simple Start program.[40] This proved particularly helpful because it guided initial setup. The software allowed for tracking of multiple business entities and it worked well while our business structures were sole proprietorships. Formation of an S corp and growth in scope meant we needed a more sophisticated system, and with the help of a bookkeeper we were able to migrate from QuickBooks Simple Start to the QuickBooks Pro platform, adding accounting items as needed.[41] This worked well in conjunction with a hard-copy filing system for items associated with accounting entries. The hard-copy files were aligned with the electronic system to provide information that would be needed for audit purposes. We were able to export relevant information from our payroll system directly into the accounting software. The accounting software provided instantaneous access to current financial information, the ability to send electronic invoices from the system, and reports that provided information needed for submitting annual tax returns. We used the software for both the sole proprietorship that we retained to cover our practice working with individuals and for the S corp that covered our organizational work and had evolved from an earlier sole proprietorship. The software could link to credit card processing, although we did not use that feature. The software could be configured on a cash or accrual basis. For the cash basis, income and expenses are shown at the time a payment is received or a bill is paid. With accrual accounting, income or expenses are shown at the time a product is shipped, a service delivered, or a purchase is received. The cash basis worked well, particularly given the importance of managing cash flow in a small enterprise. You may also find that beginning with a simpler accounting program and then migrating to a more sophisticated platform helpful as business needs dictate.

Tax return preparation. Closely connected to the accounting system is tax return preparation. During our sole proprietorship phase, when we had a federal tax return, a single state tax return, and no corporate returns, I was able to prepare annual tax returns using Intuit's TurboTax software.[42] With a nontraditional career, it is necessary to submit estimated quarterly tax payments, and the software helps with the calculation and provides the associated vouchers. When we expanded to operations in several states and formed the S corp, we needed to submit multiple state tax returns, both personal and corporate, as well as federal personal and corporate tax returns, and provide estimated tax payments to more than one state. At this point we engaged a tax accountant. This proved effective and straightforward, once we established the initial basis. It brought considerable expertise to our tax reporting process and the ability to readily address questions that surfaced. You may also find it beneficial to gradually evolve to greater levels of tax return support as business needs grow.

Financial considerations. Meeting cash flow needs from start-up through established presence is fundamental to a viable nontraditional career path. This is affected by the rate and nature of entry, whether immediate and full immersion or a paced entry. In both cases it is helpful to build a contingency fund to support the likelihood of changing income levels, particularly in the early stages.[43] It is also important to set up a one-participant 401(k) plan and/or an Individual Retirement Account to address long-term saving needs and benefit from the tax advantages of such approaches. The 401(k) plan becomes more complex when you add employees, so consulting a tax accountant, and reexamining the approach at that point is appropriate.[44]

In the earlier stages of a nontraditional career, which likely entail a sole proprietorship business structure, resources needed to assist in cash flow management will probably be bank (or credit union) accounts dedicated to the business entity. This will also provide a credit card to simplify purchasing activities. As business activity grows, cash flows will increase and additional accounts may be necessary, for example, to handle cash inflow and outflow for payroll purposes. Formation of an incorporated business entity such as an S corp will also require separate and dedicated bank accounts. We used a large national bank for our accounts, primarily for convenience, such as ease of online banking and multiple branch locations. This provided access to lines of credit if needed, though we did not use them. Lines of credit or other sources of financing, such as loans, may be needed if the approach is one of rapid, full immersion with the associated capital requirements. The Small Business Administration 7(a) loan program may be of help should you have initial funding needs.[45] It is a good idea to explore various banking relationships when starting a nontraditional career, particularly local banks, which may offer preferred services and rates.

Information technology. Information technology can provide increased operational efficiency and extend capabilities. Staying current with rapidly evolving technology is easier for a single individual or a small enterprise than it is for a large organization facing cost and upgrade compatibility issues. Core technology needs include basic computing and printing hardware, backup capabilities (for example, with an external hard drive or using a web-based backup service such as Mozy), and software for word processing, presentations, and numerical manipulation.[46] The standard Microsoft Office suite of Word, PowerPoint, and Excel will likely suffice for these latter functions. E-mail remains ubiquitous, and one of any number of platforms, including Microsoft Outlook and Gmail, will work well. We also found use of the web-based fax service eFax to be helpful.[47] It is beneficial to use a web hosting service, which will provide a platform for your website and multiple e-mail accounts that will be particularly useful as employees join. We used AT&T for this service. Likewise, smartphone capability provides good communication access to e-mail, instant messaging, social media platforms, and the Internet.

There may be specific hardware and software offerings relevant to your career components that are helpful. For example, a small, portable projector to connect to my computer proved invaluable for presentations. We also needed the ability to gather data from disparate locations, aggregate this into useful information, and create presentations and reports from this information. This need arose in multiple settings, for example, in gathering exit interview feedback, tracking time utilization to meet contractual obligations, tracking client appointments and outcomes, gathering client feedback and workshop feedback, and creating repositories of shared information for teams. We also needed a system that was secure, easy to use, remotely accessible, reliable, scalable, and could be customized for many different situations. We used Intuit's QuickBase platform, which proved excellent for these purposes.[48] Depending on your specific needs, there may be similar applications that add value. Given the absence of an information technology department, we have found it helpful to purchase extended maintenance agreements for selected hardware and software products so assistance is readily at hand.

Customer-Facing Considerations

We will examine two primary customer-facing considerations:

- Pricing
- Outreach

Pricing. Initial pricing decisions will cascade through the years as a nontraditional career grows, since it can be difficult to make subsequent adjustments with large institutional purchasers. Extended contracts with customer organizations are based on the premise that you bring experience and expertise, and initial pricing should reflect this. Pricing decisions are critical to the financial viability of a nontraditional career. It is important to know typical pricing levels for products or services similar to yours. This knowledge can come, for example, from your prior work experience, as it did in our case, or from customers. It will help ensure that your pricing is neither so low it will jeopardize returns nor so high it will jeopardize securing opportunities. Adjust your pricing to reflect the particular attributes and differentiation that you bring, being sure that proposed pricing generates an acceptable return. If considering billing on an hourly rate as a sole proprietor, be sure to allow at least half of your time for marketing and other activities so that billable time accounts for only half of your time and billing rates are set accordingly.

Pricing to supply services or products to an organization can be structured on a fixed (or lump-sum) basis, a variable basis (for example, linked to time spent or activities completed), or a hybrid basis, with a fixed component and a variable component. A fixed basis offers the benefit of guaranteed revenue, though it comes with a risk of loss should costs exceed those anticipated. A variable basis removes this risk element but may limit returns if scope is subsequently reduced. A hybrid approach may work well, bringing the positive attributes of the other approaches together with a fixed-base fee, which may be payable at regular intervals, and an additional variable fee based on activity.

Sometimes larger contracts for products or services for organizations first proceed through a request for proposal (RFP) stage. This brings a degree of rigor to the vendor selection process, though it may devolve into a price emphasis. Building an established relationship with a customer prior to the RFP stage is desirable, so that you fully understand the requirements and may be able to provide input that results in a more effective RFP structure.

There are some situations in which significant pricing adjustments are possible over time. For example, as new individual clients engaged with our practice, we gradually adjusted pricing up to market levels. Our initial pricing was based on entering the field after recently completing educational programs. Initial pricing reflected our level of early experience, and subsequent adjustments were made as our level of experience grew.

Outreach. Sustaining a nontraditional career requires continually reaching out to potential customers and clients. This is particularly important in the early stages since referrals will likely grow as a source of customers and clients over time. While the form outreach takes will vary according to specific nontraditional career components, there are some aspects that are likely of general relevance. Creation of a website is important primarily as a repository of information about your capabilities for referral purposes. We created our own website for our initial practice and then expanded it as we added additional components.[49] Anna Domek, my daughter and a member of our team, was so helpful in rebuilding the website using CoffeeCup software as the content evolved.[50] A website is also an excellent repository for other outreach elements, such as electronic newsletters.[51] We have been sending out a newsletter for almost ten years to a broad and evolving group of interested individuals. Currently it is issued quarterly and typically includes career or workforce content as well as content related to social justice and social responsibility. MailChimp has proven an excellent platform for distributing the newsletter.[52] Another important outreach area is that of social media, and Anna spearheaded our presence primarily on LinkedIn, though also with Facebook and Twitter accounts. We found the LinkedIn presence particularly helpful also as a recruiting platform when seeking new team members.[53] Exploring different social media platforms and how they can contribute to your outreach activities is an important step.

While social media are of growing significance, in-person contact with potential customers is critical to building relationships. The form this takes will depend on the nature of your nontraditional career components. In our case, presentations to professional organizations based on proprietary content proved to be one effective means of making initial connections. It may also prove to be an effective channel for you. Business cards are useful in these and other in-person settings. They are readily available through cost-effective, on-line services such as Vistaprint.[54] Advertising is an area that has not been a focus for us, other than small placements in the Yellow Pages in the early days of our practice. The relevance of advertising to you will depend on the nature of your career components.

Table 9.1 summarizes the various nuts-and-bolts items we have reviewed with an indication of those that are primarily relevant to a sole proprietorship and those that are primarily relevant to an LLC or S corp, recognizing that there may be some variations in specific situations and that this is an overview and not all inclusive.

Having thus far explored what we mean by a nontraditional career, having looked at strategic factors and at some practical aspects, in the final chapter we will highlight key questions to consider for moving forward.

Table 9.1. Checklist for Nuts-and-Bolts Aspects.

Item	Sole Proprietorship	LLC/S Corp
Build contingency fund	x	x
Select primary location	x	x
Implement hardware and software information technology infrastructure	x	x
Determine appropriate business structure	x	x
Consider engaging attorney specializing in business structure and employment law		x
File articles of incorporation with selected state		x
Apply for fictitious business name (doing business as)	x	x
Obtain business license (local)	x	x
Apply for employer identification number (EIN)		x
Obtain state seller's permit (for products)	x	x
Register for state unemployment insurance (SUI) payments		x
Establish banking relationship	x	x
Establish individual retirement account (IRA)	x	x
Establish one-participant 401(k) plan	x	x (reexamine basis if adding employees)
Engage payroll service provider and determine payroll frequency		x
Secure workers' compensation insurance		x
Secure general liability, umbrella, business automobile, and professional liability insurance as appropriate	x	x
Set up accounting software	x	x
Set up tax return preparation software	x	
Consider engaging tax accountant		x
Order business cards	x	x
Create letter of understanding for contractors	x	x
Determine employment attributes to offer: full- or part-time, exempt or non-exempt, compensation and benefits		x

(continued)

Item	Sole Proprietorship	LLC/S Corp
Create job descriptions for employees		x
Create employee engagement letters		x
Request that prospective contractors complete form W-9, request for taxpayer identification number and certification	x	x
Request that prospective employees complete form I-9 eligibility to work and W-4 withholding allowance form		x
Establish pricing basis for initial offerings	x	x
Establish outreach infrastructure, for example, website, in-person, newsletter, social media	x	x
Pay quarterly estimated taxes, federal and state	x	x
File Schedule C with annual 1040 tax return	x	
File corporate taxes		x
File and send 1099 forms for contractors at appropriate time	x	x
File and send W-2 forms for employees (and W-3 summary form) at appropriate time		x

CHAPTER 10

Moving Forward

EXHILARATION ECLIPSED A feeling of anxiety, though both emotions were there. Now was the time to decide about fully embracing a nontraditional path. Seeds sown over several years had led to some nontraditional career components taking shape. Now, with a changing employment setting, there was an opportunity to cast completely free of conventional employment into nontraditional waters. I talked about it with my family and with colleagues. It was clear the time was right. So with a slight nudge from the world of conventional employment, a fully fledged nontraditional career was born. In this, our final chapter, we will explore moving forward for such a launch and for the trajectory this nontraditional career path then takes. After a reminder about the influence of the changing work environment, we will explore various stages that can unfold along the path and questions that, when addressed, can lead to this becoming fulfilling and rewarding.

Let us bring together observations about the changing work environment that point to the growing attraction of a nontraditional career path. This work environment is characterized by the following:

- Economic disruption due to changing global and national economies speaking to the need for career agility
- Sustained high unemployment for those at an early career stage, and challenges securing conventional employment at a late career stage, highlighting the importance of nontraditional paths
- High levels of satisfaction for people engaged in nontraditional career paths
- Reduced appeal of conventional employment due to
  - stagnant wages;
  - growing gross compensation inequities favoring those at the top of many large, private sector companies;

- systematic reduction in defined benefit pension plans; and
- low levels of employee engagement
- Improved access to health-care insurance for those in nontraditional career paths due to the Affordable Care Act
- Ready access to the latest communication and analysis tools benefitting individuals and small enterprises

These factors in combination speak to an increasingly turbulent economic environment, a continuing decline in the attractiveness of conventional employment, and the growing attractiveness of a nontraditional path. As we observed in chapter 1, the sweet spot for this nontraditional path is at the intersection of passion/interests, skills, and external needs. It is here that we can affirm our values and create fulfilling work. Keeping this in mind, we will look at questions to consider at each of the following stages on the trajectory of a nontraditional career path:

- Initial entry decision
- Launch
- Growth and evolution
- Completion

Initial Entry Decision

We will look at four aspects and associated questions related to the initial entry decision:

- Critical mass of skills and attributes
- Work/life purpose
- Components to include
- Relationship strength

Critical mass of skills and attributes. Do I have what is needed to succeed in a nontraditional path? This question looms large at the start. Some suggest that only a few are cut out to be their own boss, while others suggest that it's a fallacy that only a few can take this path successfully.[1] As we have seen, there are many different paths to a nontraditional career that people have successfully negotiated and there are many options to tailor a path to individual needs. Indeed, there is historical precedent that what appears nontraditional today is closer to our natural relationship with work than that of employment in large organizations, which has come to dominate our thinking since the early 1900s. Our view about what constitutes traditional work is an artifact of this more recent past. Given this, given the flexibility to tailor a nontraditional path to individual needs, and given a changing work environment that is moving to favoring the nontraditional route, a nontraditional path is likely broadly accessible to many. In chapter 7 we explored the skills and attributes needed to be successful, and in chapter 8 we looked at how partnering can fill gaps. While some skills involve specific content for a given discipline and career component, many skills and attributes are generic, such as interpersonal skills, customer-facing skills, and support service management capabilities. Consider the following questions to help in deciding whether to pursue a nontraditional path:

- What content expertise do I possess or could I develop that would have value to customers?
- What strengths do I have in business management, time management, interpersonal skills, and customer-facing and support service management areas? What strengths do I bring to the personal characteristics of integrity, tenacity, self-awareness, empathy, and comfort with ambiguity? What gaps do I see and what development plan would be appropriate to provide the foundation of skills and personal characteristics I need?
- What skill areas could I enhance through partnering? What partnering opportunities would I consider?

Work/life purpose. Fundamental to successful engagement in a nontraditional career path is strong alignment with personal work/life purpose and support from those connected to us. We addressed this in chapter 1 and in exploring balance in chapter 5. Here are some questions to consider that can help in reflecting on this area:

- What are my aspirations and what do they mean for a nontraditional career path?
- What do my values, personality preferences, interests, and past experiences mean for a nontraditional career path?
- How is my time spent today in areas of life that are significant to me and how would I like this to be in three to five years? What might this mean for a nontraditional career path?
- What priorities of family and those close to me, and others connected to me, would I include when evaluating a nontraditional career path? Whose support is important to me and what form of support do I anticipate needing? What might this mean for a nontraditional career path?

Components to include. A key area to address when considering whether to proceed on a nontraditional path is identifying what career components to include and how they might be differentiated for customers. We have seen examples of various approaches to nontraditional careers in our book. Others have commented on the wide range of possible options, varying from consulting to web businesses, from specialty foods to handyman services, from freelancing to interior design.[2] We explored differentiation in chapter 4. Drawing on that foundation, here are some questions to consider when making a decision about components to include in a nontraditional path and whether to proceed with that path:

- What excites me in my work and interests and what might this suggest for nontraditional career components I could develop?
- How might these components be valuable to others and what approaches could I use to create differentiation?
- How might I combine more than one component to create additional value?

Relationship strength. Relationships are a cornerstone of a nontraditional career. They include customer and client relationships that initially inform the path forward and then lead to project engagements. They can include partnering relationships that broaden the scope of activities or enhance skills. They can include supplier relationships that strengthen infrastructure or contribute to product or service offerings. They can also include mentoring relationships to assist in scoping opportunities, designing the launch process, and providing ongoing support.[3] Given the significance of relationships, the following questions are helpful when considering whether to take a nontraditional path:

- What relationships will I need in order to successfully launch a nontraditional career path in areas of interest to me?
- How strong are current relationships?
- How could I deepen these relationships and identify new connections?
- What insights can my current connections offer about my prospective nontraditional path?
- Who might provide mentoring support?

Launch

Once the decision to proceed is made, the next step is launching a nontraditional career. We will examine the following five aspects:

- Planning and strategy
- Structure
- Timing

- Customers
- Infrastructure

Planning and strategy. When I was in the planning and economics department of a large company in the 1980s, it took weeks, many meetings, and much staff time to create and integrate multiyear business plans annually for the multiple business units in the company. Developing a planning structure and financial projection tools that could be aggregated at the company level helped make this manageable. The business plans were used to guide capital spending within the company and by the parent organization. However, the main benefit of the activity was not in the final documents that emerged but in the thinking and consensus building that took place. This was a complex process designed to minimize the risk of generating low returns from capital investment decisions that typically had time horizons of ten years or more. It was not designed for flexibility or rapid change. Sometimes it worked well. When I began my nontraditional career path, I had no formal business plan; rather, I had a foundation of values, clarity of purpose, implicit personal and financial objectives, and various ideas that evolved and grew as opportunities and needs emerged. This path developed by experiencing situations, by trying and exploring approaches, and by modifying them based on feedback from customers, clients, and team members. While the main career components began to take shape conceptually before they were implemented, and each step now fits in retrospect, it would have been hard to predict the detailed path in advance. Mostly this approach worked well.

Here we see two examples of planning, in one case a formalized approach, with much structure and limited flexibility, and in another case an informal approach with little structure and much flexibility. Which approach is best to use? Or would the best approach be something in between? Let me suggest that the appropriate approach depends on your personality preferences—for example, greater comfort with structure or with flexibility—and practical considerations, such as whether there is a need to secure external funding that may require a formal business plan.[4]

One benefit of a small enterprise and a nontraditional career path is the opportunity afforded to rapidly test ideas and adjust accordingly. As a result, it is possible and desirable to readily validate market concepts based on feedback from customers.[5] Take advantage of this process of learning by experience in crafting and evolving your nontraditional path. In addition, tools are available to simplify the planning process so that it is manageable and responsive.[6] Again, the primary value is in the thinking involved rather than the brief document that may be created. Regardless of the approach adopted for planning, it is helpful to establish personal and business objectives at the outset. These objectives can include both intangible aspects such as personal fulfillment and social contribution and tangible aspects such as specific financial targets. Some questions to consider:

- What balance between formal planning and learning by testing and experience will I seek?
- What personal and business objectives are appropriate for the first year? For three years in the future? For five years in the future? For a longer-term time horizon that I define?
- What market segments are attractive initial targets for each of my potential career components?
- How might I rapidly test potential business ideas with customers?

Structure. Two aspects of structure are important to the launch phase, the interrelationship of nontraditional career components and the form of business structure adopted. We reviewed the latter aspect in chapter 9, including how different business structures can fit at different stages in the evolution of a nontraditional career path. Determining the initial business structure is an important first step. We addressed the interrelationship of career components in chapter 3 and how to find an appropriate balance in their contributions in chapter 5. Determining whether to build a nontraditional path with linked components that are interdependent or with separate, independent components is a core strategic issue, as is the selection of components that provide needed balance between scale of contribution and time to build a position. Here are some questions to consider in the launch phase, related to both the business structure and components you may choose to develop as part of your nontraditional career:

- What form of business structure will be most appropriate for launching my nontraditional career?
- What are the pros and cons of connecting career components in my nontraditional career? If I choose to build links among the components, what connecting principles might apply, and if I choose not to build links, what partnership approaches may be appropriate?
- In projecting the financial trajectory into the future, are there potential gaps and, if so, how can these gaps be filled by adjusting the balance of career components?

Timing. Moving along a nontraditional path likely means letting go of conventional employment at some point. In chapter 6 we explored the implications of doing this gradually or abruptly. Timing of this step is an important consideration in the launch phase and central to implementation. Here are some questions to consider:

- In considering my purpose for embarking on a nontraditional path and my sense of urgency, what might this imply for pace of entry?
- What are the pros and cons of gradual entry versus rapid, full immersion in my nontraditional career path, and what are the financial implication of each approach?
- How might I accelerate my pace on entry?

Customers. Success in launching and sustaining a nontraditional career is dependent on securing and maintaining a sound customer base. The form and nature of this customer base will vary based on the attributes of specific career components, for example, whether they are organizational customers, individual clients, or a combination of both. A solid customer base is built on strong relationships, understanding customer needs, communicating and delivering value, and demonstrating integrity and commitment. Some questions to consider in connection with customers during the launch phase:

- What customer relationships are currently in place that could lead to initial projects? How can I best nurture those relationships?
- How might I reach additional customers and build those relationships?
- How concentrated is my likely initial distribution of sales by customer? What distribution am I seeking in the future to balance limiting dependency on only a few customers with the need to build in-depth customer relationships?
- What approach to pricing will I adopt to secure adequate returns while staying competitive?

Infrastructure. Practical operational issues related to infrastructure surface in the launch phase. Location is a primary question, and depending upon the nature of nontraditional career components, working from home may be a viable option that offers time and cost benefits. Alternatively, a co-working site, shared with others, may suffice.[7] Regardless, select an initial location that is easily accessible and financially attractive so as not to deplete resources. In chapter 9 we reviewed the nuts-and-bolts items that need to be addressed in the launch phase and summarized them in table 9.1. These items vary according to the nature of the business structure adopted and the complexity of operations, for example, engaging employees creates significant additional infrastructure requirements. The nuts-and-bolts items also include financial considerations such as building a contingency fund to address income variations that may accompany a non-traditional career and establishing a one-participant 401(k) plan and/or an individual retirement account. It will be helpful to review the items in table 9.1 and create your own checklist building on this foundation. Examples of further resources that might be helpful include the following:

- A U.S. government website that contains a compilation of many resources for those who are self-employed, including a section on starting up, which includes links to Small Business Administration resources such as a checklist of items to consider when starting a business[8]
- Resources for those beginning entrepreneurship at different life stages, for example, resources addressing encore entrepreneurship at later life stages, including a readiness self-assessment for starting a business (applicable at any life stage) and an encore entrepreneurship course, and resources for an earlier life stage, including a guide and course for young entrepreneurs starting a business[9]

Some questions that surface related to infrastructure during the launch phase are as follows:

- What location will I select as my primary base?
- What items will I include in my nuts-and-bolts checklist and what is my timing to address these items?
- How will I build a contingency fund and what funding level am I seeking?
- If I need outside funding to start, what sources of funding will I explore?
- How will I establish a one-participant 401(k) plan and/or an individual retirement account to address long-term savings?

Growth and Evolution

Once your nontraditional career has been launched, what does it take to sustain growth and evolution? Growth and evolution occur over an extended period of time, in contrast to the short launch phase that characterizes the abrupt start sometimes chosen for a nontraditional career. When gradual development is the chosen entry path into a nontraditional career, then the extended launch phase blends into that of growth and evolution. Indeed, this may then be viewed as a single phase punctuated by a decision to fully engage in the nontraditional path at some point. Given the extended time for growth and evolution, actions to sustain a vibrant nontraditional career during this phase may be prompted by changes in the external environment or by changing internal aspirations.

Staying current with emerging business and competitive trends, technology enhancements, and evolving customer needs is central to identifying external factors that may prompt changes in operating practices, services, or products. This means doing the following:

- Allocating time with customers to explore new needs
- Allocating time to participate in trade and professional organizations and to review relevant publications to understand evolving practice
- Participating in appropriate social media special interest groups, such as those associated with LinkedIn, to contribute thinking and stay current with topical issues
- Reviewing emerging offerings from suppliers that may suggest enhancements to operational effectiveness or efficiency

By staying current in this way, it is possible to identify emerging opportunities early and to spot potential challenges so that mitigating approaches can be developed.

Internally initiated steps to adjust to evolving career aspirations may include adding new career components that contribute to the vitality of a nontraditional career path or infrastructure changes that support and strengthen the approach. Infrastructure changes could include adopting a more complex business structure to accommodate growth and an expansion of capabilities. Reviewing the pros and cons of alternative business structures as summarized in chapter 9 can help, so that when significant shifts occur in career component scope, the business structure can be adjusted accordingly. A related aspect warranting continued exploration during growth and evolution is that of partnering, either to strengthen component offerings, to strengthen infrastructure, or to enhance skills, as reviewed in chapter 8. In addition to opportunities that arise from the external environment, a regular review of potential extensions from existing career components can provide insights into possible new opportunities. These extensions may be based, for example, on leveraging content expertise or broadening products or services to existing customers, as mentioned in chapter 3.

A further aspect of the growth and evolution phase is that of assessing progress and refining direction based on that assessment. Such assessment is most helpful when it addresses both personal and business perspectives, including intangible and tangible factors. Some assessment questions to consider from a personal perspective:

- How fulfilled am I on a daily basis with my nontraditional career path?
- How well aligned are my work activities with my values?

- To what extent am I using skills where I have a strong capability and that I enjoy?
- To what extent am I realizing my full potential?
- How does my nontraditional career path honor the needs of those people who are close to me and the needs of others with whom I am connected?
- What changes, if any, would be appropriate to better align what I do in my nontraditional career path with my aspirations?
- What development steps, if any, should I consider to enhance my sense of fulfillment or capabilities?

Some assessment questions to consider from a business perspective are as follows:

- How robust is my combination of career components to address possible changes in the external environment and to secure potential new opportunities? What changes, if any, are appropriate?
- How do products or services meet customer expectations, and what changes, if any, are appropriate?
- How does the financial profile of income generated, profit margins from product or service offerings, and income growth match my objectives, and what changes might be needed?
- What infrastructure changes should I consider to build on emerging technology?

Completion

The final phase is that of completion. For some, with a goal of creating a sizable enterprise, this may consist of a business transaction to maximize financial value through sale or initial public offering. For others, it may be a process of transferring assets to family members. For many of us, this is a phase of gently winding down career components after a successful extended nontraditional path before proceeding to another life stage. This completion phase can be gradual, just as the launch process can be gradual, exiting from career components over an extended time. The phase can also be more abrupt for those choosing a rapid exit. There are psychological factors to consider, such as a sense of loss of identity, which can be softened by engaging in other fulfilling activities, such as volunteer work that may be a natural extension of one or more career components. There are also practical issues to address when closing a business, such as filing dissolution documents, canceling registrations, permits, licenses, and business names, complying with employment and labor laws, addressing financial obligations, and maintaining records.[10] Completion is an important stage in which to honor the legacy that you have created. Some questions related to this phase:

- What is my preferred timing for withdrawing from each of my nontraditional career components?
- Am I seeking to sell the assets (tangible or intangible) associated with my career components? If so, what approaches should I consider for such a sale?
- Am I seeking to transfer any assets to others? If so, how might this be structured?
- For those components I will be winding down, what items need to be addressed to secure an orderly closure?
- What steps will I take to ease my transition into my next life phase?

The stages in a nontraditional career path and key factors associated with each stage in moving forward are summarized in figure 10.1.

We see interplay of personal purpose with communal contribution and a combination of giving and receiving. We see the creation and realization of aspirations brought forth in the practical world of everyday life. We see individual initiative and community participation woven through a tapestry of life. It is in finding such expression that these words of Irenaeus of Lyons from the second century CE come to life: "The glory of God is the human person fully alive."[11] These words have special meaning for me, may they also have special meaning for you.

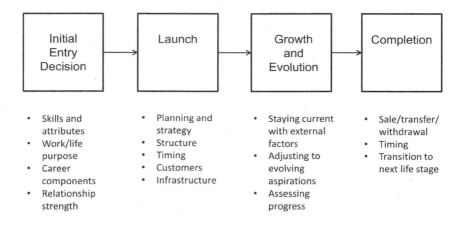

Moving Forward

Initial Entry Decision	→	Launch	→	Growth and Evolution	→	Completion
• Skills and attributes • Work/life purpose • Career components • Relationship strength		• Planning and strategy • Structure • Timing • Customers • Infrastructure		• Staying current with external factors • Adjusting to evolving aspirations • Assessing progress		• Sale/transfer/withdrawal • Timing • Transition to next life stage

Figure 10.1

Appendix: Selected Templates

Table A.1. Template of a Weighted Index of Well-Being for Assessing Career Component Contribution.

Index Element	Estimated Value for a Given Career Component on a 7-Point Scale (7 Is the Most Positive)	Weight (%)	Score (Value x Weight)
Anticipated degree of personal fulfillment			
Alignment with values			
Fit with interests			
Opportunity to use skills			
Fit with other career components			
Geographic proximity			
Low initial capital investment			
Strong financial contribution			
Low variability or seasonality			
Likelihood of extended duration of the component			
Total			

Table A.2. Template for a Grid to Guide a Decision about the Pace of Entry into a Nontraditional Career.

Category	Factor	Immediate Full-Time Engagement	Gradual Launch
Personal preferences	Personal purpose		
Personal preferences	Sense of urgency		
Personal preferences	Development		
Personal preferences	Risk tolerance		
Practical considerations	Financial base		
Practical considerations	Infrastructure needs		
Practical considerations	Geographic aspects		
Support from others	Family and friends		
Support from others	Customers		
Support from others	Current employer		

Notes

Preface

1. Ron Elsdon, *Affiliation in the Workplace: Value Creation in the New Organization* (Westport, Conn.: Praeger, 2003).

2. Zeth Ajemian, Mark Malcolm, and Ron Elsdon, "Example Case Studies: Healthcare and High-Technology," in *Building Workforce Strength: Creating Value through Workforce and Career Development,* ed. Ron Elsdon (Santa Barbara, Calif.: Praeger, 2010), 169–91.

3. Ron Elsdon, ed., *Business Behaving Well: Social Responsibility, from Learning to Doing* (Washington, D.C.: Potomac Books, 2013), 6, 7; Sarah Anderson, Scott Klinger, and Sam Pizzigati, "Executive Excess 2013: Bailed Out, Booted, and Busted," Institute for Policy Studies, August 28, 2013, 4, accessed August 29, 2013, http://www.ips-dc.org/reports/executive-excess-2013/.

4. Elsdon, *Business Behaving Well,* 2–6.

5. Zeth Ajemian, "Labor Management Partnerships: A Collaborative Path toward Improved Business Results and Employee Satisfaction," in Elsdon, *Business Behaving Well*, 151–53.

6. Bureau of Labor Statistics, "Union Members—2012," news release USDL-13-0105, January 23, 2013, accessed July 28, 2013, http://www.bls.gov/news.release/archives/union2_01232013.pdf; Public Purpose, "U.S. Private Sector Trade Union Market Share Falls Again in 2000," accessed July 28, 2013, http://www.publicpurpose.com/lm-unn2000.htm.

7. Sylvia A. Allegretto and Steven C. Pitts, "To Work with Dignity: The Unfinished March Toward a Decent Minimum Wage," Economic Policy Institute, August 26, 2013, accessed April 29, 2014, http://www.epi.org/publication/work-dignity-unfinished-march-decent-minimum/.

8. Ron Elsdon and Rita Erickson, "What Does Workforce Strength Look Like and Why Does It Matter?," in *Building Workforce Strength: Creating Value through*

Workforce and Career Development, ed. Ron Elsdon (Santa Barbara, Calif.: Praeger, 2010), 13, 14.

Introduction

1. Robert Bucholz, *Foundations of Western Civilization II: A History of the Modern Western World* (Chantilly, Va.: Great Courses, 2006), 135.

2. Zeth Ajemian, "Labor Management Partnerships: A Collaborative Path toward Improved Business Results and Employee Satisfaction," in *Business Behaving Well: Social Responsibility, from Learning to Doing*, ed. Ron Elsdon (Washington, D.C.: Potomac Books, 2013), 151.

3. Campbell Gibson, *American Demographic History Chartbook: 1790 to 2000*, chaps. 15 and 17, accessed July 30, 2013, http://www.demographicchartbook.com/Chartbook/, which is primarily based on census data from household surveys.

4. Richard Henderson, "Employment Outlook: 2010–2020: Industry Employment and Output Projections to 2020," *Monthly Labor Review*, January 2012, 66, accessed July 31, 2013, http://www.bls.gov/opub/mlr/2012/01/art4full.pdf, based on the Current Employment Statistics Survey, which counts jobs based on surveys of establishments, for 2000 sector breakdown; Gibson, *American Demographic History Chartbook*, chaps. 15 and 17 for the 2000 labor force and 1960 data.

5. Gibson, *American Demographic History Chartbook*, chap. 15.

6. The data for employment by sector up to 1900 are from Gibson, *American Demographic History Chartbook*, figure 17–2; employment by sector from 1920 to 1990 is from figure 17–1 in the same source; the data for employment by sector in 2000 and 2010 and the projection for 2020 are from Henderson, "Employment Outlook: 2010–2020," 66. It is challenging obtaining consistent historical data on self-employment, and the information shown is intended to reflect trends. Self-employment information for 2010 is based on self-employed numbers from Brian Headd, "Small Business Trending Up," *Small Business Quarterly Bulletin*, First Quarter 2013, accessed July 31, 2013, http://www.sba.gov/sites/default/files/files/SBQB_2013q1.pdf, which uses the Current Population Survey based on household data, while data for the civilian labor force for 2010 is from Bureau of Labor Statistics, Current Population Survey, accessed August 1, 2013, http://www.bls.gov/cps/cpsaat01.pdf; self-employment data for 1990 and 2000 is from Steven Hipple, "Self-Employment in the United States," *Monthly Labor Review*, September 2010, accessed July 30, 2013, http://www.bls.gov/opub/mlr/2010/09/art2full.pdf; self-employment information for 1950–1980, which only includes self-employment in unincorporated entities, is from Steven Hipple, "Self-Employment in the United States: An Update," *Monthly Labor Review*, July 2004, accessed July 31, 2013, http://www.bls.gov/opub/mlr/2004/07/art2full.pdf, 14; self-employment information for 1880–1940 is from Joseph D. Phillips, *The Self-Employed in the United States* (Urbana: University of Illinois, 1962), using the average of the different sources for 1880–1920 from table 1 on page 11 and the value for 1940 from table 3 on page 13.

7. Phillips, *Self-Employed in the United States*, 11; Ron Elsdon, *Affiliation in the Workplace: Value Creation in the New Organization* (Westport, Conn.: Praeger, 2003), 21, 22.

8. Hipple, "Self-Employment in the United States," 19, 20, which is based on the Current Population Survey, using household data.

9. In 1967 those who were both self-employed and in incorporated entities were classified as wage and salary workers in the Current Population Survey, leading to an apparent drop in self-employment. Hipple shows data from 1989 for those who were self-employed in incorporated entities so they can be included in a count of self-employment, leading to an apparent increase. In addition, there were changes in survey methodology at various times as described by Hipple, "Self Employment in the United States," 18.

10. John Schmitt and Nathan Lane, "An International Comparison of Small Business Employment," Center for Economic and Policy Research, August 2009, 1, 4, 7, accessed July 30, 2013, http://www.cepr.net/documents/publications/small-business -2009-08.pdf; Jordan Weissmann, "Think We're the Most Entrepreneurial Country in the World? Not So Fast," *Atlantic*, October 2, 2012, accessed August 18, 2013, http:// www.theatlantic.com/business/archive/2012/10/think-were-the-most -entrepreneurial-country-in-the-world-not-so-fast/263102/.

11. Schmitt and Lane, "International Comparison of Small Business Employment," 11; Deborah LeVeen provides an analysis of the evolution of health care in the United States, the strengths and challenges of the Affordable Care Act, and options moving forward in "Healthcare as Social Responsibility: Implications for Business," in *Business Behaving Well: Social Responsibility, from Learning to Doing,* ed. Ron Elsdon (Washington, D.C.: Potomac Books, 2013), 107–37.

12. Robert W. Fairlie, *Kauffman Index of Entrepreneurial Activity: 1996–2013* (Kansas City, Mo.: Ewing Marion Kauffman Foundation, 2014), 5.

13. Uschi Schreiber and Maria Pinelli, "The Power of Three: Together, Governments, Entrepreneurs and Corporations Can Spur Growth Across the G20, EY G20 Entrepreneurship Barometer 2013," Ernst & Young, 2013, accessed September 4, 2013, http:// www.ey.com/Publication/vwLUAssets/EY_G20_Entrepreneurship_Barometer_2013 _UK/$FILE/EY_G20_Entrepreneurship_Barometer_2013_UK.pdf/; observations about the Ernst & Young survey are provided by Robb Mendelbaum, "In New Study, Entrepreneurs See the Glass Nearly Empty," *You're the Boss* (blog), *New York Times*, September 3, 2013, accessed September 4, 2013, http://boss.blogs.nytimes .com/2013/09/03/in-new-study-entrepreneurs-see-the-glass-nearly-empty/.

14. Donna J. Kelley, Abdul Ali, Candida Brush, Andrew C. Corbett, Mahdi Majbouri, Edward G. Rogoff, Babson College, and Baruch College, *Global Entrepreneurship Monitor: 2012 United States Report,* 2012, 7, accessed October 30, 2013, http:// www.babson.edu/Academics/centers/blank-center/global-research/gem/Documents/ GEM%20US%202012%20Report%20FINAL.pdf.

15. Bureau of Labor Statistics, Labor Force Statistics from the Current Population Survey, accessed August 1, 2013, http://www.bls.gov/webapps/legacy/ cpsatab9.htm/.

16. MBO Partners, "The State of Independence in America," Third Annual Independent Workforce Report, September 2013, 3, 5, accessed April 29, 2014, http://www.mbopartners.com/state-of-independence/independent-workforce -index.html.

17. Wendell Cox, "Toward a Self Employed Nation," *NewGeography.com*, accessed July 31, 2013, http://www.newgeography.com/content/003761-toward-a-self-employed-nation/; Headd, "Small Business Trending Up."

18. Hipple, "Self-Employment in the United States," 24, 29.

19. Ibid.

20. Ibid.

21. David Christian, *Big History: The Big Bang, Life on Earth, and the Rise of Humanity* (Chantilly, Va.: Great Courses, 2008), 189–90.

22. The World Bank, "GDP Growth (annual %)," accessed August 6, 2013, http://data.worldbank.org/indicator/NY.GDP.MKTP.KD.ZG/.

23. An overview is provided in Richard Dobbs, Jaana Remes, Sven Smit, James Manyika, Jonathan Woetzel, and Yaw Agyenim-Boateng, "Urban World: The Shifting Global Business Landscape," McKinsey and Company, October 2013, accessed October 23, 2013, http://www.mckinsey.com/insights/urbanization/urban_world_the_shifting_global_business_landscape/.

24. World Economic Forum, *Outlook on the Global Agenda 2014*, accessed November 16, 2013, http://www3.weforum.org/docs/WEF_GAC_GlobalAgendaOutlook_2014.pdf/.

25. Ron Elsdon, ed., *Business Behaving Well: Social Responsibility, from Learning to Doing* (Washington, D.C.: Potomac Books, 2013), 187.

26. Ibid., 6.

27. Ibid.

28. David Hole, Le Zhong, and Jeff Schwartz, "Talking About Whose Generation," *Deloitte Review*, Issue 10, 2010, accessed August 6, 2013, http://www.deloitte.com/view/en_US/us/Insights/Browse-by-Content-Type/deloitte-review/5d6e2bb18ef26210VgnVCM100000ba42f00aRCRD.htm/.

29. Erica Dhawan, "Gen-Y Workforce and Workplace Are Out of Sync," *Forbes*, January 23, 2012, accessed August 6, 2013, http://www.forbes.com/sites/85broads/2012/01/23/gen-y-workforce-and-workplace-are-out-of-sync/; Kyra Friedell, Katrina Puskala, Morgan Smith, and Nicole Villa, "Hiring, Promotion, and Progress: Millennials' Expectations in the Workplace," St. Olaf College, Northfield, Minn., accessed August 6, 2013, http://wp.stolaf.edu/sociology/files/2013/06/Hiring-Promotion-and-Progress.pdf; Emily Esfahani Smith and Jennifer L. Aaker, "Millennial Searchers," *New York Times*, November 30, 2013, accessed April 21, 2014, http://www.nytimes.com/2013/12/01/opinion/sunday/millennial-searchers.html.

30. Jessica Brack, *Maximizing Millennials in the Workplace*, 2012, 2, University of North Carolina Kenan-Flagler Business School, accessed August 6, 2013, http://www.kenan-flagler.unc.edu/executive-development/custom-programs/~/media/DF1C11C056874DDA8097271A1ED48662.ashx/.

31. Elsdon, *Affiliation in the Workplace*, 2.

32. Richard W. Johnson and Barbara A. Butrica, "Age Disparities in Unemployment and Reemployment during the Great Recession and Recovery," Urban Institute, Brief 03, May 2012, 3, accessed August 6, 2013, http://www.urban.org/publications/412574.html; Steven Greenhouse, "The Gray Jobs Enigma," *New York Times*, March 12, 2014, accessed April 21, 2014, http://www.nytimes.com/2014/03/13/business/retirementspecial/the-gray-jobs-enigma.html.

33. Andrew Schaefer, "The Long-Term Unemployed in the Wake of the Great Recession," Carsey Institute, National Issue Brief #68, Winter 2014, accessed April 21, 2014, 3, http://carseyinstitute.unh.edu/sites/carseyinstitute.unh.edu/files/publications/IB-Schaefer-Long-Term-Unemployment-web.pdf.

34. Nona Willis Aronowitz, "Over 50 and Out of Work: Program Seeks to Help Long-Term Unemployed," *NBCNews.com*, November 16, 2013, accessed November 16, 2013, http://inplainsight.nbcnews.com/_news/2013/11/16/21266839-over-50-and-out-of-work-program-seeks-to-help-long-term-unemployed?gt1=43001/.

35. Bureau of Labor Statistics, "Labor Force Statistics from the Current Population Survey: Household Data Seasonally Adjusted, A-10, Unemployment Rates by Age, Sex, and Marital Status, Seasonally Adjusted," accessed June, 6, 2014, http://www.bls.gov/web/empsit/cpseea10.htm/.

36. Liz Alderman, "Young and Educated in Europe, but Desperate for Jobs," *New York Times*, November 15, 2013, accessed November 16, 2013, http://www.nytimes.com/2013/11/16/world/europe/youth-unemployement-in-europe.html?_r=0/.

37. Floyd Norris, "For Some, Joblessness Is Not a Temporary Problem," *New York Times*, October 25, 2013, accessed October 25, 2013, http://www.nytimes.com/2013/10/26/business/economy/for-some-joblessness-is-not-a-temporary-problem.html?_r=0/.

38. Ibid.

39. Ibid.

40. LeVeen, "Healthcare as Social Responsibility," 109, 110.

41. Liz Weston, "Obamacare Could Free Workers from 'Job Lock,'" *MSN Money*, accessed September 6, 2013, http://money.msn.com/personal-finance/obamacare-could-free-workers-from-job-lock/.

42. Randstad, "ACA Removes Barriers to Temporary Work," August 27, 2013, accessed September 4, 2013, http://www.randstadusa.com/workforce360/workforce-insights/aca-removes-barriers-to-temporary-work/124/.

43. Jerry Geisel, "Fewer Employers Offering Defined Benefit Pension Plans to New Salaried Employees," *Workforce*, October 3, 2012, accessed August 6, 2013, http://www.workforce.com/articles/fewer-employers-offering-defined-benefit-pension-plans-to-new-salaried-employees/.

44. "The Personal Computer," *History Learning Site*, accessed August 6, 2013, http://www.historylearningsite.co.uk/personal_computer.htm/.

45. "Gartner Says Worldwide PC, Tablet and Mobile Phone Combined Shipments to Reach 2.4 Billion Units in 2013," Gartner, Inc., press release, April 4, 2013, accessed August 6, 2013, http://www.gartner.com/newsroom/id/2408515.

46. Susan N. Houseman and Machiko Osawa, eds., *Nonstandard Work in Developed Economies: Causes and Consequences* (Kalamazoo, Mich.: W. E. Upjohn Institute for Employment Research, 2003), introduction, accessed August 1, 2013, http://research.upjohn.org/up_bookchapters/307/.

47. Michael Grabell, "The Expendables: How the Temps Who Power Corporate Giants Are Getting Crushed," *ProPublica*, June 27, 2013, accessed August 1, 2013, http://www.propublica.org/article/the-expendables-how-the-temps-who-power-corporate-giants-are-getting-crushe/.

48. Lawrence M. Fisher, "The Paradox of Charles Handy," *Strategy+Business* 32 (Fall 2003), accessed August 5, 2013, http://www.strategy-business.com/article/ 03309?gko=f3861/.

49. Rachel Pickering, "Portfolio Careers," *Triple Helix* (Summer 2006): 14, 15, accessed August 5, 2013, http://admin.cmf.org.uk/pdf/helix/sum06/36portfolio _careers.pdf.

50. Mary Mallon, "From Managerial Career to Portfolio Career: Making Sense of the Transition" (PhD diss., Sheffield Hallam University, 1998), accessed August 5, 2013, http://shura.shu.ac.uk/3093/.

51. Elsdon, *Business Behaving Well*, 2–7; James Manyika, Michael Chui, Jacques Bughin, Richard Dobbs, Peter Bisson, and Alex Marrs, "Disruptive Technologies: Advances that Will Transform Life, Business, and the Global Economy," McKinsey Global Institute, May 2013, accessed August 20, 2013, http://www.mckinsey.com/insights/ business_technology/disruptive_technologies/.

52. William Bridges, *Transitions: Making Sense of Life's Changes*, Revised 25th Anniversary Edition (Cambridge, Mass.: Da Capo Press, 2004), 4.

53. Elsdon, *Affiliation in the Workplace*, 24.

54. Marcel Boyer and Sebastien Boyer, "The Main Challenge of Our Times: A Population Growing Younger," C. D. Howe Institute, July 24, 2013, accessed August 20, 2013, http://www.cdhowe.org/the-main-challenge-of-our-times-a-population-growing -younger/22311/.

Chapter 1

1. Lauren Hargrave, "The Road Less Traveled: Navigating a Non-Traditional Career Path," *Forbes,* May 21, 2012, accessed August 12, 2013, http://www.forbes.com/sites/ dailymuse/2012/05/21/the-road-less-traveled-navigating-a-non-traditional-career-path/.

2. Rachel Nelken, "Portfolio Careers in the Arts: One Part Pain, Two Parts Pleasure," *Guardian,* March 5, 2013, accessed August 12, 2013, http://www.theguardian .com/culture-professionals-network/culture-professionals-blog/2013/mar/05/ arts-portfolio-careers-top-tips/.

3. Barrie Hopson and Katie Ledger, *And What Do You Do? 10 Steps to Creating a Portfolio Career* (London: A&C Black, 2009), 6–7.

4. Claire Adler, "Mix and Match," *Guardian,* November 4, 2006, accessed August 27, 2013, http://www.theguardian.com/money/2006/nov/04/careers.graduates1/.

5. Gerald Klickstein, "Preparing for Portfolio Careers," *Musician's Way Blog,* January 3, 2012, accessed August 12, 2013, http://musiciansway.com/blog/2012/01/ preparing-for-portfolio-careers/.

6. Claire Hope, "Portfolio Careers for Busy Mums," accessed August 12, 2013, http://www.mum-life-balance.com/flexible-work-life-balance-with-a-portfolio -career.php/.

7. Shirley Rowe and Elizabeth Craig, "Guiding Clients into Encore Careers," *Career Developments* 29, no. 3 (Summer 2013): 20, 21; Kerry Hannon, "For Many Older Americans, an Entrepreneurial Path," *New York Times,* February 7, 2014, accessed April 21, 2014, http://www.nytimes.com/2014/02/08/your-money/for-many-older -americans-an-entrepreneurial-path.html; Kerry Hannon, "An Aging Population Also

Poses Opportunities for Retirement Careers," *New York* Times, March 7, 2014, accessed April 21, 2014, http://www.nytimes.com/2014/03/08/your-money/an-aging-population-also-poses-opportunities-for-retirement-careers.html; Kerry Hannon, "Childhood Dreams Can Inspire Rewarding Second Careers," *New York Times,* April 11, 2014, accessed April 21, 2014, http://www.nytimes.com/2014/04/12/your-money/childhood-dreams-can-inspire-rewarding-second-careers-for-retirees.html.

8. Adler, "Mix and Match"; Chris Carmichael, "Etsy Artisans Reach Retailers," *New York Times* Video, April 14, 2014, accessed April 21, 2014, http://www.nytimes.com/video/business/100000002824797/etsy-goes-wholesale.html.

9. Carmichael, "Etsy Artisans Reach Retailers."

10. Leslie Kaufman, "Actors Don't Just Read for the Part. Reading IS the Part," *New York Times,* June 29, 2013, accessed August 8, 2013, http://www.nytimes.com/2013/06/30/business/media/actors-today-dont-just-read-for-the-part-reading-is-the-part.html?pagewanted=all&_r=0/.

11. Harriet Swain, "Portfolio Careers," *Times Higher Education,* April 10, 2008, accessed August 12, 2013, http://www.timeshighereducation.co.uk/401437.article/.

12. Fiona Pathiraja and Marie-Claire Wilson, "The Rise and Rise of the Portfolio Career," *BMJ Careers,* January 19, 2011, accessed August 27, 2013, http://www.careers.bmj.com/careers/advice/view-article.html?id=20001807.

13. Stacy Julien, "Want to Start a Business? Entrepreneurs Who Found Success at 50+ Offer Their Advice," AARP, May 15, 2013, accessed August 12, 2013, http://www.aarp.org/work/on-the-job/info-05-2013/advice-tips-starting-your-own-business.html/; Anika Anand, "Law Grads Going Solo and Loving It," June 20, 2011, *NBCNews.com,* accessed August 12, 2013, http://www.nbcnews.com/id/43442917/ns/business-careers/t/law-grads-going-solo-loving-it/#.UglqQX_3O2E/; Alex Williams, "Maybe It's Time for Plan C," *New York Times,* August 12, 2011, accessed August 27, 2013, http://www.nytimes.com/2011/08/14/fashion/maybe-its-time-for-plan-c.html?pagewanted=all/; Hannah Seligson, "Job Jugglers, on the Tightrope," *New York Times,* June 25, 2011, accessed August 27, 2013, http://www.nytimes.com/2011/06/26/business/26work.html?pagewanted=all&_r=0/; Alison Coleman, "How Home Business Brings Freedom for Entrepreneurs with Disabilities," *Guardian,* Home Business Hub, accessed September 19, 2013, http://www.theguardian.com/small-business-network/2013/sep/16/home-business-freedom-entrepreneurs-disability/; John Boudreau, "New Generation of Japanese Entrepreneurs Sets Sights on Silicon Valley," *Contra Costa Times,* October 12, 2012, accessed August 12, 2013, http://www.contracostatimes.com/business/ci_21707825/new-generation-japanese-entrepreneurs-sets-sights-silicon-valley?source=rss/.

14. Oscar Wilde quotation, Goodreads, accessed August 13, 2013, http://www.goodreads.com/quotes/679860-i-m-exhausted-i-spent-all-morning-putting-in-a-comma/.

15. Ron Elsdon and Seema Iyer, "Creating Value and Enhancing Retention through Employee Development: The Sun Microsystems Experience," *Human Resource Planning* 22, no. 2 (1999): 39–47; Ron Elsdon, *Affiliation in the Workplace: Value Creation in the New Organization* (Westport, Conn.: Praeger, 2003), xiv, 147–57.

16. Philip Yancey, *Soul Survivor: How Thirteen Unlikely Mentors Helped My Faith Survive the Church* (New York: Galilee Doubleday, 2003), 246.

17. Ron Elsdon and Rita Erickson, "What Does Workforce Strength Look Like and Why Does It Matter?" in *Building Workforce Strength: Creating Value through Workforce and Career Development,* ed. Ron Elsdon (Santa Barbara, Calif.: Praeger, 2010), 13, 15.

18. Elsdon, *Affiliation in the Workplace,* 24, outlines the framework of relationship to work created by Betsy Brewer.

19. Zeth Ajemian, Mark Malcolm, and Ron Elsdon, "Example Case Studies: Healthcare and High-Technology," in *Building Workforce Strength: Creating Value through Workforce and Career Development,* ed. Ron Elsdon (Santa Barbara, Calif.: Praeger, 2010), 169–91.

20. Laura Whitworth, Henry Kimsey-House, and Phil Sandahl, *Co-Active Coaching: New Skills for Coaching People Toward Success in Work and Life* (Palo Alto, Calif.: Davies-Black, 1998), 20, 182.

21. Ron Elsdon, "How Can You Grow Your Practice with Purpose," in *Starting and Growing a Business in the New Economy,* ed. Sally Gelardin (Broken Arrow, Okla.: National Career Development Association, 2007), 19–22.

22. Edgar H. Schein and John Van Maanen, *Career Anchors Self-Assessment* (San Francisco, Calif.: Wiley, 2013).

23. Rita Erickson and Ron Elsdon, "Designing, Developing and Measuring Effective Career Development Processes and Systems," in *Building Workforce Strength: Creating Value through Workforce and Career Development,* ed. Ron Elsdon (Santa Barbara, Calif.: Praeger, 2010), 83–85.

24. Schein, *Career Anchors Self-Assessment,* 1–7.

25. Isabel Briggs Myers, *Introduction to Type: A Guide to Understanding Your Results on the MBTI Instrument* (Mountain View, Calif.: CPP, 1998).

26. See CPP, Inc., MBTI Complete, accessed August 8, 2013, https://www.mbticomplete.com/en/index.aspx/.

27. Fred Borgen and Judith Grutter, *Where Do I Go Next? Using Your Strong Results to Manage Your Career* (Mountain View, Calif.: CPP, 2012).

28. Department of Labor, o*net Interest Profiler, accessed May 1, 2014, http://www.mynextmove.org/explore/ip.

29. SkillScan, accessed August 9, 2013, http://www.skillscan.com/.

30. James Flaherty, *Coaching: Evoking Excellence in Others* (Woburn, Mass.: Butterworth-Heinemann, 1999), 132.

31. Cliff Oxford, "What I Learned About Business as a South Georgia Bottom Fisher," *You're the Boss* (blog), *New York Times,* April 14, 2014, accessed April 21, 2014, http://boss.blogs.nytimes.com/2014/04/14/what-i-learned-about-business-as-a-south-georgia-bottom-fisher/.

32. Linda Faucheux, Jennifer Earls, and Deepesh Faucheux, "Contemplative Career Counseling: Utilizing Mindfulness to Navigate a Portfolio Career," *Career Developments,* 30, no. 2 (Spring 2014): 22, 23.

Chapter 2

1. Innosight, *Executive Briefing: Creative Destruction Whips through Corporate America,* Winter 2012, 4, accessed August 18, 2013, http://www.innosight.com/innovation-resources/strategy-innovation/upload/creative-destruction-whips-through-corporate-america_final2012.pdf.

2. Arie de Geus, *The Living Company: Habits for Survival in a Turbulent Business Environment* (Boston: Harvard Business School Press, 2002), 198.

3. Bureau of Labor Statistics, "Job Openings and Labor Turnover Survey Highlights, March 2014," May 9, 2014, accessed May 16, 2014, http://www.bls.gov/jlt/jlt_labstatgraphs _March2014.pdf.

4. Manpower Group, "Employers Advised to Connect Engagement to Performance to Retain Top Talent in the New Year," November 19, 2013, accessed April 21, 2014, http://www.manpowergroup.com/wps/wcm/connect/manpowergroup-en/home/newsroom/news-releases/most+employees+plan+to+pursue+new+job+opportunities +in+2014+reveals+right+management+poll#.U1WonaIvmSo.

5. Small Business Administration, Office of Advocacy, "Frequently Asked Questions about Small Business, September 2012," 3, accessed August 18, 2013, http://www.sba.gov/sites/default/files/FAQ_Sept_2012.pdf; additional perspectives on measures of small firm survival rates are provided in Ron Cook, Diane Campbell, and Caroline Kelly, "Survival Rates of New Firms: An Exploratory Study," *Small Business Institute Journal* 8, no. 2 (2012): 35–42, accessed August 18, 2013, http://www.sbij .ecu.edu/index.php/SBIJ/article/viewFile/147/91/.

6. Industry Canada, "Key Small Business Statistics—July 2012," 15, accessed August 18, 2013, http://www.ic.gc.ca/eic/site/061.nsf/eng/h_02711.html/.

7. David Blanchflower and Andrew Oswald, "Measuring Latent Entrepreneurship Across Nations," abstract, January 2000, 1–7, accessed August 19, 2013, http://www .dartmouth.edu/~blnchflr/papers/EntrepLeague.pdf.

8. Pew Research Center, "Take This Job and Love It," Pew Research Social and Demographic Trends, September 17, 2009, accessed August 19, 2013, http://www .pewsocialtrends.org/2009/09/17/take-this-job-and-love-it/.

9. Ibid.

10. Ibid.

11. Ibid.

12. Ibid.

13. MBO Partners, "The State of Independence in America," Third Annual Independent Workforce Report, September 2013, 4, 12, 17, accessed April 21, 2014, http://www .mbopartners.com/state-of-independence/independent-workforce-index.html.

14. Independent Direction Directors Advisory Service (IDDAS), "Portfolio Working—Plural Careers Prove to Be a Singular Success," 2003, IDDAS, accessed April 26, 2014, http://thought-leadership.top-consultant.com/UK/FullArticle .aspx?ID=899.

15. Ibid.

16. Ibid.

17. IPro Index 2013, Entity Solutions study conducted by Monash University, Melbourne, Australia, 2013, accessed May 31, 2014, http://www.entitysolutions .com.au/entitysolutions/resourcecentre/iproindex/.

18. Ibid.

19. Ibid.

20. Ibid.

21. BlessingWhite, "Employee Engagement Research Update—Jan 2013," 7, accessed August 19, 2013, http://www.blessingwhite.com/EEE__report.asp/.

22. Gretchen Gavett, "Ten Charts Show We've All Got a Case of the Mondays," *Harvard Business Review Blog,* June 14, 2013, referencing Gallup's 2013 State of the American Workplace report, accessed August 19, 2013, http://blogs.hbr.org/hbr/ hbreditors/2013/06/ten_charts_that_show_weve_all_got_a_case_of_the_mondays .html; Kelli B. Grant, "Americans Hate Their Jobs Even with Perks," *USA Today,* June 30, 2013, accessed August 20, 2013, http://www.usatoday.com/story/money/business/2013/ 06/30/americans-hate-jobs-office-perks/2457089/.

23. IPro Index 2013.

24. Ron Elsdon, *Affiliation in the Workplace: Value Creation in the New Organiza- tion* (Westport, Conn.: Praeger, 2003), 77.

25. Ibid., 88; this figure was also confirmed with additional exit interviews through 2009 based on Ron Elsdon unpublished studies, for a total of more than twenty-five hundred exit interviews in nineteen studies, including those referenced in *Affiliation in the Workplace.*

26. Ron Elsdon, unpublished exit interview studies.

27. Hiroko Tabuchi, "Layoff Taboo, Japan Workers Are Sent to the Boredom Room," *New York Times,* August 16, 2013, accessed August 19, 2013, http://www.nytimes .com/2013/08/17/business/global/layoffs-illegal-japan-workers-are-sent-to-the -boredom-room.html?pagewanted=all&_r=0/.

28. Gavett, "Ten Charts"; Joshua Brustein, "Yahoo's Latest HR Disaster: Ranking Workers on a Curve," *Bloomberg Businessweek,* accessed November 13, 2013, http:// www.businessweek.com/articles/2013-11-12/yahoos-latest-hr-disaster-ranking -workers-on-a-curve/; Michael Winerip, "Pushed Out of a Job Early," *New York Times,* December 6, 2013, accessed April 21, 2014, http://www.nytimes.com/2013/12/07/ booming/pushed-out-of-a-job-early.html.

29. Ron Elsdon and Rita Erickson, "What Does Workforce Strength Look Like and Why Does It Matter?" in *Building Workforce Strength: Creating Value through Workforce and Career Development,* ed. Ron Elsdon (Santa Barbara, Calif.: Praeger, 2010), 13.

30. IDDAS, "Portfolio Working."

31. IPro Index 2013.

32. Ron Elsdon, ed., *Business Behaving Well: Social Responsibility, from Learning to Doing* (Washington, D.C.: Potomac Books, 2013), 17, 21.

33. IPro Index 2013.

34. Pew Research Center, "Take This Job and Love It."

35. Lawrence Mishel and Heidi Shierholz, *A Decade of Flat Wages: The Key Barrier to Shared Prosperity and a Rising Middle Class,* Economic Policy Institute, Briefing

Paper 365, August 21, 2013, 4, 10, accessed August 23, 2013, http://www.epi.org/files/2013/BP365.pdf.

36. Uma V. Sridharan, Lori Dickes, and W. Royce Cains, "The Social Impact of Business Failure: Enron," *American Journal of Business* 17, no. 2 (Fall 2002), accessed August 21, 2013, http://www.bsu.edu/mcobwin/ajb/?p=199/.

37. Ron Elsdon, ed., *Building Workforce Strength: Creating Value through Workforce and Career Development* (Santa Barbara, Calif.: Praeger, 2010), xxii–xxiv.

38. Elsdon and Erickson, "What Does Workforce Strength Look Like," 13.

39. Elsdon, *Building Workforce Strength,* xxii–xxiv.

40. Sophie Egan, "The Health Tradeoffs of Entrepreneurship," *New York Times,* August 8, 2013, accessed August 25, 2013, http://well.blogs.nytimes.com/2013/08/08/the-health-tradeoffs-of-entrepreneurship/?_r=0/.

41. Federal Reserve Bank of St. Louis, "The Stock Market: Risk vs. Uncertainty," *Inside the Vault,* Fall 2002, accessed August 23, 2013, http://www.stlouisfed.org/publications/itv/articles/?id=1185.

42. Peter Dizikes, "Explained: Knightian Uncertainty," *MIT News,* June 1, 2010, accessed November 15, 2013, http://web.mit.edu/newsoffice/2010/explained-knightian-0602.html/.

43. Ibid.

44. Steven Greenhouse, "Tackling Concerns of Independent Workers," *New York Times,* March 23, 2013, accessed August 25, 2013, http://www.nytimes.com/2013/03/24/business/freelancers-union-tackles-concerns-of-independent-workers.html?pagewanted=all/.

45. Lauren Hargrave, "The Road Less Traveled: Navigating a Non-Traditional Career Path," *Forbes,* May 21, 2012, accessed August 12, 2013, http://www.forbes.com/sites/dailymuse/2012/05/21/the-road-less-traveled-navigating-a-non-traditional-career-path/.

46. Frederick Buechner, Goodreads, accessed August 26, 2013, http://www.goodreads.com/quotes/496215-vocation-is-the-place-where-our-deep-gladness-meets-the/.

Chapter 3

1. Rachel Nelken, "Portfolio Careers in the Arts: One Part Pain, Two Parts Pleasure," *Guardian,* March 5, 2013, accessed August 12, 2013, http://www.theguardian.com/culture-professionals-network/culture-professionals-blog/2013/mar/05/arts-portfolio-careers-top-tips/.

2. Claire Adler, "Mix and Match," *Guardian,* November 4, 2006, accessed August 27, 2013, http://www.theguardian.com/money/2006/nov/04/careers.graduates1/.

3. Mind Tools, "The Boston Matrix: Focusing Effort to Give the Greatest Returns," accessed August 28, 2013, http://www.mindtools.com/pages/article/newTED_97.htm/; Ron Elsdon, *Affiliation in the Workplace: Value Creation in the New Organization* (Westport, Conn.: Praeger, 2003), 46, 47; Robert G. Cooper, *Winning at New Products: Accelerating the Process from Idea to Launch* (Reading, Mass.: Addison-Wesley, 1993), 185.

4. Lesah Beckhusen, *Skills-Focused Career Development: Facilitator's Manual* (Orinda, Calif.: SkillScan Professional Pack, 1993), 3–6.

5. Ibid.

6. Ralph Biggadike, "The Risky Business of Diversification," *Harvard Business Review,* May 1979, accessed August 30, 2013, http://hbr.org/1979/05/the-risky-business-of-diversification/ar/.

Chapter 4

1. Michael E. Porter, *Competitive Advantage: Creating and Sustaining Superior Performance* (New York: Free Press, 1985), 5.

2. Ibid., 11, for the classical view of mutually exclusive strategies, while the benefits of simultaneously executing low-cost and differentiation strategies are shown in Peter Wright, Mark Kroll, Ben Kedia, and Charles Pringle, "Strategic Profiles, Market Share, and Business Performance," *Industrial Management,* May 1, 1990, accessed September 6, 2013, http://www.thefreelibrary.com/Strategic+profiles,+market+share,+and+business+performance.-a09306235/.

3. Ron Elsdon and Rita Erickson, "What Does Workforce Strength Look Like and Why Does It Matter?" in *Building Workforce Strength: Creating Value through Workforce and Career Development,* ed. Ron Elsdon (Santa Barbara, Calif.: Praeger, 2010), 13.

4. Ron Elsdon, ed., *Business Behaving Well: Social Responsibility, from Learning to Doing* (Washington, D.C.: Potomac Books, 2013), 17, 21.

5. DuPont, "A World Leader in Market-Driven Innovation and Science," accessed September 4, 2013, http://www.dupont.com/corporate-functions/our-company.html/; Ron Elsdon, *Affiliation in the Workplace: Value Creation in the New Organization* (Westport, Conn.: Praeger, 2003), 30.

6. Alexander H. Tullo, "DuPont May Sell Chemical Units," *Chemical and Engineering News,* July 24, 2013, accessed September 4, 2013, http://cen.acs.org/articles/91/web/2013/07/DuPont-Sell-Chemical-Units.html/.

7. Environmental Protection Agency, "Lean Thinking and Methods," accessed September 5, 2013, http://www.epa.gov/lean/environment/methods/index.htm.

8. Leonard L. Berry, *On Great Service: A Framework for Action* (New York: Free Press, 1995), 5.

9. Ibid., 8.

10. Porter, *Competitive Advantage,* 159.

11. Rita Erickson and Ron Elsdon, "Designing, Developing and Measuring Effective Career Development Processes and Systems," in *Building Workforce Strength: Creating Value through Workforce and Career Development,* ed. Ron Elsdon (Santa Barbara, Calif.: Praeger, 2010), 96–103.

12. Ibid.

Chapter 5

1. Robert G. Cooper, *Winning at New Products: Accelerating the Process from Idea to Launch* (Reading, Mass.: Addison-Wesley, 1993), 185.

2. Christine Benz, "Find the Right Stock/Bond Mix," *Morningstar,* October 18, 2011, accessed September 8, 2013, http://news.morningstar.com/articlenet/article.aspx?id=397904.

3. Census Bureau, *Statistical Abstract of the United States: 2012* (Washington, D.C.: Government Printing Office, 2013), table 1207, p. 749, accessed September 9, 2013, http://www.census.gov/compendia/statab/2012/tables/12s1207.pdf.

4. Aaron Hurst, "Pro Bono Service: Driving Social Impact with Professional Skills," 81–94, and Barbara Langsdale, "Partnerships between the For-Profit and Nonprofit Sectors," 56–67, in *Business Behaving Well: Social Responsibility, from Learning to Doing,* ed. Ron Elsdon (Washington, D.C.: Potomac Books, 2013).

5. Donna J. Kelley, Abdul Ali, Candida Brush, Andrew C. Corbett, Mahdi Majbouri, Edward G. Rogoff, Babson College, and Baruch College, *Global Entrepreneurship Monitor: 2012 United States Report,* 2012, 19, accessed October 30, 2013, http://www.babson.edu/Academics/centers/blank-center/global-research/gem/Documents/GEM%20US%202012%20Report%20FINAL.pdf.

6. Ron Elsdon, *Affiliation in the Workplace: Value Creation in the New Organization* (Westport, Conn.: Praeger, 2003), 74–83.

Chapter 6

1. Eric Ries, *The Lean Startup: How Today's Entrepreneurs Use Continuous Innovation to Create Radically Successful Businesses* (New York: Crown Business, 2011), 92–113.

2. William Bridges, *Transitions: Making Sense of Life's Changes,* Revised 25th Anniversary Edition (Cambridge, Mass.: Da Capo Press, 2004), 4.

Chapter 7

1. Examples of individual skill descriptions are at SkillScan's website, accessed August 9, 2013, http://www.skillscan.com/; Michael M. Lombardo and Robert W. Eichinger provide insights into needed skills and developmental activities in *FYI: For Your Improvement: A Guide for Development and Coaching for Learners, Managers, Mentors, and Feedback Givers* (Minneapolis: Korn/Ferry International, 2009); examples of various approaches to organizational leadership and related skills are summarized in Ron Elsdon, ed., *Business Behaving Well: Social Responsibility, from Learning to Doing* (Washington, D.C.: Potomac Books, 2013), 22–24; examples of skills related to small business leadership are summarized in Small Business Administration, "Being a Leader," *SBA.gov,* accessed September 23, 2013, http://www.sba.gov/content/being-leader/.

2. Lombardo and Eichinger, *FYI: For Your Improvement.*

3. SkillScan, http://www.skillscan.com/.

4. Daniel Goleman, Richard Boyatzis, and Annie McKee, *Primal Leadership: Realizing the Power of Emotional Intelligence* (Boston: Harvard Business School Press, 2002), 39; Larry C. Spears, "Practicing Servant Leadership," *Leader to Leader* 34 (Fall 2004): 7–11; Sam M. Intrator and Megan Scribner, *Leading from Within: Poetry that Sustains the Courage to Lead* (San Francisco: Jossey-Bass, 2007); Ron Elsdon, *Affiliation in the Workplace: Value Creation in the New Organization* (Westport, Conn.: Praeger, 2003), 169–76.

5. Lesah Beckhusen, *Skills-Focused Career Development: Facilitator's Manual* (Orinda, Calif.: SkillScan Professional Pack, 1993), 3–6.

6. PwC, Annual Global CEO Survey, "15th Annual Global CEO survey 2012, Delivering Results: Growth and Value in a Volatile World," 3, accessed September 24, 2013, http://www.pwc.com/gx/en/ceo-survey/2012/ceo-profiles/index.jhtml/; PwC, Annual Global CEO Survey, "16th Annual Global CEO Survey, Dealing with Disruption: Adapting to Survive and Thrive," 2013, accessed September 24, 3013, http://www.pwc.com/gx/en/ceo-survey/index.jhtml/.

7. See the Small Business Administration's website, accessed September 23, 2013, http://www.sba.gov/, and for SCORE, see http://www.sba.gov/content/score/.

8. The Time Mastery Profile is available from many distributors. One example is the Center for Internal Change, accessed April 26, 2014, http://www.internalchange.com/disc_profile_store/mall/timemasteryonline.htm.

9. Robert Bolton, *People Skills: How to Assert Yourself, Listen to Others, and Resolve Conflicts* (New York: Touchstone, Simon and Schuster, 1979), 78.

10. Examples of continuing education courses in communication through the University of California—Berkeley Extension, accessed September 23, 2013, http://extension.berkeley.edu/public/category/programStream.do?method=load&selectedProgramAreaId=11463&selectedProgramStreamId=15569#/, and through the American Management Association, accessed September 23, 2013, http://www.amanet.org/training/seminars/communication-skills-training.aspx/; Bolton's book *People Skills* provides excellent insights; Bridget Weston Pollack provides additional insights in "Communication Converts Leads," *SBA.gov,* accessed September 23, 2013, http://www.sba.gov/community/blogs/communication-converts-leads/.

11. Bolton, *People Skills.*

12. David Freedman, "How to Become a First-Class Negotiator," *Huthwaite International,* May 21, 2013, accessed September 24, 2013, http://www.huthwaite.co.uk/resource-centre/articles/how-to-become-a-first-class-negotiator/.

13. For examples of negotiation training, see negotiation training program, *Huthwaite International*, accessed September 24, 2013, http://www.huthwaite.co.uk/solutions/negotiation-training/; and Business Negotiating course (BUS ADM X451.3), University of California–Berkeley Extension, accessed September 24, 2013, http://extension.berkeley.edu/search/publicCourseSearchDetails.do;jsessionid=4558F30B72035956CB4F08C304BCD785?method=load&courseId=40301. The Small Business Administration (http://www.sba.gov/) provides resources on negotiating skills.

14. Independent Direction Directors Advisory Service (IDDAS), "Portfolio Working—Plural Careers Prove to Be a Singular Success," 2003, accessed April 26, 2014, http://thought-leadership.top-consultant.com/UK/FullArticle.aspx?ID=899.

15. Donna Fisher and Sandy Vilas, *Power Networking: 55 Secrets for Personal and Professional Success* (Austin, Tex.: Mountain Harbour Publications, 1992), 22.

16. Ibid., 14.

17. See the following: seminars from the American Management Association, for example, http://www.amanet.org/training/webcasts/Relationship-Networking.aspx; and resources through the Small Business Administration, accessed September 24, 2013, http://www.sba.gov/community/blogs/power-face-face-networking/.

18. See LinkedIn's website, accessed November 17, 2013, http://www.linkedin.com/.

19. Small Business Administration, "Marketing," *SBA.gov,* accessed September 24, 2013, http://www.sba.gov/category/navigation-structure/starting-managing-business/managing-business/running-business/marketing/.

20. The American Marketing Association is a professional body dedicated to the discipline of marketing. American Marketing Association's website, accessed September 24, 2013, http://www.marketingpower.com/; the American Management Association offers seminars on marketing: AMA, Marketing Seminars and Courses, accessed September 24, 2013, http://www.amanet.org/training/seminars/Marketing -training.aspx/; an example of a continuing education program in marketing is the University of California–Berkeley Extension certificate program in marketing, accessed September 24, 2013, http://extension.berkeley.edu/public/category/course CategoryCertificate Profile.do?method=load&certificateId=17111.

21. Professional organizations in the consulting field include the Institute of Management Consultants USA, accessed September 25, 2013, http://www.imcusa.org/; Professional Consultants Association, accessed September 25, 2013, http://www.professionalconsultantsassociation.org/; and Society of Professional Consultants, accessed September 25, 2013, http://www.spconsultants.org/.

22. Jim Lomac, Management Research Group, Portland, Maine, private communication, 2011, internal report analyzing results from 242 participants completing sales performance and leadership effectiveness analysis assessments.

23. Ibid.

24. Ibid.

25. Management Research Group, Portland, Maine, private communication, 2009, "Leadership Best Practices for Sales Managers."

26. Examples include Raven House, which specializes in customizing sales training, accessed September 25, 2013, http://www.ravenhouseint.com/documents/RavenHouse -ReinvetingYourSalesOrganization.pdf; sales seminars from the American Management Association, accessed September 25, 2013, http://www.amanet.org/training/seminars/Sales-training.aspx/; Small Business Administration, "Sales Strategy," SBA.gov, accessed September 25, 2013, http://www.sba.gov/content/sales-strategy/.

27. For example, we have used Automatic Data Processing (ADP) for payroll processing, see ADP's website, http://www.adp.com/; see also the Society of Human Resource Managers (SHRM) website, http://www.shrm.org/.

28. Max DePree, *Leadership Jazz* (New York: Dell Trade Paperback, 1992), 3.

29. Marc Santora and Jess Wisloski, "8 Aboard Rescued in Another East River Helicopter Crash," *New York Times,* June 18, 2005, accessed September 29, 2013, http://www.nytimes.com/2005/06/18/nyregion/18chopper.html?pagewanted=all/.

30. Andrew Ross Sorkin, "Ex-Chief and Aide Guilty of Looting Millions at Tyco," *New York Times,* June 18, 2005, accessed September 29, 2013, http://www.nytimes.com/2005/06/18/business/18tyco.html?pagewanted=all/.

31. Carmen DeNavas-Walt, Bernadette D. Proctor, and Jessica C. Smith, U.S. Census Bureau Current Population Reports, P60-245, *Income, Poverty, and Health Insurance Coverage in the United States: 2012* (Washington, D.C.: Government Printing Office, 2013), 33, accessed September 26, 2013, http://www.census.gov/prod/2013pubs/p60-245.pdf.

32. Rob Kabacoff, "Ten Years of Best Practice Studies: What Have We Learned?," webinar presented on behalf of the Management Research Group, Portland, Maine, September 26, 2013.

33. Bolton, *People Skills,* 93; Dictionary.com, accessed September 26, 2013, http://dictionary.reference.com/browse/empathy?s=t/.

34. Lombardo and Eichinger, *FYI: For Your Improvement,* 7.

35. Ibid., 7–12.

Chapter 8

1. Ron Elsdon and Seema Iyer, "Creating Value and Enhancing Retention through Employee Development: The Sun Microsystems Experience," *Human Resource Planning* 22, no. 2 (1999): 39–47; Ron Elsdon, *Affiliation in the Workplace: Value Creation in the New Organization* (Westport, Conn.: Praeger, 2003), xiv, 147–57.

2. Ron Elsdon, ed., *Business Behaving Well: Social Responsibility, from Learning to Doing* (Washington, D.C.: Potomac Books, 2013).

3. Ron Elsdon, ed., *Building Workforce Strength: Creating Value through Workforce and Career Development* (Santa Barbara, Calif.: Praeger, 2010).

4. See the Taproot Foundation's website, http://www.taprootfoundation.org/.

5. Elsdon, *Affiliation in the Workplace.*

6. Rita Erickson and Ron Elsdon, "Designing, Developing and Measuring Effective Career Development Processes and Systems," in Elsdon, *Building Workforce Strength,* 96–103.

7. Zeth Ajemian, Mark Malcolm, and Ron Elsdon, "Example Case Studies: Healthcare and High-Technology," in Elsdon, *Building Workforce Strength,* 169–91.

8. Small Business Administration, "Hire a Contractor or an Employee?" SBA.gov, accessed September 28, 2013, http://www.sba.gov/content/hire-contractor-or-employee/.

9. Elsdon, *Affiliation in the Workplace,* 205–21.

10. See the following in Elsdon, *Business Behaving Well*: Barbara Langsdale, "Partnerships between the For-Profit and Nonprofit Sectors," 56–67; Andrew Domek, "Government as a Business Partner in the Pursuit of Social Responsibility," 68–77; Aaron Hurst, "Pro-Bono Service: Driving Social Impact with Professional Skills," 81–94; Linda Williams, "Public/Private Sector Collaboration," 95–106; and Zeth Ajemian, "Labor-Management Partnerships: A Collaborative Path toward Improved Business Results and Employee Satisfaction," 150–67.

11. Elsdon, *Affiliation in the Workplace,* 215.

12. Inscape Publishing, Team Dimensions Profile, Team Innovation Seminar, Facilitator's Kit, Volume 2 (2003, Version 3.0), 43.

13. Michael Watkins, "What Is Organizational Culture? And Why Should We Care?" *HBR Blog Network,* accessed September 29, 2013, http://blogs.hbr.org/2013/05/what-is-organizational-culture/.

14. For examples of organizational culture dimensions, see Kevin Johnston, "Ten Dimensions of an Organizational Culture," *Houston Chronicle,* accessed September 30, 2013, http://smallbusiness.chron.com/ten-dimensions-organizational-culture-25955.html; Nathalie Delobbe, Robert R. Haccoun, and Christian Vandenberghe, "Measuring

Core Dimensions of Organizational Culture: A Review of Research and Development of a New Instrument," unpublished manuscript, 2002, accessed September 30, 2013, http://www.uclouvain.be/cps/ucl/doc/iag/documents/WP_53_Delobbe.pdf; Michael A. Diamond, "Dimensions of Organizational Culture and Beyond," *Political Psychology* 12, no. 3 (September 1991): 509–22; Hofstede Centre, "Dimensions," *Hofstede Centre* website, accessed September 30, 2013, http://geert-hofstede.com/organisational-culture-dimensions.html; Elsdon, *Affiliation in the Workplace,* 27–29, 217; and Rita Erickson and Ron Elsdon, "Designing, Developing and Measuring Effective Career Development Processes and Systems," in Elsdon, *Building Workforce Strength,* 82–86.

15. Elsdon, *Business Behaving Well,* 16–22.

16. Niels van Quaquebeke, Daniel C. Henrich, and Tilman Ecklof, "'It's Not Tolerance I'm Asking for, It's Respect!' A Conceptual Framework to Differentiate between Tolerance, Acceptance and (Two Types of) Respect," *Gruppendynamik und Organisationsberatung* 38, no. 2 (2007): 185–200, accessed September 30, 2013, http://www.respectresearchgroup.org/rrg/files/pdf/Articles/van%20Quaquebeke%2C%20N.%2C%20Henrich%2C%20D.%20C.%2C%20Eckloff%2C%20T.%20%282007%29.%20A%20conceptual%20framework%20to%20differentiate%20between%20tolerance%2C%20acceptance%20and%20respect.pdf.

17. Carl R. Rogers, *On Becoming a Person: A Therapist's View of Psychotherapy* (New York: Houghton Mifflin, 1995).

18. Elsdon, *Affiliation in the Workplace,* 220–21.

Chapter 9

1. Investor Words, "Limited Liability," accessed October 8, 2013, http://www.investorwords.com/2816/limited_liability.html.

2. See, for example, Internal Revenue Service, "Exemption Requirements—Section 501(c)(3), Organizations," accessed October 4, 2013, http://www.irs.gov/Charities-&-Non-Profits/Charitable-Organizations/Exemption-Requirements-Section-501%28c%29%283%29-Organizations/; Minnesota Council on Foundations, "Create a Private Foundation," accessed April 29, 2014, http://www.mcf.org/donors/private.

3. Small Business Administration, "Sole Proprietorship," accessed October 4, 2013, http://www.sba.gov/content/sole-proprietorship-0/.

4. Internal Revenue Service, "Sole Proprietorships," accessed October 5, 2013, http://www.irs.gov/Businesses/Small-Businesses-&-Self-Employed/Sole-Proprietorships/.

5. State of California, Franchise Tax Board, "Sole Proprietorship," accessed October 8, 2013, https://www.ftb.ca.gov/businesses/bus_structures/soleprop.shtml/.

6. Small Business Administration, "Partnership," accessed October 4, 2013, http://www.sba.gov/content/partnership/.

7. Internal Revenue Service, "Partnerships," accessed October 5, 2013, http://www.irs.gov/Businesses/Small-Businesses-&-Self-Employed/Partnerships/.

8. State of California, Franchise Tax Board, "Withholding on Partnerships and Limited Liability Companies," accessed October 8, 2013, https://www.ftb.ca.gov/individuals/wsc/partnerships_and_limited_liability_companies.shtml/.

9. Small Business Administration, "Cooperative," accessed October 4, 2013, http://www.sba.gov/content/cooperative/; Michael Wilson, "Reality Intrudes at a Utopian Market in Brooklyn, with Light Fingers," *New York Times,* November 29, 2013, accessed April 21, 2014, http://www.nytimes.com/2013/11/30/nyregion/reality-intrudes-at-a-utopian-market-in-brooklyn-with-light-fingers.html.

10. Shaila Dewan, "Who Needs a Boss?" *New York Times,* March 25, 2014, accessed April 21, 2014, http://www.nytimes.com/2014/03/30/magazine/who-needs-a-boss.html; Catherine Lacey, "A Way for Artists to Live," *New York Times,* April 19, 2014, accessed April 21, 2014, http://www.nytimes.com/2014/04/20/opinion/sunday/a-way-for-artists-to-live.html.

11. Small Business Administration, "Limited Liability Company," accessed October 4, 2013, http://www.sba.gov/content/limited-liability-company-llc/.

12. Internal Revenue Service, "Limited Liability Company (LLC)," accessed April 29, 2014,http://www.irs.gov/Businesses/Small-Businesses-&-Self-Employed/Limited-Liability-Company-LLC.

13. Kevin Finck, *How to Form a California Corporation or LLC from Any State* (Irvine, Calif.: Entrepreneur Press, 2005), 7.

14. Internal Revenue Service, "S Corporations," accessed October 5, 2013, http://www.irs.gov/Businesses/Small-Businesses-&-Self-Employed/S-Corporations/.

15. Small Business Administration, "S Corporation," accessed October 4, 2013, http://www.sba.gov/content/s-corporation/.

16. Small Business Administration, "Corporation," accessed October 4, 2013, http://www.sba.gov/content/corporation/.

17. Internal Revenue Service, "Corporations," accessed October 5, 2013, http://www.irs.gov/Businesses/Small-Businesses-&-Self-Employed/Corporations/.

18. Internal Revenue Service, "Employer ID Numbers (EINs)," accessed October 5, 2013, http://www.irs.gov/Businesses/Small-Businesses-&-Self-Employed/Employer-ID-Numbers-EINs/.

19. Internal Revenue Service, "Do You Need an EIN?" accessed October 5, 2013, http://www.irs.gov/Businesses/Small-Businesses-&-Self-Employed/Do-You-Need-an-EIN/.

20. Internal Revenue Service, "State Government Websites," accessed October 5, 2013, http://www.irs.gov/Businesses/Small-Businesses-&-Self-Employed/State-Links-1/.

21. Internal Revenue Service, "Topic 762—Independent Contractor vs. Employee," accessed October 5, 2013, http://www.irs.gov/taxtopics/tc762.html.

22. Internal Revenue Service, Form 1099-MISC, Miscellaneous Income, accessed October 5, 2013, http://www.irs.gov/uac/Form-1099-MISC,-Miscellaneous-Income/.

23. Internal Revenue Service, "Forms and Associated Taxes for Independent Contractors," accessed November 17, 2013, http://www.irs.gov/Businesses/Small-Businesses-%26-Self-Employed/Forms-and-Associated-Taxes-for-Independent-Contractors/.

24. Internal Revenue Service, Form W-2, Wage and Tax Statement, accessed October 5, 2013, http://www.irs.gov/uac/Form-W-2,-Wage-and-Tax-Statement/.

25. Ron Elsdon, ed., *Business Behaving Well: Social Responsibility, from Learning to Doing* (Washington, D.C.: Potomac Books, 2013), 16–17.

26. Automatic Data Processing, Inc., *Guide to FLSA Rules Regarding Exempt/Non-exempt Employees,* accessed October 5, 2013, https://www.adp.com/pdf/979_686_ct-03-173-0971.pdf.

27. Tom Janz, Lowell Hellervik, and David C. Gilmore, *Behavior Description Interviewing: New, Accurate, Cost Effective* (Newton, Mass.: Allyn and Bacon, 1986), 31.

28. See, for example, Salary.com, accessed October 5, 2013, http://www.salary.com/.

29. Department of Homeland Security, "Instructions for Employment Eligibility Verification," USCIS Form I-9, accessed October 5, 2013, http://www.uscis.gov/files/form/i-9.pdf/.

30 Internal Revenue Service, Form W-4 (2014), accessed April 29, 2014, http://www.irs.gov/pub/irs-pdf/fw4.pdf.

31. Healthcare Providers Service Organization's website, accessed October 6, 2013, http://www.hpso.com/.

32. Travelers Insurance Company's website, accessed October 6, 2013, https://www.travelers.com/.

33. An example of workers' compensation requirements in California is summarized at State of California, Department of Industrial Relations, accessed October 6, 2013, http://www.dir.ca.gov/dwc/faqs.html/.

34. Department of Labor, "State Unemployment Insurance Benefits," accessed October 6, 2013, http://workforcesecurity.doleta.gov/unemploy/uifactsheet.asp.

35. State and Local Government on the Net, "State Unemployment Insurance Sites," accessed October 7, http://www.statelocalgov.net/50states-unemployment.cfm.

36. Internal Revenue Service, "Federal Unemployment Tax," accessed October 6, 2013, http://www.irs.gov/Individuals/International-Taxpayers/Federal-Unemployment-Tax/.

37. State of California, Employment Development Department, "Overview—State Disability Insurance," accessed October 7,2013, http://www.edd.ca.gov/disability/.

38. HealthCare.gov, accessed October 6, 2013, https://www.healthcare.gov/.

39. Deborah LeVeen, "Healthcare as Social Responsibility: Implications for Business," in Elsdon, *Business Behaving Well,* 107–37.

40. Intuit, "QuickBooks Products," accessed October 7, 2013, http://quickbooks.intuit.com/products.

41. Ibid.

42. Intuit, "TurboTax," accessed October 7, 2013, https://turbotax.intuit.com/.

43. Vanguard Group, Inc., "Your Investing Life: Starting Your Own Business," accessed November 4, 2013, https://personal.vanguard.com/us/insights/article/starting-business-investing-092013/.

44. Internal Revenue Service, "One-Participant 401(k) Plans," accessed November 24,2013, http://www.irs.gov/Retirement-Plans/One-Participant-401%28k%29-Plans/.

45. Small Business Administration, "SBA's 7(a) Loan Program Explained," accessed November 3, 2013, http://www.sba.gov/community/blogs/sbas-7a-loan-program-explained/.

46. Mozy's website, accessed April 26, 2014, http://www.mozy.com.

47. See eFax's website, accessed April 26, 2014, http://www.efax.com.

48. Intuit, "QuickBase," accessed October 8, 2013, http://quickbase.intuit.com/.

49. Elsdon Organizational Renewal and New Beginnings Career and College Guidance, "Welcome to Elsdon Organizational Renewal and New Beginnings," accessed October 8, 2013, http://www.elsdon.com.

50. CoffeeCup, accessed October 8, 2013, http://www.coffeecup.com/.

51. Elsdon Organizational Renewal, "Newsletters," accessed October 8, 2013, http://www.elsdon.com/EOR_Newsletters.html.

52. MailChimp, "Send Better Email," accessed October 8, 2013, http://mailchimp.com/.

53. Elsdon, Inc., accessed October 8, 2013, http://www.linkedin.com/company/elsdon-inc.

54. Vistaprint's website, accessed April 26, 2014, http://www.vistaprint.com.

Chapter 10

1. For the view that only a few are suited to self-employment, see Don McNay, "Are You Ready to Jump into the World of Self-Employment," *Huffington Post, Small Business America,* August 10, 2011, accessed October 22, 2013, http://www.huffington post.com/don-mcnay/are-you-ready-to-jump-int_b_922926.html/; and for the view that this is accessible to many, see Leo Babauta, "The Get-Started-Now Guide to Becoming Self-Employed," *Zenhabits,* accessed October 22, 2013, http://zenhabits.net/the-get-started-now-guide-to-becoming-self-employed/.

2. Anita Campbell, "10 Businesses You Can Start with Little Capital," Small Business Administration website, accessed November 4, 2013, http://www.sba.gov/community/blogs/guest-blogs/industry-word/10-businesses-you-can-start-with-little-capital/; Caron Beesley, "7 Inspiring Home Business Ideas for Stay-at-Home Moms (or Dads)," Small Business Administration website, accessed November 4, 2013, http://www.sba.gov/community/blogs/community-blogs/small-business-matters/7-inspiring-home-business-ideas-stay-home-mom/.

3. Mark Oppenheimer, "Taste-Testing a Second Career, with a Mentor," *New York Times,* September 14, 2013, accessed October 22, 2103, http://www.nytimes.com/2013/09/15/business/taste-testing-a-second-career.html?_r=0/.

4. The Myers-Briggs Type Indicator mentioned in chapter 1 provides insights into personal preferences for structure or spontaneity.

5. Eric Ries, *The Lean Startup: How Today's Entrepreneurs Use Continuous Innovation to Create Radically Successful Businesses* (New York: Crown Business, 2011), 56–72.

6. One approach is described in Jim Horan, *The One Page Business Plan for the Creative Entrepreneur* (Berkeley, Calif.: One Page Business Plan, 2004); various one-page business plan books are available at: http://www.onepagebusinessplan.com/books2.htm, accessed October 23, 2013.

7. David Zweig, "A Communal Space, but Still My Own," *New York Times,* March 29, 2014, http://www.nytimes.com/2014/03/30/jobs/a-communal-space-but-still-my-own.html.

8. Usa.gov, "USA.gov for the Self-Employed," accessed October 23, 2013, http://www.usa.gov/Business/Self-Employed.shtml; Small Business Administration, "10 Steps to Starting a Business," accessed October 23, 2013, http://www.sba.gov/content/follow-these-steps-starting-business/.

9. Small Business Administration, "50+ Entrepreneurs," accessed October 23, 2013, http://www.sba.gov/encore/; Small Business Administration, "Encore Entrepreneurs: An Introduction to Starting Your Own Business," accessed October 23, 2013, http://www.sba.gov/tools/sba-learning-center/training/encore-entrepreneurs-introduction-starting-your-own-business/; Small Business Administration, "Young Entrepreneurs," accessed October 23, 2013, http://www.sba.gov/content/young-entrepreneurs/; Small Business Administration, "Young Entrepreneurs, An Essential Guide to Starting Your Own Business," accessed October 23, 2013, http://www.sba.gov/tools/sba-learning-center/training/young-entrepreneurs-essential-guide-starting-your-own-business/.

10. Internal Revenue Service, "Closing a Business Checklist," accessed October 23, 2013, http://www.irs.gov/Businesses/Small-Businesses-&-Self-Employed/Closing-a-Business-Checklist/; Small Business Administration, "Steps to Closing a Business," accessed October 23, 2013, http://www.sba.gov/content/steps-closing-business/.

11. Irenaeus of Lyons, Goodreads, accessed October 24, 2013, http://www.goodreads.com/quotes/172267-the-glory-of-god-is-the-human-person-fully-alive/.

Selected Bibliography

Berry, Leonard L. *On Great Service: A Framework for Action*. New York: Free Press, 1995.

Bolton, Robert. *People Skills: How to Assert Yourself, Listen to Others, and Resolve Conflicts*. New York: Touchstone, Simon and Schuster, 1979.

Bridges, William. *Transitions: Making Sense of Life's Changes*. Revised 25th Anniversary Edition. Cambridge, Mass.: Da Capo Press, 2004.

Briggs Myers, Isabel. *Introduction to Type: A Guide to Understanding Your Results on the MBTI Instrument*. Mountain View, Calif.: CPP, 1998.

Cooper, Robert G. *Winning at New Products: Accelerating the Process from Idea to Launch*. Reading, Mass.: Addison-Wesley, 1993.

Dobbs, Richard, Jaana Remes, Sven Smit, James Manyika, Jonathan Woetzel, and Yaw Agyenim-Boateng. "Urban World: The Shifting Global Business Landscape."

Elsdon, Ron. *Affiliation in the Workplace: Value Creation in the New Organization*. Westport, Conn.: Praeger, 2003.

Elsdon, Ron, ed. *Building Workforce Strength: Creating Value through Workforce and Career Development*. Santa Barbara, Calif.: Praeger, 2010.

Elsdon, Ron. *Business Behaving Well: Social Responsibility, from Learning to Doing*. Washington, D.C.: Potomac Books, 2013.

Gelardin, Sally, ed. *Starting and Growing a Business in the New Economy*. Broken Arrow, Okla.: National Career Development Association, 2007.

Goleman, Daniel, Richard Boyatzis, and Annie McKee. *Primal Leadership: Realizing the Power of Emotional Intelligence*. Boston: Harvard Business School Press, 2002.

Hopson, Barrie, and Katie Ledger. *And What Do You Do? 10 Steps to Creating a Portfolio Career*. London: A&C Black, 2009.

Horan, Jim. *The One Page Business Plan for the Creative Entrepreneur*. Berkeley, Calif.: One Page Business Plan, 2004.

Intrator, Sam M., and Megan Scribner. *Leading from Within: Poetry that Sustains the Courage to Lead*. San Francisco: Jossey-Bass, 2007.

IPro Index 2013. Entity Solutions sponsored study conducted by Monash University, Melbourne, Australia, 2013. Accessed May 31, 2014. http://www.entitysolutions .com.au/entitysolutions/resourcecentre/iproindex/.

Lombardo, Michael M., and Robert W. Eichinger. *FYI: For Your Improvement: A Guide for Development and Coaching for Learners, Managers, Mentors, and Feedback Givers.* Minneapolis: Korn/Ferry International, 2009.

McKinsey and Company, October 2012. Accessed October 23, 2013. http://www .mckinsey.com/insights/urbanization/urban_world_the_shifting_global _business_landscape/.

Phillips, Joseph D. *The Self-Employed in the United States.* University of Illinois Bulletin 88. Urbana: University of Illinois, 1962.

Porter, Michael E. *Competitive Advantage: Creating and Sustaining Superior Performance.* New York: Free Press, 1985.

Ries, Eric. *The Lean Startup: How Today's Entrepreneurs Use Continuous Innovation to Create Radically Successful Businesses.* New York: Crown Business, 2011.

Schein, Edgar H., and John Van Maanen. *Career Anchors Self-Assessment.* San Francisco: Wiley, 2013.

U.S. Small Business Administration. Website. http://www.sba.gov/.

Yancey, Philip. *Soul Survivor: How Thirteen Unlikely Mentors Helped My Faith Survive the Church.* New York: Galilee Doubleday, 2003.

Index

About the Author

Ron Elsdon's nontraditional career path includes founding organizations that specialize in career and workforce development, where he has worked for many years with individuals as well as with the for-profit, nonprofit, and public sectors. He also has leadership experience in various organizations and has been an adjunct faculty member at, or affiliated with, several universities. He is the editor of *Business Behaving Well: Social Responsibility, from Learning to Doing,* the editor of *Building Workforce Strength: Creating Value through Workforce and Career Development,* and the author of *Affiliation in the Workplace: Value Creation in the New Organization.* Ron holds a PhD in chemical engineering from Cambridge University, a master's in career development from John F. Kennedy University, and a first-class honors degree in chemical engineering from Leeds University. He lives in Danville, California, and can be reached at renewal@ elsdon.com.